C. W. GROETSCH J. THOMAS KING

University of Cincinnati

MATRIX METHODS AND APPLICATIONS

PRENTICE HALL, Englewood Cliffs, New Jersey 07632

Library of Congress Cataloging-in-Publication Data

Groetsch, C. W.
Matrix methods and applications: an introduction to linear algebra/C. W. Groetsch and J. Thomas King.
p. cm.
Includes bibliographies and index.
ISBN 0-13-565565-X
1. Matrices. 2. Algebras, Linear. I. King, J. Thomas.
II. Title.
QA188.G76 1988 87–17833
512.9′434—dc19 CIP

To Sandra, Kurt, and Heidi
To Susan, Kathy, and Alex

Editorial/production supervision and interior design: *Eleanor Henshaw Hiatt*
Cover illustration: *Kathleen King*
Cover design: *Ben Santora*
Manufacturing buyer: *Paula Benevento*

Printed in the United States of America

10 9 8 7 6 5 4 3 2 1

ISBN 0-13-565565-X 01

PRENTICE-HALL INTERNATIONAL (UK) LIMITED, *London*
PRENTICE-HALL OF AUSTRALIA PTY. LIMITED, *Sydney*
PRENTICE-HALL CANADA INC., *Toronto*
PRENTICE-HALL HISPANOAMERICANA, S.A., *Mexico*
PRENTICE-HALL OF INDIA PRIVATE LIMITED, *New Delhi*
PRENTICE-HALL OF JAPAN, INC., *Tokyo*
SIMON & SCHUSTER ASIA PTE. LTD., *Singapore*
EDITORA PRENTICE-HALL DO BRASIL, LTDA., *Rio de Janeiro*

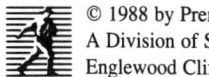

Contents

Preface

This book developed from a course in matrix methods that we have taught for more than fifteen years to students in the sciences, engineering, and business, who need to use matrices in their further course work. Our primary concern in this course is to provide the mathematical tools necessary for problem solving by matrix techniques while at the same time supplying the theoretical framework in linear algebra requisite for more advanced courses in the quantitative sciences and engineering. In this book we attempt to attain these goals within the framework of a short course (one semester or one/two quarters) at the sophomore level that is suitable for students from a variety of backgrounds and disciplines.

We have tried in our treatment to (1) present an elementary exposition of the essential concepts of matrix algebra that are routinely used in problem solving, (2) give an informal and concrete account of basic linear algebra, (3) provide simple and effective methods for the solution of problems that arise in applications, and (4) equip the student with basic mathematical tools necessary to solve significant linear problems on a modern digital computer. As a consequence our book is considerably different from more traditional introductions to linear algebra. For example, such topics as determinants and the Jordan canonical form are given little or no coverage; on the other hand, the Gaussian elimination algorithm assumes a central role in our treatment of both theory and applications. We also stress factorization methods, such as Gaussian elimination interpreted as the LU factorization and the Gram–Schmidt process in the guise of the QR factorization, in a way that is unusual in elementary texts. Another novel feature for an elementary text is the systematic use of coordinate transformations for solving systems of difference and differential equations.

Far too many books present methods that are suitable only for solving small matrix problems using pencil and paper. Although some methods and problems of this sort are necessary for beginning students, we are convinced of the necessity to point out the shortcomings of such methods and to suggest how such difficulties can be overcome. By way of illustration we mention that the student's needs are not well served if he or she is led to believe that Cramer's rule is a realistic tool for solving linear systems.

To be more accessible to sophomore students we have adopted a casual writing style wherein only the most important results are identified as theorems. The formal Theorem/Proof format is avoided, but in most cases the essential ideas of a proof are given in the textual exposition itself. Similarly, we do not give formal definitions but, rather, feel free to identify important terminology, using boldface type, in the middle of a discussion. As a further aid to the student each chapter contains a glossary of important terms that have appeared in the chapter.

Each chapter develops an essential concept to a conclusion appropriate to the intended audience, and at least one application is presented. We have not attempted to be comprehensive in our coverage of applications, nor have we tried to develop each application fully. Rather our aim is to *introduce* some typical applications thereby stirring the reader's interest in other applications that arise in various disciplines. Although the coverage of applications is necessarily limited in scope, we feel that enough examples are provided to satisfy the intended readership. Moreover we are convinced, for example, that a computer science student can and does benefit by solving a problem that arises in economics.

The list of exercises following each section forms an integral part of the text. In our opinion little is gained by solving many exercises each of which is a slight variation of a single problem. Therefore we have designed a lengthy and varied set of exercises some of which require the application of matrix techniques to a problem arising outside of mathematics. Each chapter culminates with a set of notes and comments, which refer the reader to literature for further study. At the end of the book we provide answers to selected odd-numbered exercises.

We have used this text at the University of Cincinnati in one-quarter courses aimed at two different audiences. One group consisted primarily of science or engineering students all of whom had completed a full year of calculus with many having been introduced to elementary differential equations. In this course we cover all the mathematical material in Chapters 1–3, with one of the applications in Chapter 1, Sections 4.1–4.3, 5.1–5.3, 6.1–6.3, 7.1, 7.2, and 7.4. A one-semester course to a similar audience could be based on all the first seven chapters, or the material outlined above plus selections from Chapter 8. The other group to whom we have taught a one-quarter course consisted mainly of students of business or the liberal arts. These students typically had a much more modest mathematical background and were more interested in applications to economics and the social sciences. In this course we covered Chapters 1–3, with selected applications, and parts of Sections 4.1–4.2, 6.1–6.3. There is more than enough material in the entire book for use in a two-quarter course.

We would like to thank the following reviewers: Carl C. Cowen, Purdue University; Herbert E. Kasube, Bradley University; John Brillhart, University of Arizona; Larry C. Grove, University of Arizona; Richard T. Bumby, Rutgers University at New Brunswick; Carl D. Meyer, North Carolina State University; Joaquin Bustoz, Arizona State University; Jack R. Porter, University of Kansas; G.W. Stewart, University of Maryland; James E. Hall, Westminster College; William Coppage, Wright State University; John Ewing, Indiana University; and Vincent Giambalvo, University of Connecticut.

C.W.G.
J.T.K.

1

Matrices and Linear Equations

Matrices play a pervasive role in mathematics and its applications. Their uses range from simple tabular representation of numerical data (such data structures are called *arrays* in computer science) to sophisticated applications in physics and economics. However, the central use of matrices is in the study of systems of linear equations.

In this chapter we introduce matrices, study their basic properties, and relate them to systems of linear equations. We also illustrate, by some simple examples, a few diverse uses of matrices in fields outside mathematics.

1.1 Matrices and Linear Equations

Simply stated, a **matrix** is a rectangular array of numbers. These numbers are arranged horizontally in **rows** and vertically in **columns**. If the matrix has m rows and n columns, we say that it is of **size** m by n (denoted $m \times n$). Consider, for example, the 2×3 matrix

$$\mathbf{A} = \begin{bmatrix} -1 & 4 & 0 \\ 1 & 6 & -3 \end{bmatrix}.$$

This matrix is uniquely determined by specifying its **entries**, that is, the numbers that occur in each row and column. A matrix entry is determined by specifying both its numerical value and the row and column in which it appears. The entries of a matrix will be denoted by the corresponding lowercase letter with two subscripts, the first subscript indicating the row of the entry and the second subscript indicating its column. In the example above we have

$$\begin{array}{lll} a_{11} = -1 & a_{12} = 4 & a_{13} = 0 \\ a_{21} = 1 & a_{22} = 6 & a_{23} = -3. \end{array}$$

In general, an $m \times n$ matrix has the form

$$\mathbf{A} = \begin{bmatrix} a_{11} & a_{12} & \cdots & a_{1n} \\ a_{21} & a_{22} & \cdots & a_{2n} \\ \vdots & \vdots & & \vdots \\ a_{m1} & a_{m2} & \cdots & a_{mn} \end{bmatrix}.$$

Matrices that have only a single column are called column **vectors** and a column vector having n entries is called an n-vector. For example,

$$\begin{bmatrix} 1 \\ 0 \end{bmatrix}, \quad \begin{bmatrix} 0 \\ 1 \\ 3 \end{bmatrix}, \quad \text{and} \quad \begin{bmatrix} 1 \\ -1 \\ 0 \\ 0 \end{bmatrix}$$

are all column vectors; the first is a 2-vector, the second a 3-vector, and the third a 4-vector. We will usually designate a column vector by a lowercase letter; for example, the general form of an m-vector $\mathbf{b}$ is

$$\mathbf{b} = \begin{bmatrix} b_1 \\ b_2 \\ \vdots \\ b_m \end{bmatrix}.$$

The number b_i is commonly called the ith **component** of the column vector $\mathbf{b}$. A special notation will be used when referring to the column vectors of a matrix. The jth column vector of a given matrix $\mathbf{A}$ will be denoted $\mathbf{A}_j$. In the previous example we have

$$\mathbf{A}_1 = \begin{bmatrix} -1 \\ 1 \end{bmatrix}, \quad \mathbf{A}_2 = \begin{bmatrix} 4 \\ 6 \end{bmatrix}, \quad \text{and} \quad \mathbf{A}_3 = \begin{bmatrix} 0 \\ -3 \end{bmatrix}.$$

Later we will find it convenient to refer to column vectors simply as "vectors." If we wish to specify the size of the vector, we use the term m-vector to indicate a column vector having m entries.

There is a corresponding notion of row vectors. A row vector having m entries takes the form

$$\mathbf{b} = [b_1 \quad b_2 \quad \cdots \quad b_m].$$

We will not introduce a special notation for row vectors since we will not use them frequently.

The great importance of matrices resides in the fact that they provide a powerful and very compact notation for systems of linear equations. By an "$m \times n$ system of linear equations" we mean a system of equations of the form

$$\begin{aligned} a_{11}x_1 + a_{12}x_2 + \cdots + a_{1n}x_n &= b_1 \\ a_{21}x_1 + a_{22}x_2 + \cdots + a_{2n}x_n &= b_2 \\ \vdots \qquad\qquad\qquad\quad \vdots \quad & \quad \vdots \\ a_{m1}x_1 + a_{m2}x_2 + \cdots + a_{mn}x_n &= b_m, \end{aligned} \tag{1}$$

where the $a_{ij}, 1 \le i \le m$, $1 \le j \le n$, are given numbers, called the *coefficients* of the system; the b_i, $1 \le i \le m$ are given numbers, called the *right-hand sides*; and x_j, $1 \le j \le n$ are the *unknowns*. To write (1) more economically, we collect the coefficients into a coefficient matrix,

$$\mathbf{A} = \begin{bmatrix} a_{11} & a_{12} & \cdots & a_{1n} \\ a_{21} & a_{22} & \cdots & a_{2n} \\ \vdots & \vdots & & \vdots \\ a_{m1} & a_{m2} & \cdots & a_{mn} \end{bmatrix}$$

and similarly we collect the unknowns and right-hand sides into vectors,

$$\mathbf{x} = \begin{bmatrix} x_1 \\ x_2 \\ \vdots \\ x_n \end{bmatrix} \quad \text{and} \quad \mathbf{b} = \begin{bmatrix} b_1 \\ b_2 \\ \vdots \\ b_m \end{bmatrix}.$$

In Section 1.2 (see p. 10) we define matrix multiplication, which will enable us to write (1) as follows:

$$\begin{bmatrix} a_{11} & a_{12} & \cdots & a_{1n} \\ a_{21} & a_{22} & \cdots & a_{2n} \\ \vdots & \vdots & & \vdots \\ a_{m1} & a_{m2} & \cdots & a_{mn} \end{bmatrix} \begin{bmatrix} x_1 \\ x_2 \\ \vdots \\ x_n \end{bmatrix} = \begin{bmatrix} b_1 \\ b_2 \\ \vdots \\ b_m \end{bmatrix} \tag{2}$$

or even more compactly as

$$\mathbf{Ax} = \mathbf{b}.$$

We will have much more to say about solutions of linear systems in later chapters. However, at this stage we would like to point out that not all linear systems have unique solutions. A given $m \times n$ linear system is said to be **overdetermined** if $m > n$, for in this case more conditions (i.e., m equations) are imposed than there are unknowns (n). In such a case we are ordinarily asking too much of the unknowns and should generally expect that no solution exists. Consider, for example, the simple 2×1 overdetermined system

$$\begin{aligned} x_1 &= 1 \\ 2x_1 &= 22 \end{aligned}$$

which obviously has no solution.

An $m \times n$ linear system is called **underdetermined** if $m < n$. In this case we are demanding relatively little of the system since there are fewer conditions than unknowns and hence we should naturally expect many solutions. For example, the 1×2 underdetermined system

$$x_1 + x_2 = 1$$

has solutions

$$\begin{bmatrix} 1 \\ 0 \end{bmatrix}, \quad \begin{bmatrix} 0 \\ 1 \end{bmatrix}, \quad \begin{bmatrix} -1 \\ 2 \end{bmatrix}.$$

Indeed, for any number t, a vector of the form

$$\begin{bmatrix} t \\ 1 - t \end{bmatrix}$$

is a solution. A system of equations that has a solution (one or many) is called a **consistent** system.

Overdetermined systems typically occur in situations where there is an overabundance of empirical data, while underdetermined systems result from insufficient data. Overdetermined systems occur more frequently than underdetermined systems and they are dealt with further in Chapter 8.

The upshot of the previous discussion is that if we expect to have a unique solution for each right-hand side, the coefficient matrix should be **square**, that is, it should have the same number of rows as columns. We investigate the theory of linear systems, including nonsquare systems, in much greater detail in Chapter 4.

EXERCISES 1.1

1. Write the following system in matrix form.

$$\begin{aligned} -x_1 \qquad\quad + 7x_3 &= 1 \\ 2x_1 - 3x_2 + x_3 &= 7 \\ x_2 - 2x_3 &= -4 \end{aligned}$$

2. One number is 3 greater than twice another number. The sum of the two numbers is 7. Write this problem as a linear system in the unknown numbers. Convert this system into matrix form.

3. Verify by direct substitution that

$$\begin{bmatrix} x_1 \\ x_2 \end{bmatrix} = \begin{bmatrix} -1 \\ 1 \end{bmatrix}$$

is a solution of the linear system

$$\begin{aligned} -2x_1 + 3x_2 &= 5 \\ 7x_2 &= 7 \\ 2x_1 + 6x_2 &= 4. \end{aligned}$$

4. Suppose that

$$\mathbf{A} = \begin{bmatrix} 1 & s & 3 \\ 2t & -1 & 1 \\ 3 & s & 2 \end{bmatrix}$$

and that $a_{ij} = a_{ji}$, $1 \le i, j \le 3$ (i.e., for all indices i and j between 1 and 3). Find s and t.

5. Solve the linear system

$$\begin{aligned} 3x_1 \qquad\quad &= 6 \\ -x_1 + 2x_2 &= -7. \end{aligned}$$

(*Hint:* Start with the first equation.)

6. Solve

$$\begin{aligned} 2x_1 + x_2 &= 5 \\ 3x_2 &= 9. \end{aligned}$$

7. Solve

$$\begin{aligned} x_1 - x_2 + 2x_3 &= 5 \\ 2x_2 - x_3 + x_4 &= 5 \\ -2x_3 - x_4 &= -10 \\ 2x_4 &= 8. \end{aligned}$$

8. What is the common feature of the systems in Exercises 5–7 which made solving them straightforward?

9. Show that the system

$$\begin{aligned} 4x_1 - 2x_2 &= 3 \\ 2x_1 + 3x_2 &= 5 \\ x_2 &= 1 \end{aligned}$$

is inconsistent, that is, it has no solution.

10. Is the system

$$\begin{aligned} x_1 + 2x_2 - x_3 + 4x_4 &= 6 \\ x_3 - 2x_4 &= 1 \\ -2x_3 + 6x_4 &= 4 \\ 3x_4 &= 3 \end{aligned}$$

consistent? That is, does it have a solution?

11. Show that for any number t the vector

$$\begin{bmatrix} -t \\ t \\ 1 - t \end{bmatrix}$$

is a solution of the linear system

$$\begin{aligned} x_1 + x_2 &= 0 \\ x_2 + x_3 &= 1. \end{aligned}$$

12. **(a)** Verify that

$$\begin{bmatrix} 1 \\ 1 \\ 1 \end{bmatrix} \quad \text{and} \quad \begin{bmatrix} \frac{10}{3} \\ \frac{1}{3} \\ 0 \end{bmatrix}$$

are both solutions of the system

$$\begin{aligned} x_1 + 2x_2 + x_3 &= 4 \\ -x_1 + x_2 - 3x_3 &= -3 \\ x_1 + 5x_2 - x_3 &= 5. \end{aligned}$$

(b) Let

$$\begin{bmatrix} x_1 \\ x_2 \\ x_3 \end{bmatrix} = \begin{bmatrix} 1 - \frac{10}{3} \\ 1 - \frac{1}{3} \\ 1 - 0 \end{bmatrix} = \begin{bmatrix} -\frac{7}{3} \\ \frac{2}{3} \\ 1 \end{bmatrix}.$$

Fill in the right-hand side of the equations

$$\begin{aligned} x_1 + 2x_2 + x_3 &= ? \\ -x_1 + x_2 - 3x_3 &= ? \\ x_1 + 5x_2 - x_3 &= ? \end{aligned}$$

and compare with part (a).

13. The vector

$$\begin{bmatrix} 2 \\ -3 \\ 3 \end{bmatrix}$$

is a solution of the system

$$\begin{aligned} 3x_1 + 2x_2 \qquad\quad &= 0 \\ 6x_1 + x_2 - 3x_3 &= 0 \\ -x_2 - x_3 &= 0 \end{aligned}$$

(Check it!) Verify that

$$\begin{bmatrix} 2t \\ -3t \\ 3t \end{bmatrix}$$

is also a solution for any number t.

14. The vector

$$\begin{bmatrix} 1 \\ 1 \end{bmatrix}$$

is a solution of the system

$$\begin{aligned} 2x_1 + x_2 &= 3 \\ x_1 + 3x_2 &= 4 \end{aligned}$$

and the vector

$$\begin{bmatrix} 2 \\ 3 \end{bmatrix}$$

is a solution of the system

$$\begin{aligned} 2x_1 + x_2 &= 7 \\ x_1 + 3x_2 &= 11 \end{aligned}$$

which has the same coefficient matrix but different right-hand sides. Verify that

$$\begin{bmatrix} 3 \\ 4 \end{bmatrix} = \begin{bmatrix} 1 + 2 \\ 1 + 3 \end{bmatrix}$$

is a solution of the system

$$\begin{aligned} 2x_1 + x_2 &= 10 \\ x_1 + 3x_2 &= 15 \end{aligned}$$

which has the same coefficient matrix but whose right-hand sides are the sums of the right-hand sides of the previous two systems.

1.2 Matrix Operations

In this section we show that with a couple of notable exceptions, the usual arithmetic operations for numbers can be extended to matrices in a natural way. First we point out the meaning of equality for matrices. Two matrices **A** and **B** are equal if they have the same number of rows and columns, respectively, and have identical entries, that is, $a_{ij} = b_{ij}$ for each i and j.

The simplest arithmetic operation is addition. If **A** and **B** are two $m \times n$ matrices, then their *sum* $\mathbf{A} + \mathbf{B}$ is defined to be the $m \times n$ matrix **C** which satisfies

$$c_{ij} = a_{ij} + b_{ij}, \qquad 1 \le i \le m, \quad 1 \le j \le n.$$

That is, **C** is the matrix that results when the corresponding entries of **A** and **B** are added.

Example 1.1

$$\begin{bmatrix} 2 & 1 \\ 0 & -1 \\ 7 & 6 \end{bmatrix} + \begin{bmatrix} 3 & 4 \\ -2 & 2 \\ -3 & 3 \end{bmatrix} = \begin{bmatrix} 2+3 & 1+4 \\ 0-2 & -1+2 \\ 7-3 & 6+3 \end{bmatrix} = \begin{bmatrix} 5 & 5 \\ -2 & 1 \\ 4 & 9 \end{bmatrix}$$

■

Example 1.2

Let

$$\mathbf{A} = \begin{bmatrix} 1 & 2 \\ -2 & 3 \end{bmatrix}, \qquad \mathbf{B} = \begin{bmatrix} 4 & -1 \\ 0 & 2 \end{bmatrix}.$$

Then

$$\mathbf{A} + \mathbf{B} = \begin{bmatrix} 1+4 & 2-1 \\ -2+0 & 3+2 \end{bmatrix} = \begin{bmatrix} 5 & 1 \\ -2 & 5 \end{bmatrix}$$

and

$$\mathbf{B} + \mathbf{A} = \begin{bmatrix} 4+1 & -1+2 \\ 0-2 & 2+3 \end{bmatrix} = \begin{bmatrix} 5 & 1 \\ -2 & 5 \end{bmatrix}.$$

■

It should be clear from this example that in general $\mathbf{A} + \mathbf{B} = \mathbf{B} + \mathbf{A}$ if **A** and **B** are $m \times n$ matrices. Similarly, since $a + (b + c) = (a + b) + c$ if a, b, and c are numbers, it follows that, in general, $\mathbf{A} + (\mathbf{B} + \mathbf{C}) = (\mathbf{A} + \mathbf{B}) + \mathbf{C}$ for $m \times n$ matrices **A**, **B**, and **C**.

In a similar way we can define the *difference* of matrices, that is, $\mathbf{C} = \mathbf{A} - \mathbf{B}$, if

$$c_{ij} = a_{ij} - b_{ij}.$$

The negative of matrix **A**, denoted $-\mathbf{A}$, is the matrix whose i,j-entry is $-a_{ij}$. The $m \times n$ matrix all of whose entries are zero is called the $m \times n$ **zero matrix**. This matrix will be denoted by **O**, regardless of its size. Note that

$$\mathbf{A} + \mathbf{O} = \mathbf{A} = \mathbf{O} + \mathbf{A}$$

for any $m \times n$ matrix **A**. A **zero vector** is denoted by $\boldsymbol{\theta}$, regardless of its size.

We now define the notion of scalar multiplication. In matrix algebra it is traditional to use the term "scalar" to refer to a number. Given a scalar α and an $m \times n$ matrix **A**, we define the multiplication of matrix **A** by α in a natural way: $\mathbf{B} = \alpha\mathbf{A}$ if

$$b_{ij} = \alpha a_{ij} \qquad \text{for } 1 \le i \le m \text{ and } 1 \le j \le n.$$

That is, $\mathbf{B} = \alpha\mathbf{A}$ results when each entry of **A** is multiplied by α.

Example 1.3

$$3\begin{bmatrix} 1 & 2 & 6 \\ 0 & -1 & 4 \end{bmatrix} = \begin{bmatrix} 3 & 6 & 18 \\ 0 & -3 & 12 \end{bmatrix}$$

■

Example 1.4

Let

$$\mathbf{A} = \begin{bmatrix} 2 & 1 \\ 0 & -1 \end{bmatrix}, \qquad \mathbf{B} = \begin{bmatrix} 1 & 3 \\ 1 & 1 \end{bmatrix}.$$

Then

$$5\mathbf{A} = \begin{bmatrix} 10 & 5 \\ 0 & -5 \end{bmatrix}, \qquad 5\mathbf{B} = \begin{bmatrix} 5 & 15 \\ 5 & 5 \end{bmatrix}$$

and hence

$$5\mathbf{A} + 5\mathbf{B} = \begin{bmatrix} 10 & 5 \\ 0 & -5 \end{bmatrix} + \begin{bmatrix} 5 & 15 \\ 5 & 5 \end{bmatrix} = \begin{bmatrix} 15 & 20 \\ 5 & 0 \end{bmatrix}.$$

Also,

$$\mathbf{A} + \mathbf{B} = \begin{bmatrix} 3 & 4 \\ 1 & 0 \end{bmatrix},$$

and hence

$$5(\mathbf{A} + \mathbf{B}) = 5\begin{bmatrix} 3 & 4 \\ 1 & 0 \end{bmatrix} = \begin{bmatrix} 15 & 20 \\ 5 & 0 \end{bmatrix},$$

that is, $5(\mathbf{A} + \mathbf{B}) = 5\mathbf{A} + 5\mathbf{B}$.

■

Examples 1.1 through 1.4 and familiar properties of numbers lead us to state the following theorem, in which some simple and readily verified properties of matrix addition and scalar multiplication are listed.

Theorem 1.1. If **A**, **B**, and **C** are $m \times n$ matrices and α and β are scalars, then

(a) $\mathbf{A} + \mathbf{B} = \mathbf{B} + \mathbf{A}$
(b) $\mathbf{A} + (\mathbf{B} + \mathbf{C}) = (\mathbf{A} + \mathbf{B}) + \mathbf{C}$
(c) $\alpha(\beta\mathbf{A}) = (\alpha\beta)\mathbf{A}$
(d) $(\alpha + \beta)\mathbf{A} = \alpha\mathbf{A} + \beta\mathbf{A}$
(e) $\alpha(\mathbf{A} + \mathbf{B}) = \alpha\mathbf{A} + \alpha\mathbf{B}$
(f) $(-1)\mathbf{A} = -\mathbf{A}$
(g) $-(-\mathbf{A}) = \mathbf{A}$
(h) $0\mathbf{A} = \mathbf{O}$
(i) $\alpha\mathbf{O} = \mathbf{O}$

Simply put, this theorem says that *as far as addition, subtraction, and multiplication by scalars are concerned, the arithmetic of matrices is exactly like the arithmetic of numbers.*

The situation for matrix products is, as we shall see, strikingly different. The product of two matrices is not defined as the matrix obtained by multiplying the respective entries. Rather, the matrix product is defined by the following rule, which at first sight seems somewhat strange:

If **A** is an $m \times n$ matrix and **B** is an $n \times p$ matrix, then **AB** is defined to be the $m \times p$ matrix **C** satisfying

$$c_{ij} = \sum_{k=1}^{n} a_{ik}b_{kj}, \qquad 1 \le i \le m, \quad 1 \le j \le p.$$

Notice that in order for **C** to be defined, it is necessary for **A** to have the same number of columns as **B** has rows.

Example 1.5

Suppose that

$$\mathbf{A} = \begin{bmatrix} 1 & -1 & 2 \\ 0 & 4 & 3 \end{bmatrix} \quad \text{and} \quad \mathbf{B} = \begin{bmatrix} 0 & 6 \\ 1 & -1 \\ 2 & 0 \end{bmatrix},$$

then $\mathbf{C} = \mathbf{AB}$ is computed as follows:

$$\begin{bmatrix} 1 & -1 & 2 \\ 0 & 4 & 3 \end{bmatrix}\begin{bmatrix} 0 & 6 \\ 1 & -1 \\ 2 & 0 \end{bmatrix} = \begin{bmatrix} 1\cdot 0 + (-1)\cdot 1 + 2\cdot 2 & 1\cdot 6 + (-1)\cdot(-1) + 2\cdot 0 \\ 0\cdot 0 + \quad 4\cdot 1 + 3\cdot 2 & 0\cdot 6 + \quad 4\cdot(-1) + 3\cdot 0 \end{bmatrix}$$

$$= \begin{bmatrix} 3 & 7 \\ 10 & -4 \end{bmatrix} = \mathbf{C}.$$

■

Note that the i,j-entry of the matrix product **AB** is found by summing the products of the entries in the ith row of **A** with the corresponding entries in the jth column of **B**. For hand computation, the student may find the following schematic illustration of the process useful:

$$\begin{bmatrix} b_{11} & \cdots & b_{1j} & \cdots & b_{1p} \\ & & b_{2j} & & \\ \vdots & & \vdots & & \vdots \\ b_{n1} & \cdots & b_{nj} & & b_{np} \end{bmatrix}$$

$$\begin{bmatrix} a_{11} & \cdots & a_{1n} \\ \vdots & & \vdots \\ a_{i1} & a_{i2} \cdots & a_{in} \\ \vdots & & \vdots \\ a_{m1} & \cdots & a_{mn} \end{bmatrix} \begin{bmatrix} c_{11} & \cdots & c_{1j} & \cdots & c_{1p} \\ & & \vdots & & \\ \vdots & & c_{ij} & & \vdots \\ & & \vdots & & \\ c_{m1} & \cdots & c_{mj} & & c_{mp} \end{bmatrix}$$

To motivate the matrix product, we consider the following.

Example 1.6

An electronics manufacturer produces two types of circuit boards, whose major components are diodes, transistors, and resistors. Let y_1, y_2, and y_3 denote the unit cost of diodes, transistors, and resistors, respectively. Then the production cost of the circuit boards, where x_i = cost of circuit board i ($i = 1, 2$), is given by

$$x_1 = b_{11}y_1 + b_{12}y_2 + b_{13}y_3$$
$$x_2 = b_{21}y_1 + b_{22}y_2 + b_{23}y_3,$$

where b_{ij} gives the number of components j used to produce circuit board i. For example, b_{13} gives the number of resistors needed to build a circuit board of type 1. We associate with the system above the component cost vector **y**, the circuit board cost vector **x**, and the quantity matrix **B**, that is,

$$\mathbf{y} = \begin{bmatrix} y_1 \\ y_2 \\ y_3 \end{bmatrix}, \quad \mathbf{x} = \begin{bmatrix} x_1 \\ x_2 \end{bmatrix}, \quad \mathbf{B} = \begin{bmatrix} b_{11} & b_{12} & b_{13} \\ b_{21} & b_{22} & b_{23} \end{bmatrix}.$$

Then

$$\begin{bmatrix} x_1 \\ x_2 \end{bmatrix} = \begin{bmatrix} b_{11}y_1 + b_{12}y_2 + b_{13}y_3 \\ b_{21}y_1 + b_{22}y_2 + b_{23}y_3 \end{bmatrix}. \tag{3}$$

Given **B** and **y**, **x** is determined by (3). Moreover, (3) suggests how to define the product of vector **y** by matrix **B**, that is, **By**, in order that **By** = **x**. According to (3), **By** should be the 2-vector whose first and second components are

$$b_{11}y_1 + b_{12}y_2 + b_{13}y_3 = \sum_{k=1}^{3} b_{1k}y_k$$

and

$$b_{21}y_1 + b_{22}y_2 + b_{23}y_3 = \sum_{k=1}^{3} b_{2k}y_k,$$

respectively.

If **B** is an $m \times n$ matrix and **y** is an n-vector, we define **By** to be the m-vector **x** given by

$$x_i = b_{i1}y_1 + b_{i2}y_2 + \cdots + b_{in}y_n, \qquad 1 \le i \le m,$$

or in more compact notation,

$$x_i = \sum_{k=1}^{n} b_{ik}y_k, \qquad 1 \le i \le m.$$

Since **y** is an $n \times 1$ matrix we see that this is just a restatement of the matrix product $\mathbf{x} = \mathbf{By}$.

Example 1.7

For

$$\mathbf{B} = \begin{bmatrix} 1 & -1 & 2 \\ 3 & 0 & 1 \end{bmatrix}, \qquad \mathbf{y} = \begin{bmatrix} 1 \\ 2 \\ 3 \end{bmatrix},$$

we have

$$\mathbf{By} = \begin{bmatrix} 1 \cdot 1 + (-1) \cdot 2 + 2 \cdot 3 \\ 3 \cdot 1 + \quad 0 \cdot 2 + 1 \cdot 3 \end{bmatrix} = \begin{bmatrix} 5 \\ 6 \end{bmatrix}.$$

■

Note that in order that **By** be defined, **y** must have the same number of components (rows) as **B** has columns. Thus if **B** is an $m \times n$ matrix and **y** is an r-vector, then **By** is defined only for $r = n$.

Example 1.8

We continue the problem of Example 1.6. The manufacturer uses the circuit boards in the production of stereo sets, of which there are three models: standard, superior, and deluxe. Let z_1, z_2, and z_3 denote the costs of producing the standard, superior, and deluxe models, respectively. Then

$$\begin{aligned} z_1 &= a_{11}x_1 + a_{12}x_2 \\ z_2 &= a_{21}x_1 + a_{22}x_2 \\ z_3 &= a_{31}x_1 + a_{32}x_2, \end{aligned} \qquad (4)$$

where a_{ij} gives the number of circuit boards of type j used to produce stereo model i; for example, the deluxe stereo (model 3) requires a_{32} circuit boards of type 2. It then follows that (4) can be written as

$$\mathbf{z} = \mathbf{Ax},$$

where

$$\mathbf{z} = \begin{bmatrix} z_1 \\ z_2 \\ z_3 \end{bmatrix} \quad \text{and} \quad \mathbf{A} = \begin{bmatrix} a_{11} & a_{12} \\ a_{21} & a_{22} \\ a_{31} & a_{32} \end{bmatrix}.$$

The cost of the stereo sets depends on the cost of the circuit boards, which in turn depends on the cost of the components. We now show that there is a matrix **C** such that

$$\begin{aligned} z_1 &= c_{11}y_1 + c_{12}y_2 + c_{13}y_3 \\ z_2 &= c_{21}y_1 + c_{22}y_2 + c_{23}y_3 \\ z_3 &= c_{31}y_1 + c_{32}y_2 + c_{33}y_3, \end{aligned} \tag{5}$$

where c_{ij} = the number of components j needed to produce stereo model i. It should be clear that the entries of **C** are determined by those of **A** and **B**. To relate **A**, **B**, and **C** we substitute equations (3) into (4). For the standard model we substitute (3) into the first equation of (4) to get

$$z_1 = a_{11}(b_{11}y_1 + b_{12}y_2 + b_{13}y_3) + a_{12}(b_{21}y_1 + b_{22}y_2 + b_{23}y_3),$$

or equivalently,

$$z_1 = (a_{11}b_{11} + a_{12}b_{21})y_1 + (a_{11}b_{12} + a_{12}b_{22})y_2 + (a_{11}b_{13} + a_{12}b_{23})y_3.$$

Similarly, for the superior and deluxe models, we have

$$\begin{aligned} z_2 &= (a_{21}b_{11} + a_{22}b_{21})y_1 + (a_{21}b_{12} + a_{22}b_{22})y_2 + (a_{21}b_{13} + a_{22}b_{23})y_3 \\ z_3 &= (a_{31}b_{11} + a_{32}b_{21})y_1 + (a_{31}b_{12} + a_{32}b_{22})y_2 + (a_{31}b_{13} + a_{32}b_{23})y_3. \end{aligned}$$

By comparing the last three equations with (5) we obtain the required relationship among **A**, **B**, and **C**. For example,

$$\begin{aligned} c_{12} &= a_{11}b_{12} + a_{12}b_{22} \\ c_{32} &= a_{31}b_{12} + a_{32}b_{22}, \end{aligned}$$

and in general

$$c_{ij} = a_{i1}b_{1j} + a_{i2}b_{2j}, \qquad 1 \le i, j \le 3. \tag{6}$$

But we recall that

$$\mathbf{By} = \mathbf{x} \quad \text{and} \quad \mathbf{z} = \mathbf{Ax},$$

and hence we require a definition of matrix multiplication which gives

$$\mathbf{z} = \mathbf{Ax} = \mathbf{A(By)} = \mathbf{Cy}$$

with **C** = **AB**. That is, we need to define the matrix product **AB** in order that **C** = **AB** is given by (6).

■

Note that if **A** is an $m \times n$ matrix and **x** is an $n \times 1$ matrix (i.e., an n-vector), then

$$\mathbf{Ax} = \begin{bmatrix} a_{11} & a_{12} & \cdots & a_{1n} \\ a_{21} & a_{22} & \cdots & a_{2n} \\ \vdots & \vdots & & \vdots \\ a_{m1} & a_{m2} & \cdots & a_{mn} \end{bmatrix} \begin{bmatrix} x_1 \\ x_2 \\ \vdots \\ x_n \end{bmatrix} = \begin{bmatrix} a_{11}x_1 + a_{12}x_2 + \cdots + a_{1n}x_n \\ a_{21}x_1 + a_{22}x_2 + \cdots + a_{2n}x_n \\ \vdots \\ a_{m1}x_1 + a_{m2}x_2 + \cdots + a_{mn}x_n \end{bmatrix},$$

which shows that our notion of matrix product agrees with the matrix representation of linear systems given in (1) and (2).

Example 1.9

Let

$$\mathbf{A} = \begin{bmatrix} -2 & 1 & 4 \\ 3 & 0 & 1 \\ 2 & 4 & 1 \end{bmatrix} \quad \text{and} \quad \mathbf{x} = \begin{bmatrix} 1 \\ 2 \\ 1 \end{bmatrix}.$$

Then

$$\mathbf{Ax} = \begin{bmatrix} -2 & 1 & 4 \\ 3 & 0 & 1 \\ 2 & 4 & 1 \end{bmatrix} \begin{bmatrix} 1 \\ 2 \\ 1 \end{bmatrix} = \begin{bmatrix} 4 \\ 4 \\ 11 \end{bmatrix},$$

that is, **x** is a solution of the linear system

$$\begin{aligned} -2x_1 + x_2 + 4x_3 &= 4 \\ 3x_1 + x_3 &= 4 \\ 2x_1 + 4x_2 + x_3 &= 11. \end{aligned}$$

■

The next example shows that even for square matrices, the matrix product has some unexpected properties.

Example 1.10

Suppose that

$$\mathbf{A} = \begin{bmatrix} 1 & 1 \\ 1 & 1 \end{bmatrix} \quad \text{and} \quad \mathbf{B} = \begin{bmatrix} 1 & -1 \\ 2 & -2 \end{bmatrix}.$$

Then

$$\mathbf{AB} = \begin{bmatrix} 1 & 1 \\ 1 & 1 \end{bmatrix} \begin{bmatrix} 1 & -1 \\ 2 & -2 \end{bmatrix} = \begin{bmatrix} 3 & -3 \\ 3 & -3 \end{bmatrix}$$

but

$$\mathbf{BA} = \begin{bmatrix} 1 & -1 \\ 2 & -2 \end{bmatrix} \begin{bmatrix} 1 & 1 \\ 1 & 1 \end{bmatrix} = \begin{bmatrix} 0 & 0 \\ 0 & 0 \end{bmatrix}.$$

■

In Example 1.10, $\mathbf{AB} \neq \mathbf{BA}$, which shows that *matrix multiplication is not a commutative operation*. Moreover, in this example $\mathbf{BA} = \mathbf{O}$, while neither $\mathbf{A}$ nor $\mathbf{B}$ is the zero matrix. One serious consequence of this fact is the failure of the cancellation law for matrix multiplication. If α, β, and γ are numbers and $\alpha \neq 0$, then $\alpha\beta = \alpha\gamma$ implies that $\beta = \gamma$. Indeed, since $\alpha\beta = \alpha\gamma$, we have $\alpha(\beta - \gamma) = 0$ and hence $\beta - \gamma = 0$, since $\alpha \neq 0$. This is the cancellation law for numbers. In the example above we have $\mathbf{BA} = \mathbf{O} = \mathbf{BO}$ and $\mathbf{B} \neq \mathbf{O}$, yet $\mathbf{A} \neq \mathbf{O}$: The cancellation law fails for matrices.

Fortunately, not all properties of numerical multiplication fail for matrices, as the following theorem shows.

Theorem 1.2. Suppose that $\mathbf{A}$, $\mathbf{B}$, and $\mathbf{C}$ are matrices for which the following products are defined and α is a scalar. Then:

(a) $(\mathbf{AB})\mathbf{C} = \mathbf{A}(\mathbf{BC})$.
(b) $\mathbf{A}(\alpha\mathbf{B}) = (\alpha\mathbf{A})\mathbf{B} = \alpha(\mathbf{AB})$.
(c) $\mathbf{A}(\mathbf{B} + \mathbf{C}) = \mathbf{AB} + \mathbf{AC}$.
(d) $(\mathbf{A} + \mathbf{B})\mathbf{C} = \mathbf{AC} + \mathbf{BC}$.

■

We illustrate parts (a) and (d) in the next examples.

Example 1.11

Suppose that

$$\mathbf{A} = \begin{bmatrix} 1 & -1 & 3 \\ 2 & 0 & 4 \end{bmatrix}, \quad \mathbf{B} = \begin{bmatrix} 1 & 2 \\ -1 & 1 \\ 3 & 0 \end{bmatrix}, \quad \text{and} \quad \mathbf{C} = \begin{bmatrix} 4 & -3 \\ 1 & 5 \end{bmatrix}.$$

Then

$$\mathbf{AB} = \begin{bmatrix} 1 & -1 & 3 \\ 2 & 0 & 4 \end{bmatrix} \begin{bmatrix} 1 & 2 \\ -1 & 1 \\ 3 & 0 \end{bmatrix} = \begin{bmatrix} 11 & 1 \\ 14 & 4 \end{bmatrix}$$

$$\mathbf{BC} = \begin{bmatrix} 1 & 2 \\ -1 & 1 \\ 3 & 0 \end{bmatrix} \begin{bmatrix} 4 & -3 \\ 1 & 5 \end{bmatrix} = \begin{bmatrix} 6 & 7 \\ -3 & 8 \\ 12 & -9 \end{bmatrix}$$

and thus

$$(\mathbf{AB})\mathbf{C} = \begin{bmatrix} 11 & 1 \\ 14 & 4 \end{bmatrix} \begin{bmatrix} 4 & -3 \\ 1 & 5 \end{bmatrix} = \begin{bmatrix} 45 & -28 \\ 60 & -22 \end{bmatrix}$$

$$\mathbf{A}(\mathbf{BC}) = \begin{bmatrix} 1 & -1 & 3 \\ 2 & 0 & 4 \end{bmatrix} \begin{bmatrix} 6 & 7 \\ -3 & 8 \\ 12 & -9 \end{bmatrix} = \begin{bmatrix} 45 & -28 \\ 60 & -22 \end{bmatrix}.$$

■

Example 1.12

Suppose that

$$\mathbf{A} = \begin{bmatrix} 5 & 0 \\ 2 & -1 \\ 6 & 3 \end{bmatrix}, \quad \mathbf{B} = \begin{bmatrix} 1 & -2 \\ 4 & 0 \\ 3 & 3 \end{bmatrix}, \quad \text{and} \quad \mathbf{C} = \begin{bmatrix} 2 & 1 \\ 1 & 3 \end{bmatrix}.$$

Then

$$\mathbf{A} + \mathbf{B} = \begin{bmatrix} 6 & -2 \\ 6 & -1 \\ 9 & 6 \end{bmatrix},$$

$$\mathbf{AC} = \begin{bmatrix} 5 & 0 \\ 2 & -1 \\ 6 & 3 \end{bmatrix}\begin{bmatrix} 2 & 1 \\ 1 & 3 \end{bmatrix} = \begin{bmatrix} 10 & 5 \\ 3 & -1 \\ 15 & 15 \end{bmatrix},$$

and

$$\mathbf{BC} = \begin{bmatrix} 1 & -2 \\ 4 & 0 \\ 3 & 3 \end{bmatrix}\begin{bmatrix} 2 & 1 \\ 1 & 3 \end{bmatrix} = \begin{bmatrix} 0 & -5 \\ 8 & 4 \\ 9 & 12 \end{bmatrix}.$$

Therefore,

$$(\mathbf{A} + \mathbf{B})\mathbf{C} = \begin{bmatrix} 6 & -2 \\ 6 & -1 \\ 9 & 6 \end{bmatrix}\begin{bmatrix} 2 & 1 \\ 1 & 3 \end{bmatrix} = \begin{bmatrix} 10 & 0 \\ 11 & 3 \\ 24 & 27 \end{bmatrix}$$

and

$$\mathbf{AC} + \mathbf{BC} = \begin{bmatrix} 10 & 0 \\ 11 & 3 \\ 24 & 27 \end{bmatrix}.$$

■

EXERCISES 1.2

1. Find numbers w, x, y, and z such that

$$\begin{bmatrix} x + 2 & 2y - 6 \\ z - 1 & 2w - 1 \end{bmatrix} = \begin{bmatrix} 0 & 2 \\ 4 & w + 1 \end{bmatrix}.$$

2. Find $\mathbf{A} + \mathbf{B}$ if

$$\mathbf{A} = \begin{bmatrix} 1 & -2 \\ 3 & 7 \end{bmatrix} \quad \text{and} \quad \mathbf{B} = \begin{bmatrix} 6 & 0 \\ -1 & 4 \end{bmatrix}.$$

3. Compute

$$\begin{bmatrix} 1 & 2 & -1 \\ 4 & 1 & 6 \\ 3 & -1 & 2 \end{bmatrix} - \begin{bmatrix} 1 & 0 & 1 \\ 0 & 1 & 0 \\ -6 & 1 & 4 \end{bmatrix} + \begin{bmatrix} 2 & -1 & 5 \\ 0 & 0 & 1 \\ 7 & 4 & 2 \end{bmatrix}.$$

4. Solve the equation

$$\mathbf{X} + \begin{bmatrix} 1 & 2 & 0 \\ 4 & 3 & -1 \\ 0 & 0 & 1 \end{bmatrix} = \begin{bmatrix} 7 & 4 & 1 \\ 0 & 2 & 1 \\ 3 & 6 & 4 \end{bmatrix}$$

for the matrix **X**.

5. Solve for **X**:

$$\begin{bmatrix} 1 & 10 \\ 7 & 6 \end{bmatrix} - \mathbf{X} = \begin{bmatrix} 2 & 1 \\ 4 & 8 \end{bmatrix} - \begin{bmatrix} 3 & 5 \\ 1 & 0 \end{bmatrix}.$$

6. Compute

$$2\begin{bmatrix} 1 & 3 \\ 0 & 1 \end{bmatrix} - 7\begin{bmatrix} 1 & 0 \\ 0 & 1 \end{bmatrix} + 4\begin{bmatrix} 1 & -1 \\ 1 & 1 \end{bmatrix}.$$

7. Suppose that

$$\mathbf{A} = \begin{bmatrix} 2 & -3 \\ 4 & 0 \end{bmatrix}, \quad \mathbf{B} = \begin{bmatrix} 1 & 0 \\ 3 & 1 \end{bmatrix}, \quad \mathbf{C} = \begin{bmatrix} 4 & 2 \\ 1 & 0 \end{bmatrix}.$$

Compute
(a) $4\mathbf{A} - 3\mathbf{B} - 2\mathbf{C}$
(b) $3\mathbf{A} - 2(\mathbf{B} - \mathbf{C})$
(c) $3\mathbf{A} - 2\mathbf{B} + 2\mathbf{C}$

8. Solve for **X**:

$$2\mathbf{X} + \frac{1}{2}\begin{bmatrix} 1 & 2 \\ 4 & 6 \end{bmatrix} = 3\left(\mathbf{X} + \begin{bmatrix} 1 & 0 \\ 0 & 1 \end{bmatrix}\right).$$

9. Let **A**, **B**, and **C** be as in Exercise 7. Compute
(a) **AB** **(b)** **BA** **(c)** **(AB)C** **(d)** **A(BC)** **(e)** **A(B − C)**
(f) **AB − AC** **(g)** **A(B + C)** **(h)** **AB + AC**

10. Suppose that **A** is 3×7, **B** is 7×3, and **C** is 7×7. What are the sizes of the following?
(a) **(AB)C** **(b)** **C(BA)** **(c)** **B(AC)** **(d)** **(AC)B**

11. Compute

$$\begin{bmatrix} 1 & 0 & 0 \\ 0 & 1 & 0 \\ 0 & 0 & 1 \end{bmatrix}\begin{bmatrix} 1 & 2 & 3 \\ 4 & 5 & 6 \\ 7 & 8 & 9 \end{bmatrix} \quad \text{and} \quad \begin{bmatrix} 1 & 2 & 3 \\ 4 & 5 & 6 \\ 7 & 8 & 9 \end{bmatrix}\begin{bmatrix} 1 & 0 & 0 \\ 0 & 1 & 0 \\ 0 & 0 & 1 \end{bmatrix}.$$

12. Let

$$\mathbf{A} = \begin{bmatrix} 1 & -1 \\ 0 & 1 \end{bmatrix}, \quad \mathbf{B} = \begin{bmatrix} 1 & 1 \\ 1 & 1 \end{bmatrix}.$$

Show that $(\mathbf{AB})(\mathbf{AB}) = \mathbf{AB}$.

13. Suppose that α, β, γ, and δ are any numbers and

$$\mathbf{A} = \begin{bmatrix} \alpha & \beta \\ -\beta & \alpha \end{bmatrix}, \qquad \mathbf{B} = \begin{bmatrix} \gamma & \delta \\ -\delta & \gamma \end{bmatrix}.$$

Show that $\mathbf{AB} = \mathbf{BA}$.

14. Suppose that α and β are any numbers and

$$\mathbf{A} = \begin{bmatrix} \alpha\beta & \beta^2 \\ -\alpha^2 & -\alpha\beta \end{bmatrix}.$$

show that $\mathbf{AA} = \mathbf{O}$.

15. Let

$$\mathbf{A} = \begin{bmatrix} 1 & 2 \\ 3 & 0 \end{bmatrix}.$$

Verify that $\mathbf{A}$ satisfies the equation

$$\mathbf{AA} - 3\mathbf{A} + 4\begin{bmatrix} -1 & 1 \\ \frac{3}{2} & -\frac{3}{2} \end{bmatrix} = \mathbf{O}.$$

16. Let

$$\mathbf{A} = \begin{bmatrix} 0 & 0 \\ 1 & 0 \end{bmatrix}.$$

Find a 2×2 matrix $\mathbf{B}$ such that $\mathbf{AB} \neq \mathbf{O}$ but $\mathbf{BA} = \mathbf{O}$.

17. Find a 2×2 matrix $\mathbf{A}$ such that

$$\begin{bmatrix} 1 & 0 \\ 2 & 1 \end{bmatrix}\mathbf{A} + \begin{bmatrix} 1 & -1 \\ 1 & 1 \end{bmatrix} = \begin{bmatrix} 2 & -1 \\ 4 & 3 \end{bmatrix}.$$

(*Hint:* Let

$$\mathbf{A} = \begin{bmatrix} a & b \\ c & d \end{bmatrix}$$

and expand the product, etc.)

18. Consider the three molecules water (H_2O), methane (CH_4), and acetic acid (CH_3COOH). We can form a "molecule–atom" matrix $\mathbf{A}$ which summarizes the atomic constituents of the molecules as follows:

	HYDROGEN	CARBON	OXYGEN
Water (H_2O)	2	0	1
Methane (CH_4)	4	1	0
Acetic acid (CH_3COOH)	4	2	2

Similarly, we can form an "atom–particle" matrix **B** for the atoms:

$$\begin{array}{l} \\ \text{Hydrogen} \\ \text{Carbon} \\ \text{Oxygen} \end{array} \begin{array}{c} \begin{array}{cc} \text{PROTONS} & \text{NEUTRONS} \end{array} \\ \begin{bmatrix} 1 & 0 \\ 6 & 6 \\ 8 & 8 \end{bmatrix} \end{array}.$$

Show that the "molecule–particle" matrix that gives the total number of protons and neutrons in each of the three molecules is the 3×2 matrix **AB**.

19. **(a)** Compute

$$\begin{bmatrix} 1 & 6 & 3 \\ -1 & 0 & 5 \\ 0 & 2 & 3 \end{bmatrix} \begin{bmatrix} x_1 \\ x_2 \\ x_3 \end{bmatrix}.$$

(b) Compute

$$\begin{bmatrix} a_{11} & a_{12} & a_{13} \\ a_{21} & a_{22} & a_{23} \end{bmatrix} \begin{bmatrix} 2 \\ 1 \\ 3 \end{bmatrix}.$$

20. Verify part (a) of Theorem 1.2 for the matrices

$$\mathbf{A} = \begin{bmatrix} 1 & 3 & -1 \\ 0 & 1 & 4 \end{bmatrix}, \quad \mathbf{B} = \begin{bmatrix} 1 & 1 & 2 & 0 \\ 2 & 3 & 0 & -1 \\ 5 & 1 & 2 & -1 \end{bmatrix}, \quad \mathbf{C} = \begin{bmatrix} 2 & 3 \\ -1 & 5 \\ 0 & 4 \\ 1 & 2 \end{bmatrix}.$$

21. Suppose that **A** is $m \times n$ and **B** is $n \times k$. Explain why mnk multiplications are required to form the matrix product **AB** using the definition of matrix multiplication given in the text.

22. Let

$$\mathbf{A} = \begin{bmatrix} 1 & 2 \\ 3 & 4 \end{bmatrix}, \quad \mathbf{B} = \begin{bmatrix} -1 \\ 3 \end{bmatrix}, \quad \mathbf{C} = [2 \quad 3].$$

(a) Compute the product (**AB**)**C** and count carefully the total number of multiplications you perform.

(b) Compute **A**(**BC**), counting the total number of multiplications you perform.

(c) Conclude that even though (**AB**)**C** and **A**(**BC**) are the same matrix, the computational work required to compute them is not the same.

23. **(a)** Suppose that **A** is $m \times n$, **B** is $n \times k$, and **C** is $k \times l$. How many multiplications are required to compute (**AB**)**C**? How many multiplications are required to compute **A**(**BC**)?

(b) Suppose that **A** is 100×1, **B** is 1×100, and **C** is 100×1. Compare the work, in terms of the number of multiplications, required to compute (**AB**)**C** and **A**(**BC**).

24. Suppose that $\mathbf{A}$ is an $m \times n$ matrix and $\mathbf{x}$ and $\mathbf{b}$ are vectors such that $\mathbf{Ax} = \mathbf{b}$. What are the sizes of the vectors $\mathbf{x}$ and $\mathbf{b}$?

***25.** Find all 2×2 matrices $\mathbf{A}$ such that

$$\mathbf{AA} = \begin{bmatrix} 1 & 0 \\ 0 & 1 \end{bmatrix}.$$

How many solutions are there?

26. (a) Show that the matrix problem

$$\begin{bmatrix} 2 & -1 & 1 & 0 \\ 6 & 0 & 2 & 4 \\ 3 & 4 & -1 & 2 \end{bmatrix} \begin{bmatrix} x_1 \\ x_2 \\ x_3 \\ x_4 \end{bmatrix} = \begin{bmatrix} -1 \\ 1 \\ 1 \end{bmatrix}$$

can also be written as the vector equation

$$x_1 \begin{bmatrix} 2 \\ 6 \\ 3 \end{bmatrix} + x_2 \begin{bmatrix} -1 \\ 0 \\ 4 \end{bmatrix} + x_3 \begin{bmatrix} 1 \\ 2 \\ -1 \end{bmatrix} + x_4 \begin{bmatrix} 0 \\ 4 \\ 2 \end{bmatrix} = \begin{bmatrix} -1 \\ 1 \\ 1 \end{bmatrix}.$$

***(b)** If $\mathbf{A}$ is $m \times n$ and $\mathbf{Ax} = \mathbf{b}$, show that

$$\sum_{k=1}^{n} x_k \mathbf{A}_k = \mathbf{b},$$

where $\mathbf{A}_k$ is the kth column of $\mathbf{A}$.

27. (a) Let

$$\mathbf{A} = \begin{bmatrix} 2 & 1 & 3 \\ 1 & 0 & 4 \end{bmatrix}, \qquad \mathbf{B} = \begin{bmatrix} 2 & 5 \\ 0 & 3 \\ 2 & 1 \end{bmatrix}.$$

Verify that

$$\mathbf{A} \begin{bmatrix} 2 \\ 0 \\ 2 \end{bmatrix}$$

is the first column vector of the product matrix $\mathbf{AB}$. What is the second column vector of $\mathbf{AB}$?

***(b)** Consider the matrix product $\mathbf{AB}$. Using the definition of matrix multiplication, show that

$$j\text{th column of } (\mathbf{AB}) = \mathbf{A} \text{ times } (j\text{th column of } \mathbf{B}).$$

Verify this on a specific product $\mathbf{AB}$.

1.3 Special Matrices

An $n \times n$ matrix $\mathbf{A}$ that has 1's along the main diagonal (i.e., $a_{ii} = 1$ for all i) and zeros for all other entries is called the $n \times n$ **identity matrix** and is denoted $\mathbf{I}_n$ (or simply $\mathbf{I}$ when the size is obvious). For example,

*Exercises of a more theoretical or challenging nature are marked with an asterisk.

$$\mathbf{I}_2 = \begin{bmatrix} 1 & 0 \\ 0 & 1 \end{bmatrix} \quad \text{and} \quad \mathbf{I}_3 = \begin{bmatrix} 1 & 0 & 0 \\ 0 & 1 & 0 \\ 0 & 0 & 1 \end{bmatrix}.$$

To put it another way, the i, j-entry of an identity matrix is given by

$$\delta_{ij} = \begin{cases} 1 & \text{if } i = j \\ 0 & \text{if } i \neq j \end{cases}$$

(the symbol δ_{ij} is called "Kronecker's delta"). We call $\mathbf{I}_n$ an identity matrix because if $\mathbf{A}$ is any $n \times n$ matrix, then $\mathbf{I}_n\mathbf{A} = \mathbf{A}\mathbf{I}_n = \mathbf{A}$. To see this, notice that the i, j-entry of $\mathbf{I}_n\mathbf{A}$ is

$$\sum_{p=1}^{n} \delta_{ip}a_{pj} = \delta_{ii}a_{ij} = a_{ij};$$

the other equality is established in the same way. For any n-vector $\mathbf{x}$ we have $\mathbf{I}_n\mathbf{x} = \mathbf{x}$.

Therefore, we see that the matrix $\mathbf{I}_n$ plays a role for $n \times n$ matrices similar to that played by 1 for numbers. We know that nonzero numbers have reciprocals; that is, if $\alpha \neq 0$, then there is a number α^{-1} such that

$$\alpha^{-1}\alpha = \alpha\alpha^{-1} = 1.$$

However, this is not always the case for matrices. Indeed, for the matrix

$$\mathbf{A} = \begin{bmatrix} 0 & 0 \\ 1 & 0 \end{bmatrix}$$

we have

$$\mathbf{AX} = \begin{bmatrix} 0 & 0 \\ 1 & 0 \end{bmatrix}\begin{bmatrix} x_{11} & x_{12} \\ x_{21} & x_{22} \end{bmatrix} = \begin{bmatrix} 0 & 0 \\ x_{11} & x_{12} \end{bmatrix} \neq \mathbf{I}_2$$

for any 2×2 matrix $\mathbf{X}$.

Still for certain $n \times n$ matrices $\mathbf{A}$ there will exist an $n \times n$ matrix $\mathbf{X}$ such that

$$\mathbf{XA} = \mathbf{AX} = \mathbf{I}_n.$$

We say that an $n \times n$ matrix $\mathbf{A}$ is **invertible** if there is an $n \times n$ matrix $\mathbf{X}$ such that $\mathbf{AX} = \mathbf{XA} = \mathbf{I}_n$. The matrix $\mathbf{X}$ is called an *inverse* of $\mathbf{A}$. Actually, either of the conditions $\mathbf{AX} = \mathbf{I}_n$ or $\mathbf{XA} = \mathbf{I}_n$ is sufficient to guarantee that $\mathbf{X}$ is an inverse of $\mathbf{A}$ (see Exercise 9 of Section 3.2).

Example 1.13

We show that

$$\mathbf{A} = \begin{bmatrix} 1 & 2 \\ 3 & 4 \end{bmatrix}$$

is invertible by demonstrating that $\mathbf{AX} = \mathbf{XA} = \mathbf{I}$, where

$$\mathbf{X} = \begin{bmatrix} -2 & 1 \\ \frac{3}{2} & -\frac{1}{2} \end{bmatrix}.$$

We have

$$\begin{bmatrix} 1 & 2 \\ 3 & 4 \end{bmatrix}\begin{bmatrix} -2 & 1 \\ \frac{3}{2} & -\frac{1}{2} \end{bmatrix} = \begin{bmatrix} 1 & 0 \\ 0 & 1 \end{bmatrix} = \begin{bmatrix} -2 & 1 \\ \frac{3}{2} & -\frac{1}{2} \end{bmatrix}\begin{bmatrix} 1 & 2 \\ 3 & 4 \end{bmatrix},$$

and hence **A** is invertible.

■

The matrix **X** in Example 1.13 can be determined by a special formula (see Exercise 1) which works for 2×2 matrices. We shall investigate the notion of an invertible matrix in more detail later in the book. For now we will be content to show that an invertible matrix has exactly one inverse. Suppose that **X** and **Y** are both inverses of **A**, that is,

$$\mathbf{AX} = \mathbf{XA} = \mathbf{I}$$

and

$$\mathbf{AY} = \mathbf{YA} = \mathbf{I}.$$

Then

$$\mathbf{X} = \mathbf{XI} = \mathbf{X}(\mathbf{AY})$$

and by part (a) of Theorem 1.2 it follows that

$$\mathbf{X} = \mathbf{X}(\mathbf{AY}) = (\mathbf{XA})\mathbf{Y} = \mathbf{IY} = \mathbf{Y}$$

and therefore **A** has only one inverse. Henceforth we denote the unique inverse of an invertible matrix **A** by $\mathbf{A}^{-1}$.

Thus if **A** is an $n \times n$ invertible matrix, then there is a unique $n \times n$ matrix $\mathbf{A}^{-1}$ satisfying

$$\mathbf{AA}^{-1} = \mathbf{A}^{-1}\mathbf{A} = \mathbf{I}_n.$$

If **A** is an invertible matrix, then we can always solve the system $\mathbf{Ax} = \mathbf{b}$. Indeed, multiplying both sides by $\mathbf{A}^{-1}$, we have

$$\mathbf{x} = \mathbf{Ix} = \mathbf{A}^{-1}\mathbf{Ax} = \mathbf{A}^{-1}\mathbf{b}.$$

There are, however, much more efficient ways of solving a linear system which do not involve computing the inverse. We have more to say about this in Chapter 2.

An important property of inverses concerns the inverse of the product of two matrices. If α and β are nonzero numbers, then

$$(\alpha\beta)^{-1} = \alpha^{-1}\beta^{-1}.$$

Does the analogous property hold for matrices? That is, for $n \times n$ matrices **A** and **B** is it true that

$$(\mathbf{AB})^{-1} = \mathbf{A}^{-1}\mathbf{B}^{-1} ?$$

Example 1.14

Let **A** be the matrix from Example 1.13 and let

$$\mathbf{B} = \begin{bmatrix} 0 & 1 \\ 1 & 0 \end{bmatrix}$$

It is easy to check that $\mathbf{BB} = \mathbf{I}$ and hence that $\mathbf{B} = \mathbf{B}^{-1}$. We have

$$\mathbf{AB} = \begin{bmatrix} 1 & 2 \\ 3 & 4 \end{bmatrix}\begin{bmatrix} 0 & 1 \\ 1 & 0 \end{bmatrix} = \begin{bmatrix} 2 & 1 \\ 4 & 3 \end{bmatrix}$$

and

$$\mathbf{A}^{-1}\mathbf{B}^{-1} = \begin{bmatrix} -2 & 1 \\ \frac{3}{2} & -\frac{1}{2} \end{bmatrix}\begin{bmatrix} 0 & 1 \\ 1 & 0 \end{bmatrix} = \begin{bmatrix} 1 & -2 \\ -\frac{1}{2} & \frac{3}{2} \end{bmatrix}$$

but

$$(\mathbf{AB})(\mathbf{A}^{-1}\mathbf{B}^{-1}) = \begin{bmatrix} 2 & 1 \\ 4 & 3 \end{bmatrix}\begin{bmatrix} 1 & -2 \\ -\frac{1}{2} & \frac{3}{2} \end{bmatrix} = \begin{bmatrix} \frac{3}{2} & -\frac{5}{2} \\ \frac{5}{2} & -\frac{7}{2} \end{bmatrix}$$

and hence $(\mathbf{AB})(\mathbf{A}^{-1}\mathbf{B}^{-1}) \neq \mathbf{I}$, that is, $(\mathbf{AB})^{-1} \neq \mathbf{A}^{-1}\mathbf{B}^{-1}$.

■

Fortunately, there is a a simple relationship between the inverse of a product and the product of inverses. The relationship, called the *reverse order rule for inverses,* says that the inverse of the product of two invertible matrices is equal to the product of the inverses, but *in reverse order*. That is,

$$(\mathbf{AB})^{-1} = \mathbf{B}^{-1}\mathbf{A}^{-1}.$$

We ask you to verify this in Exercise 11.

If $\mathbf{A}$ is a square matrix, then we can define powers of $\mathbf{A}$ in a natural way:

$$\mathbf{A}^1 = \mathbf{A},$$
$$\mathbf{A}^2 = \mathbf{AA},$$
$$\mathbf{A}^3 = \mathbf{AA}^2, \qquad \text{etc.}$$

By convention we take $\mathbf{A}^0 = \mathbf{I}$. The usual law of exponents holds in this setting:

$$\mathbf{A}^n\mathbf{A}^m = \mathbf{A}^{n+m},$$

for integers n and m.

Given an $m \times n$ matrix $\mathbf{A}$ we define an $n \times m$ matrix $\mathbf{A}^{\mathrm{T}}$, called the **transpose** of $\mathbf{A}$, by

$$a_{ij}^{\mathrm{T}} = a_{ji}.$$

Example 1.15

If

$$\mathbf{A} = \begin{bmatrix} 1 & 2 & 3 \\ 4 & 5 & 6 \end{bmatrix}, \qquad \text{then} \qquad \mathbf{A}^{\mathrm{T}} = \begin{bmatrix} 1 & 4 \\ 2 & 5 \\ 3 & 6 \end{bmatrix}.$$

■

Notice that the column vectors of $\mathbf{A}^T$ are the row vectors of $\mathbf{A}$; in fact, "transpose" refers to transposing rows and columns. Consequently, if $\mathbf{A}$ has m rows and n columns, then $\mathbf{A}^T$ has n rows and m columns.

Since the transpose is formed by interchanging the rows and columns of a matrix, if this operation is performed twice, we end up with the original matrix, that is,

$$(\mathbf{A}^T)^T = \mathbf{A}.$$

Also, since the rows (columns) of the matrix which is the sum of two other matrices are the sums of the rows (columns) of the individual matrices, it follows that the transpose of the sum is the sum of the transposes:

$$(\mathbf{A} + \mathbf{B})^T = \mathbf{A}^T + \mathbf{B}^T.$$

Example 1.16

Let $\mathbf{A}$ be the matrix of Example 1.14; then

$$(\mathbf{A}^T)^T = \begin{bmatrix} 1 & 4 \\ 2 & 5 \\ 3 & 6 \end{bmatrix}^T = \begin{bmatrix} 1 & 2 & 3 \\ 4 & 5 & 6 \end{bmatrix} = \mathbf{A}.$$

If

$$\mathbf{B} = \begin{bmatrix} 0 & 1 & -2 \\ 1 & 3 & 4 \end{bmatrix}$$

then

$$\mathbf{A} + \mathbf{B} = \begin{bmatrix} 1 & 3 & 1 \\ 5 & 8 & 10 \end{bmatrix} \quad \text{and} \quad (\mathbf{A} + \mathbf{B})^T = \begin{bmatrix} 1 & 5 \\ 3 & 8 \\ 1 & 10 \end{bmatrix}.$$

Therefore,

$$\mathbf{A}^T + \mathbf{B}^T = \begin{bmatrix} 1 & 4 \\ 2 & 5 \\ 3 & 6 \end{bmatrix} + \begin{bmatrix} 0 & 1 \\ 1 & 3 \\ -2 & 4 \end{bmatrix} = \begin{bmatrix} 1 & 5 \\ 3 & 8 \\ 1 & 10 \end{bmatrix} = (\mathbf{A} + \mathbf{B})^T.$$

■

A matrix that is equal to its transpose is called **symmetric**. Note that a symmetric matrix is necessarily square.

Example 1.17

$$\mathbf{A} = \begin{bmatrix} -1 & 2 & 3 \\ 2 & 6 & 0 \\ 3 & 0 & 4 \end{bmatrix}$$

is a symmetric matrix.

■

We now consider the relationship between transposition and matrix products. For example, is it the case that $(\mathbf{AB})^T = \mathbf{A}^T\mathbf{B}^T$? A little thought shows that this cannot be. Suppose, for example, that $\mathbf{A}$ is 3×2 and $\mathbf{B}$ is 2×2. Then $\mathbf{AB}$ is 3×2 and $(\mathbf{AB})^T$ is 2×3. However, $\mathbf{A}^T$ is 2×3 and $\mathbf{B}^T$ is 2×2 and therefore the product $\mathbf{A}^T\mathbf{B}^T$ is not even defined! But $\mathbf{B}^T\mathbf{A}^T$ is defined and in fact has the same size (3×2) as $(\mathbf{AB})^T$. This at least makes it plausible that $(\mathbf{AB})^T = \mathbf{B}^T\mathbf{A}^T$.

Example 1.18

Let

$$\mathbf{A} = \begin{bmatrix} 2 & -1 \\ 2 & 3 \end{bmatrix} \quad \text{and} \quad \mathbf{B} = \begin{bmatrix} -2 & 2 \\ 1 & 1 \end{bmatrix}.$$

Then

$$\mathbf{AB} = \begin{bmatrix} -5 & 3 \\ -1 & 7 \end{bmatrix} \quad \text{and} \quad (\mathbf{AB})^T = \begin{bmatrix} -5 & -1 \\ 3 & 7 \end{bmatrix}.$$

However,

$$\mathbf{A}^T\mathbf{B}^T = \begin{bmatrix} 2 & 2 \\ -1 & 3 \end{bmatrix}\begin{bmatrix} -2 & 1 \\ 2 & 1 \end{bmatrix} = \begin{bmatrix} 0 & 4 \\ 8 & 2 \end{bmatrix}.$$

Therefore, even if "the sizes are right," it does not follow that $(\mathbf{AB})^T$ is equal to $\mathbf{A}^T\mathbf{B}^T$.

■

The next theorem, in fact, establishes the important *reverse order rule for the transpose of a product* which is analogous to the corresponding rule for inverses.

Theorem 1.3. If $\mathbf{A}$ is $n \times m$ and $\mathbf{B}$ is $m \times k$, then $(\mathbf{AB})^T = \mathbf{B}^T\mathbf{A}^T$.

Proof. Suppose that $\mathbf{AB} = \mathbf{C}$; then the i,j-entry of $\mathbf{C}$ is given by

$$c_{ij} = \sum_{p=1}^{m} a_{ip}b_{pj}.$$

Therefore, the i,j-th entry of $\mathbf{C}^T = (\mathbf{AB})^T$ is

$$c_{ij}^T = \sum_{p=1}^{m} a_{jp}b_{pi}.$$

But since $b_{pi} = b_{ip}^T$ and $a_{jp} = a_{pj}^T$, we find that

$$c_{ij}^T = \sum_{p=1}^{m} b_{ip}^T a_{pj}^T,$$

which is the i,j-entry of $\mathbf{B}^T\mathbf{A}^T$.

■

We will have more to say about transposes, particularly concerning the crucial relationship between transpose and inner product, in Chapter 5.

Example 1.19

Suppose that:

$$\mathbf{A} = \begin{bmatrix} 0 & 1 \\ 2 & 4 \end{bmatrix} \quad \text{and} \quad \mathbf{B} = \begin{bmatrix} 1 & -1 & 1 \\ 1 & 2 & 0 \end{bmatrix}.$$

Then

$$\mathbf{AB} = \begin{bmatrix} 1 & 2 & 0 \\ 6 & 6 & 2 \end{bmatrix}, \qquad (\mathbf{AB})^{\mathrm{T}} = \begin{bmatrix} 1 & 6 \\ 2 & 6 \\ 0 & 2 \end{bmatrix},$$

$$\mathbf{B}^{\mathrm{T}} = \begin{bmatrix} 1 & 1 \\ -1 & 2 \\ 1 & 0 \end{bmatrix}, \qquad \mathbf{A}^{\mathrm{T}} = \begin{bmatrix} 0 & 2 \\ 1 & 4 \end{bmatrix},$$

and

$$\mathbf{B}^{\mathrm{T}}\mathbf{A}^{\mathrm{T}} = \begin{bmatrix} 1 & 6 \\ 2 & 6 \\ 0 & 2 \end{bmatrix} = (\mathbf{AB})^{\mathrm{T}}.$$

■

EXERCISES 1.3

1. Suppose that

$$\mathbf{A} = \begin{bmatrix} a & b \\ c & d \end{bmatrix} \quad \text{and} \quad ad - bc \neq 0.$$

Let

$$\mathbf{X} = \frac{1}{ad - bc}\begin{bmatrix} d & -b \\ -c & a \end{bmatrix}.$$

Compute the matrix products **AX** and **XA**. What do you conclude about **X**?

2. Find the inverse of the matrix

$$\begin{bmatrix} -1 & 2 \\ 3 & 2 \end{bmatrix}.$$

(*Hint:* See Exercise 1.)

3. Suppose that

$$\mathbf{A} = \begin{bmatrix} 1 & 0 & 0 \\ 0 & 1 & 0 \\ 1 & 0 & 1 \end{bmatrix}.$$

Show that

$$\mathbf{A}^{-1} = \begin{bmatrix} 1 & 0 & 0 \\ 0 & 1 & 0 \\ -1 & 0 & 1 \end{bmatrix}.$$

***4.** (**a**) Suppose that $\mathbf{A}^2 = \mathbf{I}$. Show that $\mathbf{A} = \mathbf{A}^{-1}$.
(**b**) Suppose that $\mathbf{A}^2 = \mathbf{O}$. Show that $\mathbf{A}$ is not invertible.

5. Let

$$\mathbf{A} = \begin{bmatrix} 1 & -1 \\ 0 & 1 \end{bmatrix} \quad \text{and} \quad \mathbf{B} = \begin{bmatrix} 1 & 0 \\ 1 & 1 \end{bmatrix}.$$

For these matrices check that

(**a**) $\mathbf{AB} \neq \mathbf{BA}$

(**b**) $(\mathbf{A} + \mathbf{B})^2 \neq \mathbf{A}^2 + 2\mathbf{AB} + \mathbf{B}^2$

(**c**) $(\mathbf{A} + \mathbf{B})(\mathbf{A} - \mathbf{B}) \neq \mathbf{A}^2 - \mathbf{B}^2$

(**d**) What are the general formulas for $(\mathbf{A} + \mathbf{B})^2$ and $(\mathbf{A} - \mathbf{B})^2$?

6. Suppose that t is any nonzero number and

$$\mathbf{A} = \begin{bmatrix} 0 & t \\ \frac{1}{t} & 0 \end{bmatrix}.$$

Verify that $\mathbf{A}^2 = \mathbf{I}_2$. (We might then say that $\mathbf{I}_2$ has infinitely many square roots.)

7. Let

$$\mathbf{A} = \begin{bmatrix} 2 & 3t \\ \frac{3}{t} & 2 \end{bmatrix},$$

where t is any nonzero number. Verify by direct substitution that $\mathbf{A}$ satisfies the quadratic matrix equation $\mathbf{A}^2 - 4\mathbf{A} - 5\mathbf{I}_2 = \mathbf{O}$. (That is, a quadratic matrix equation may have infinitely many solutions.)

8. Let

$$\mathbf{A} = \begin{bmatrix} 3 & 5 \\ 2 & 4 \end{bmatrix}.$$

Compute $\mathbf{B} = \mathbf{A}^{-1}$. Compute $\mathbf{B}^{-1}$.

9. Show that if $\mathbf{A}$ is invertible, then $(\mathbf{A}^{-1})^{-1} = \mathbf{A}$. [*Hint:* $(\mathbf{A}^{-1})^{-1}$ is the matrix $\mathbf{X}$ satisfying $\mathbf{XA}^{-1} = \mathbf{A}^{-1}\mathbf{X} = \mathbf{I}$.]

10. Let

$$\mathbf{A} = \begin{bmatrix} 2 & 1 \\ 3 & 2 \end{bmatrix} \quad \text{and} \quad \mathbf{B} = \begin{bmatrix} 4 & 2 \\ 3 & 2 \end{bmatrix}.$$

Compute $(\mathbf{AB})^{-1}$. Compute $\mathbf{A}^{-1}$, $\mathbf{B}^{-1}$, and $\mathbf{B}^{-1}\mathbf{A}^{-1}$. Compare $(\mathbf{AB})^{-1}$ and $\mathbf{B}^{-1}\mathbf{A}^{-1}$.

11. Show that if $\mathbf{A}$ is an $n \times n$ invertible matrix and $\mathbf{B}$ is an $n \times n$ invertible matrix, then $\mathbf{AB}$ is invertible and $(\mathbf{AB})^{-1} = \mathbf{B}^{-1}\mathbf{A}^{-1}$. [*Hint:* Let $\mathbf{X} = \mathbf{B}^{-1}\mathbf{A}^{-1}$. Show that $\mathbf{X}(\mathbf{AB}) = (\mathbf{AB})\mathbf{X} = \mathbf{I}$.]

12. Let

$$\mathbf{A} = \begin{bmatrix} 2 & -1 & 3 \\ 1 & 1 & 0 \end{bmatrix}.$$

Compute $\mathbf{A}^T$. Let $\mathbf{B} = \mathbf{A}^T$, and compute $\mathbf{B}^T$.

13. Let

$$\mathbf{A} = \begin{bmatrix} 3 & 6 \\ -1 & 0 \\ 4 & 1 \end{bmatrix} \quad \text{and} \quad \mathbf{B} = \begin{bmatrix} 0 & -1 \\ 2 & 1 \\ 1 & 3 \end{bmatrix}.$$

Verify that $(\mathbf{A} + \mathbf{B})^{\mathrm{T}} = \mathbf{A}^{\mathrm{T}} + \mathbf{B}^{\mathrm{T}}$.

14. Show that $(\alpha\mathbf{A})^{\mathrm{T}} = \alpha\mathbf{A}^{\mathrm{T}}$ for any scalar α and any matrix $\mathbf{A}$.

15. Is the product of two symmetric matrices necessarily symmetric? (*Hint:* Try two specific 2×2 matrices.)

16. Let

$$\mathbf{A} = \begin{bmatrix} 2 & 0 & 0 \\ 1 & -1 & 0 \\ 0 & 1 & 4 \end{bmatrix} \quad \text{and} \quad \mathbf{B} = \begin{bmatrix} -1 & 0 & 0 \\ 3 & 1 & 0 \\ 2 & -2 & 5 \end{bmatrix}.$$

Compute **AB** and **BA**.

17. Let

$$\mathbf{A} = \begin{bmatrix} 1 & 2 & 3 & 1 \\ 0 & 4 & 1 & 0 \\ 0 & 0 & 2 & 1 \\ 0 & 0 & 0 & 3 \end{bmatrix} \quad \text{and} \quad \mathbf{B} = \begin{bmatrix} 2 & -1 & 1 & 4 \\ 0 & -1 & 3 & 1 \\ 0 & 0 & 2 & 1 \\ 0 & 0 & 0 & 1 \end{bmatrix}.$$

Compute **AB** and **BA**.

18. A square matrix **A** is called ***lower triangular*** if a $a_{ij} = 0$ for $i < j$. **A** is called ***upper triangular*** if $a_{ij} = 0$ for $i > j$. Show that the product of two lower triangular matrices is lower triangular and that the product of two upper triangular matrices is upper triangular.

19. Let

$$\mathbf{D} = \begin{bmatrix} 1 & 0 & 0 \\ 0 & 2 & 0 \\ 0 & 0 & 3 \end{bmatrix}.$$

Compute $\mathbf{D}^2$ and $\mathbf{D}^3$.

20. Let

$$\mathbf{D} = \begin{bmatrix} a & 0 & 0 \\ 0 & b & 0 \\ 0 & 0 & c \end{bmatrix} \quad \text{and} \quad \mathbf{A} = \begin{bmatrix} -1 & 2 & 3 \\ 2 & 1 & 4 \\ -3 & 5 & 3 \end{bmatrix}.$$

Compute **DA** and **AD**.

21. A matrix that is simultaneously lower triangular and upper triangular is called ***diagonal***. An $n \times n$ diagonal matrix therefore has the form

$$\mathbf{D} = \begin{bmatrix} d_1 & & & \text{\Large 0} \\ & d_2 & & \\ & & \ddots & \\ \text{\Large 0} & & & d_n \end{bmatrix}$$

which we will sometimes write more economically as $\mathbf{D} = \mathbf{diag}(d_1, \ldots, d_n)$.

(a) Show that premultiplying an $n \times n$ matrix $\mathbf{A}$ by the diagonal matrix $\mathbf{diag}\,(d_1, \ldots, d_n)$, that is, forming the product $\mathbf{diag}\,(d_1, \ldots, d_n)\mathbf{A}$, has the effect of multiplying each row of $\mathbf{A}$ by the corresponding diagonal entry of $\mathbf{diag}\,(d_1, \ldots, d_n)$.

(b) Show that postmultiplying an $n \times n$ matrix $\mathbf{A}$ by $\mathbf{diag}\,(d_1, \ldots, d_n)$, that is, forming the product $\mathbf{A}\,\mathbf{diag}\,(d_1, \ldots, d_n)$, has the effect of multiplying each column of $\mathbf{A}$ by the corresponding entry of $\mathbf{diag}\,(d_1, \ldots, d_n)$.

22. (a) Given that $\mathbf{D} = \mathbf{diag}\,(d_1, \ldots, d_n)$. Is $\mathbf{D}^2$ a diagonal matrix? What diagonal matrix?

(b) Suppose that an $n \times n$ matrix $\mathbf{A}$ can be written as $\mathbf{A} = \mathbf{B}^{-1}\mathbf{D}\mathbf{B}$ for some invertible matrix $\mathbf{B}$. Verify that $\mathbf{A}^2 = \mathbf{B}^{-1}\mathbf{D}^2\mathbf{B}$.

23. (a) Suppose that $\mathbf{A}$ is a square matrix satisfying $\mathbf{A}^3 = \mathbf{O}$. Show that $(\mathbf{I} + \mathbf{A})^{-1} = \mathbf{I} - \mathbf{A} + \mathbf{A}^2$, by computing $(\mathbf{I} + \mathbf{A})(\mathbf{I} - \mathbf{A} + \mathbf{A}^2)$.

(b) Use part (a) to find

$$\begin{bmatrix} 1 & 1 & 1 \\ 0 & 1 & 1 \\ 0 & 0 & 1 \end{bmatrix}^{-1}.$$

(*Hint:* Let

$$\mathbf{A} = \begin{bmatrix} 0 & 1 & 1 \\ 0 & 0 & 1 \\ 0 & 0 & 0 \end{bmatrix}.)$$

24. (a) Suppose that

$$\mathbf{x} = \begin{bmatrix} 1 \\ 2 \\ -1 \end{bmatrix}.$$

Compute $\mathbf{x}^T\mathbf{x}$ and $\mathbf{x}\mathbf{x}^T$.

(b) Let

$$\mathbf{A} = \begin{bmatrix} 1 & -1 & 0 \\ 2 & 1 & 3 \\ 0 & 1 & 5 \end{bmatrix}.$$

Compute $\mathbf{A}\mathbf{x}$ and $\mathbf{x}^T\mathbf{A}\mathbf{x}$ for $\mathbf{x}$ as in part (a).

25. (a) Show that

$$\begin{bmatrix} 2 & 0 \\ 1 & 3 \end{bmatrix}^{-1} = \begin{bmatrix} \frac{1}{2} & 0 \\ -\frac{1}{6} & \frac{1}{3} \end{bmatrix}.$$

(b) Show that if

$$\mathbf{A} = \begin{bmatrix} a_{11} & 0 \\ a_{21} & a_{22} \end{bmatrix}$$

with $a_{11} \neq 0$, $a_{22} \neq 0$, then

$$\mathbf{A}^{-1} = \begin{bmatrix} a_{11}^{-1} & 0 \\ b_{21} & a_{22}^{-1} \end{bmatrix},$$

where $b_{21}a_{11} + a_{21}a_{22}^{-1} = 0$, *that is*, $b_{21} = -a_{21}/(a_{22}a_{11})$.

***(c)** Show that if **A** is a triangular matrix with no zero on its diagonal, then **A** is invertible. (*Hint:* If **A** is lower triangular, show that

$$\mathbf{A}^{-1} = \begin{bmatrix} a_{11}^{-1} & 0 & \cdots & 0 \\ b_{21} & a_{22}^{-1} & & \cdot \\ \cdot & \cdot & & \cdot \\ \cdot & \cdot & & \cdot \\ \cdot & \cdot & & \cdot \\ b_{n1} & b_{n2} & \cdots & a_{nn}^{-1} \end{bmatrix},$$

where b_{ij} can be uniquely determined in terms of the a_{ij}'s in the order b_{21}, b_{32}, b_{31}, ... A similar argument works for upper triangular matrices.)

26. Let

$$\mathbf{A} = \begin{bmatrix} -2 & 1 & 3 \\ 4 & -1 & 0 \\ 0 & 2 & 1 \end{bmatrix}.$$

Compute $\mathbf{A}^T\mathbf{A}$ and $\mathbf{A}\mathbf{A}^T$.

27. Let **A** be an $m \times n$ matrix and let $\mathbf{B} = \mathbf{A}^T\mathbf{A}$ and $\mathbf{C} = \mathbf{A}\mathbf{A}^T$. Show that **B** and **C** are symmetric. (*Hint:* Do not compute $\mathbf{A}^T\mathbf{A}$; use Theorem 1.3.)

28. If **A** is a square matrix that satisfies $\mathbf{A}^T = -\mathbf{A}$, then **A** is called **skew-symmetric.** Show that the diagonal entries of a skew-symmetric matrix **A** must be zero, that is, $a_{ii} = 0$ for all i.

***29.** Suppose that **A** is an $n \times n$ matrix. Show that **A** can be written as $\mathbf{A} = \mathbf{B} + \mathbf{C}$, where **B** is symmetric and **C** is skew-symmetric. [*Hint:* Consider $\mathbf{B} = \frac{1}{2}(\mathbf{A} + \mathbf{A}^T)$.]

1.4 Some Simple Applications

Input–Output Economics

Linear input–output models of an economy were developed by W. Leontief in the 1930s. In this analysis an economy is broken down into sectors and the interchange of goods and services among sectors is quantified in order to draw systematic conclusions on the state of the economy.

Consider a much simplified model that consists of three sectors each of which produces exactly one type of product, which may be goods or services. For example, one could think of an economy that is broken down into agricultural, manufacturing, and service sectors. We suppose that in order to produce its product, each sector buys products from the other sectors and produces just enough to provide products to the other sectors and all other consumers; that is, in no case is a surplus allowed to accumulate.

We suppose that the jth sector requires y_{ij} units of product from the ith sector and that b_i units are demanded from the ith sector by all other consumers. If we let x_i stand for number of units of output produced by the ith sector, then we have

$$x_i = y_{i1} + y_{i2} + y_{i3} + b_i, \qquad 1 \le i \le 3. \tag{7}$$

Finally, we suppose that the amount of product i needed by the jth sector to produce product j is directly proportional to the amount of product j produced. If we call the constant of proportionality a_{ij}, then

$$y_{ij} = a_{ij}x_j.$$

Substituting this into (7) we have

$$\begin{aligned} x_1 &= a_{11}x_1 + a_{12}x_2 + a_{13}x_3 + b_1 \\ x_2 &= a_{21}x_1 + a_{22}x_2 + a_{23}x_3 + b_2 \\ x_3 &= a_{31}x_1 + a_{32}x_2 + a_{33}x_3 + b_3, \end{aligned}$$

or, collecting the unknowns on the left-hand side and expressing the system in matrix form,

$$(\mathbf{I} - \mathbf{A})\mathbf{x} = \mathbf{b}.$$

Therefore, if the demands $\mathbf{b}$ are specified and the matrix $\mathbf{I} - \mathbf{A}$ is invertible, the outputs of the various sectors are given by

$$\mathbf{x} = (\mathbf{I} - \mathbf{A})^{-1}\mathbf{b}.$$

Note that a_{ij} is the number of units of the ith product needed by the jth sector to produce one unit of the jth product. Therefore, the matrix $\mathbf{A}$ obviously reflects in some way the technologies of the sectors. For example, low values for a_{11}, a_{12}, and a_{13} would mean that the first sector can produce its product while consuming relatively little. The matrix $\mathbf{A}$ is therefore called the *technology matrix.*

Example 1.20

Consider an economy which for simplicity is divided into two sectors, say manufacturing (sector 1) and service (sector 2). Suppose that the outputs of the sectors are related as follows:

$$\begin{aligned} x_1 &= .6x_1 + .2x_2 + 15 \\ x_2 &= .3x_1 + .1x_2 + 60. \end{aligned}$$

That is, the technology matrix for the economy is

$$\mathbf{A} = \begin{bmatrix} .6 & .2 \\ .3 & .1 \end{bmatrix}$$

and the outside demands are given by

$$\mathbf{b} = \begin{bmatrix} 15 \\ 60 \end{bmatrix}.$$

Then

$$\mathbf{I} - \mathbf{A} = \begin{bmatrix} .4 & -.2 \\ -.3 & .9 \end{bmatrix}$$

and by Exercise 1 of Section 1.3,

$$(\mathbf{I} - \mathbf{A})^{-1} = \frac{1}{.36 - .06}\begin{bmatrix} .9 & .2 \\ .3 & .4 \end{bmatrix}$$
$$= \begin{bmatrix} 3 & \frac{2}{3} \\ 1 & \frac{4}{3} \end{bmatrix}.$$

The output of the economy is then given by

$$\mathbf{x} = (\mathbf{I} - \mathbf{A})^{-1}\mathbf{b}$$
$$= \begin{bmatrix} 3 & \frac{2}{3} \\ 1 & \frac{4}{3} \end{bmatrix}\begin{bmatrix} 15 \\ 60 \end{bmatrix} = \begin{bmatrix} 85 \\ 95 \end{bmatrix}.$$

■

If the price of one unit of the ith product is p_i, then the cost of producing a unit of jth product is

$$a_{1j}p_1 + a_{2j}p_2 + a_{3j}p_3, \qquad 1 \le j \le 3.$$

If we call the difference between the price and production cost of one unit of the jth product the *value added* by the jth sector and denote it by v_j, then we have

$$p_j - (a_{1j}p_1 + a_{2j}p_2 + a_{3j}p_3) = v_j, \qquad 1 \le j \le 3.$$

Writing this in more detail we have

$$\begin{aligned} (1 - a_{11})p_1 - a_{21}p_2 - a_{31}p_3 &= v_1 \\ -a_{12}p_1 + (1 - a_{22})p_2 - a_{32}p_3 &= v_2 \\ -a_{13}p_1 - a_{23}p_2 + (1 - a_{33})p_3 &= v_3 \end{aligned}$$

or

$$(\mathbf{I} - \mathbf{A})^{\mathrm{T}}\mathbf{p} = \mathbf{v}.$$

Therefore, if $(\mathbf{I} - \mathbf{A})^{\mathrm{T}}$ is invertible and the values added $\mathbf{v}$ are specified, then the prices $\mathbf{p}$ are determined by

$$\mathbf{p} = [(\mathbf{I} - \mathbf{A})^{\mathrm{T}}]^{-1}\mathbf{v}.$$

It is important to realize that any realistic model of an economy would take many more factors into account than we have in our simplified discussion. For example, any realistic model would take hundreds of sectors into account, resulting in a very large technology matrix. The entries of this matrix as well as the components of the outside demand vector would naturally be subject to some uncertainties. In analyzing such a model the economist would be faced with some formidable problems of data collection, computation, and interpretation. Nonetheless, such computations are carried out routinely by the World Bank and other large international agencies. Our aim in presenting this simplified discussion is merely to point out in an uncomplicated fashion the role that matrices can play in economic analysis.

Electrical Ladder Networks

Such networks may be analyzed by splitting them into smaller basic networks and applying the rule for matrix multiplication. We may represent a network by the diagram shown in Figure 1.1, where i_1 and v_1 are the input current and voltage, respectively, and i_2 and v_2 are the output current and voltage, respectively. Inside the box are various circuit elements. We will consider *ladder networks* in which the circuit elements are all resistors wired in series or as shunts, as shown in Figure 1.2. The orientation of the voltages and currents in the circuit may be assigned in an arbitrary way.

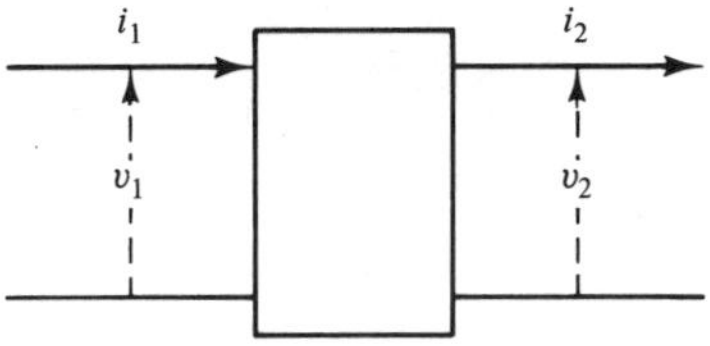

Figure 1.1

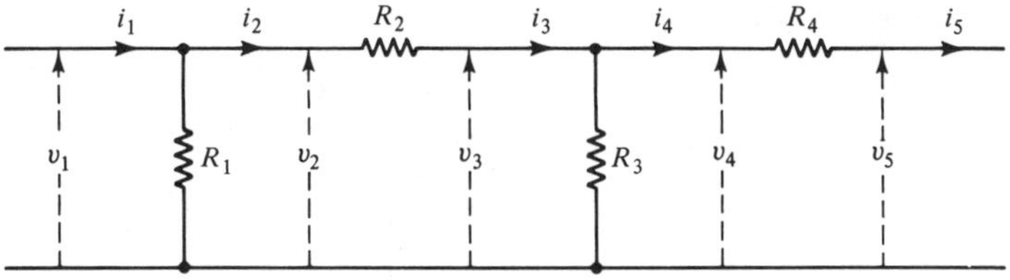

Figure 1.2

Consider first the basic shunt network, that is, a network element of the form shown in Figure 1.3. A basic law of electrical circuits, *Ohm's law*, states that the voltage drop v across a resistance R is equal to the product of R and the current i passing through R:

$$v = Ri.$$

Applying Ohm's law to the shunt network shown in Figure 1.3, we have

$$i = \frac{v_1}{R} = \frac{v_2}{R}$$

and hence

$$v_2 = v_1. \tag{8}$$

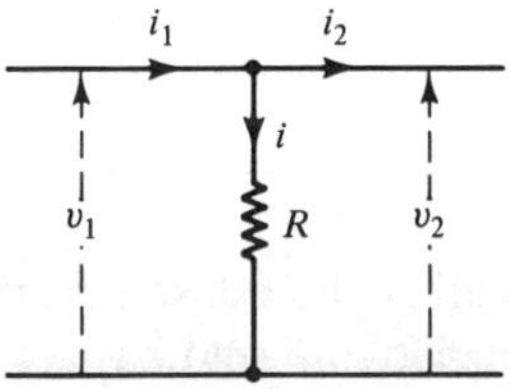

Figure 1.3

Another basic principle of circuit theory, *Kirchhoff's law*, states that the algebraic sum of all currents at a junction must be zero. Applying Kirchhoff's law to the junction above the resistance in the shunt network, we have

$$i_1 - i - i_2 = 0.$$

Combining this with Ohm's law,we have

$$i_2 = i_1 - i = -\frac{v_1}{R} + i_1. \tag{9}$$

Therefore, if we represent the input by the vector

$$\mathbf{x}^{(1)} = \begin{bmatrix} v_1 \\ i_1 \end{bmatrix}$$

and the output by

$$\mathbf{x}^{(2)} = \begin{bmatrix} v_2 \\ i_2 \end{bmatrix}$$

then by (8) and (9) we have

$$\mathbf{x}^{(2)} = \mathbf{A}\mathbf{x}^{(1)},$$

where

$$\mathbf{A} = \begin{bmatrix} 1 & 0 \\ -\frac{1}{R} & 1 \end{bmatrix}.$$

The series network is simpler (see Figure 1.4). Here $i_1 = i_2$ and by Ohm's law,

$$v_1 - v_2 = Ri_1, \quad \text{that is,} \quad v_2 = v_1 - Ri_1.$$

Therefore,

$$\mathbf{x}^{(2)} = \mathbf{A}\mathbf{x}^{(1)},$$

where

$$\mathbf{A} = \begin{bmatrix} 1 & -R \\ 0 & 1 \end{bmatrix}.$$

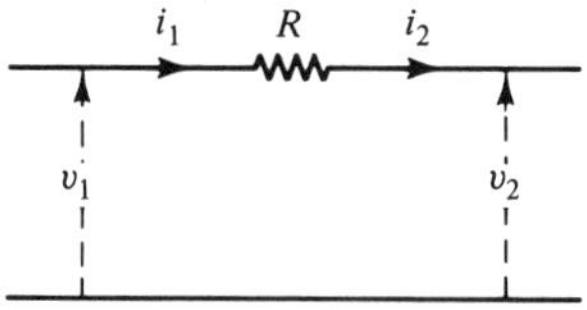

Figure 1.4

In each case we will call **A** the transfer matrix for the basic network. We may now split the original ladder network into basic shunt and series networks as shown in Figure 1.5. To obtain the transfer matrix for the entire network, note that

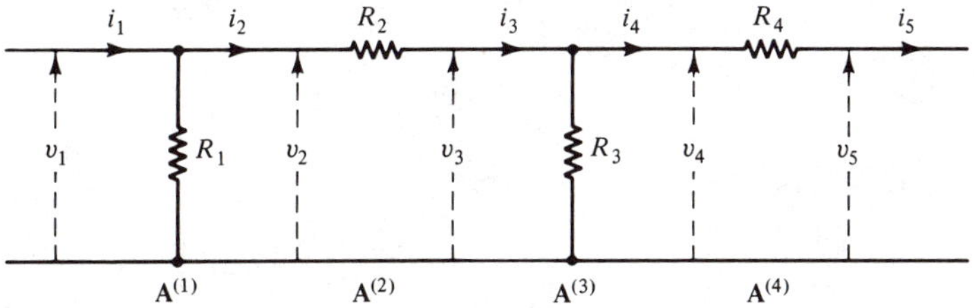

Figure 1.5

$$\begin{bmatrix} v_5 \\ i_5 \end{bmatrix} = \mathbf{A}^{(4)}\begin{bmatrix} v_4 \\ i_4 \end{bmatrix} = \mathbf{A}^{(4)}\mathbf{A}^{(3)}\begin{bmatrix} v_3 \\ i_3 \end{bmatrix} = \cdots = \mathbf{A}^{(4)}\mathbf{A}^{(3)}\mathbf{A}^{(2)}\mathbf{A}^{(1)}\begin{bmatrix} v_1 \\ i_1 \end{bmatrix}.$$

Therefore, the transfer matrix for any ladder network may be obtained by multiplying the transfer matrices of the basic shunt and series networks.

Example 1.21

Consider the simple ladder network shown in Figure 1.6. We split this into series and shunt networks as indicated. The transfer matrix for the first series network is then

$$\mathbf{A}^{(1)} = \begin{bmatrix} 1 & -R \\ 0 & 1 \end{bmatrix} = \begin{bmatrix} 1 & -1 \\ 0 & 1 \end{bmatrix}.$$

The next two subnetworks have transfer matrices

$$\mathbf{A}^{(2)} = \begin{bmatrix} 1 & 0 \\ -\dfrac{1}{R} & 1 \end{bmatrix} = \begin{bmatrix} 1 & 0 \\ -\frac{1}{2} & 1 \end{bmatrix}$$

and

$$\mathbf{A}^{(3)} = \begin{bmatrix} 1 & -R \\ 0 & 1 \end{bmatrix} = \begin{bmatrix} 1 & -3 \\ 0 & 1 \end{bmatrix},$$

respectively. The transfer matrix for the ladder network

$$\mathbf{A}^{(3)}\mathbf{A}^{(2)}\mathbf{A}^{(1)} = \begin{bmatrix} 1 & -3 \\ 0 & 1 \end{bmatrix}\begin{bmatrix} 1 & 0 \\ -\frac{1}{2} & 1 \end{bmatrix}\begin{bmatrix} 1 & -1 \\ 0 & 1 \end{bmatrix}$$
$$= \frac{1}{2}\begin{bmatrix} 5 & -11 \\ -1 & 3 \end{bmatrix}.$$

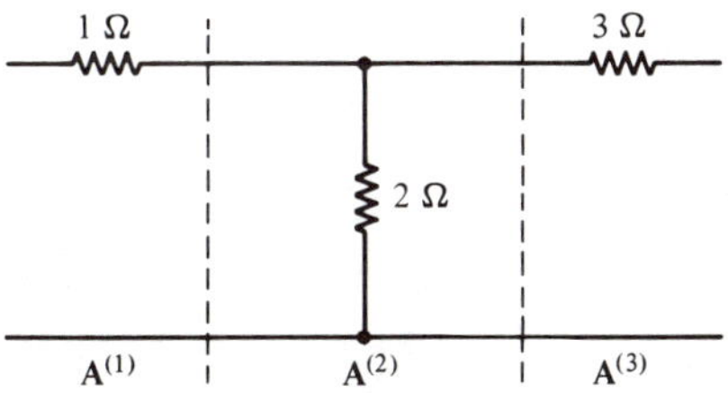

Figure 1.6

Therefore, if the inputs to the ladder network are

$$\begin{bmatrix} v \\ i \end{bmatrix} = \begin{bmatrix} 40 \\ 10 \end{bmatrix},$$

the output voltage and current would be

$$\frac{1}{2}\begin{bmatrix} 5 & -11 \\ -1 & 3 \end{bmatrix}\begin{bmatrix} 40 \\ 10 \end{bmatrix} = \frac{1}{2}\begin{bmatrix} 90 \\ -10 \end{bmatrix}.$$

■

Artificial Inbreeding

We now give a simple application of matrix powers to genetics. Inherited traits in plants are determined by a pair of genes each of which may be of two types, say G and g. A plant inherits one gene of its pair from each parent and it is assumed that these genes are selected at random.

Consider a population of plants which are reproduced by self-fertilization only. A plant of the GG type will then, of course, produce only GG plants and a gg plant will produce only gg offspring. However, a plant of the Gg type may produce plants of any of the three types. Such a plant will produce GG, Gg, and gg offspring in proportions $\frac{1}{4}$, $\frac{1}{2}$, and $\frac{1}{4}$, respectively.

Suppose that $x_1^{(0)}$, $x_2^{(0)}$, and $x_3^{(0)}$ are the proportions of plants in the original population of the types GG, Gg, and gg, respectively (note that $x_1^{(0)} + x_2^{(0)} + x_3^{(0)} = 1$). We may then represent this population by the vector

$$\mathbf{x}^{(0)} = \begin{bmatrix} x_1^{(0)} \\ x_2^{(0)} \\ x_3^{(0)} \end{bmatrix}.$$

The corresponding proportions in the next generation will be denoted by $\mathbf{x}^{(1)}$. Our assumptions then imply that $\mathbf{x}^{(0)}$ and $\mathbf{x}^{(1)}$ are related by

$$\mathbf{x}^{(1)} = \mathbf{H}\mathbf{x}^{(0)},$$

where

$$\mathbf{H} = \begin{bmatrix} 1 & \frac{1}{4} & 0 \\ 0 & \frac{1}{2} & 0 \\ 0 & \frac{1}{4} & 1 \end{bmatrix}.$$

To see this, note that, for example, a GG parent always has GG offspring, a Gg parent has GG offspring one-fourth of the time, and a gg parent never has a GG offspring. That is,

$$\begin{aligned} x_1^{(1)} &= \text{fraction of first generation of type } GG \\ &= 1 \cdot \text{fraction of original population of type } GG \\ &\quad + \tfrac{1}{4} \cdot \text{fraction of original population of type } Gg \\ &= 1 \cdot x_1^{(0)} + \tfrac{1}{4} \cdot x_2^{(0)} + 0 \cdot x_3^{(0)} \end{aligned}$$

and the other rows of $\mathbf{H}$ can be justified in a similar way. Now, in the second generation we have

$$\mathbf{x}^{(2)} = \mathbf{H}\mathbf{x}^{(1)} = \mathbf{H}^2\mathbf{x}^{(0)},$$

and in general we have

$$\mathbf{x}^{(n)} = \mathbf{H}^n\mathbf{x}^{(0)}$$

for the nth generation. Note that

$$\mathbf{H}^2 = \begin{bmatrix} 1 & \frac{1}{4} + \frac{1}{8} & 0 \\ 0 & \frac{1}{4} & 0 \\ 0 & \frac{1}{8} + \frac{1}{4} & 1 \end{bmatrix} = \begin{bmatrix} 1 & 2^{-1} - 2^{-3} & 0 \\ 0 & 2^{-2} & 0 \\ 0 & 2^{-1} - 2^{-3} & 1 \end{bmatrix}$$

and in general it is not difficult to convince oneself (see also Exercise 4) that

$$\mathbf{H}^n = \begin{bmatrix} 1 & 2^{-1} - 2^{-(n+1)} & 0 \\ 0 & 2^{-n} & 0 \\ 0 & 2^{-1} - 2^{-(n+1)} & 1 \end{bmatrix}.$$

In the nth generation we therefore find the proportions

$$\begin{aligned} x_1^{(n)} &= x_1^{(0)} + (2^{-1} - 2^{-(n+1)})x_1^{(0)} \\ x_2^{(n)} &= 2^{-n}x_2^{(0)} \\ x_3^{(n)} &= x_3^{(0)} + (2^{-1} - 2^{-(n+1)})x_2^{(0)}. \end{aligned}$$

Geneticists are often interested in *limiting populations*, that is, the values which the proportions approach as the number of generations grows without bound. In our example, as n gets arbitrarily large we therefore approach a limiting population with proportions of $x_1^{(0)} + \frac{1}{2}x_2^{(0)}$, 0, and $x_3^{(0)} + \frac{1}{2}x_2^{(0)}$ for the genotypes GG, Gg, and gg, respectively.

Computer Networks

Consider a network of computers connected by telephone lines. If computer C_i can communicate with (i.e., transmit data and commands to) computer C_j, we will write $C_i \rightarrow C_j$. In the case of two-way communication between these machines we write $C_i \leftrightarrow C_j$. The computer network may then be pictured as a graph as in Figure 1.7. The relationships between the computers in this network can also be described by a matrix. To do this we simply set $C_{ij} = 1$ if $C_i \rightarrow C_j$ and $C_{ij} = 0$ otherwise. We then obtain the 4×4 *communications matrix* (also called an incidence matrix in graph theory)

$$\mathbf{C} = \begin{bmatrix} 0 & 1 & 1 & 0 \\ 0 & 0 & 0 & 1 \\ 1 & 0 & 0 & 1 \\ 1 & 0 & 0 & 0 \end{bmatrix}.$$

Note that the diagonal entries of this matrix are zero because we are assuming that no computer communicates with itself (obviously, each computer communicates with

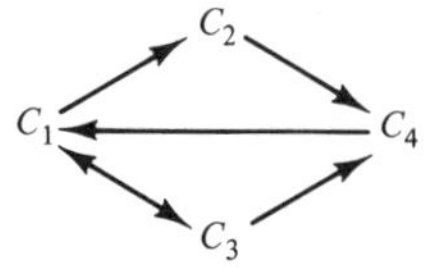

Figure 1.7

itself internally, but we are assuming that no computer talks to itself on the telephone). In the network above, computer C_1 cannot communicate directly with C_4; however, it is capable of two-stage communication with C_4 (e.g., via the link $C_1 \to C_2 \to C_4$).

We now show that such two-stage communication links may be described by use of the matrix $\mathbf{C}^2$. According to the definition of matrix multiplication, the i, j-entry of $\mathbf{C}^2$ is given by

$$c_{i1}c_{1j} + c_{i2}c_{2j} + c_{i3}c_{3j} + c_{i4}c_{4j}. \tag{10}$$

Now the term $c_{i1}c_{1j}$ is nonzero only if $c_{i1} = 1$ and $c_{1j} = 1$. That is, $c_{i1}c_{1j} = 1$ exactly when the links $C_i \to C_1$ and $C_1 \to C_j$ both exist; otherwise, $c_{i1}c_{1j} = 0$. But this is the same as saying that $c_{i1}c_{1j} = 1$ when the two-stage link $C_i \to C_1 \to C_j$ exists and $c_{i1}c_{1j} = 0$ otherwise. We therefore see that (10), that is, the i,j-entry of $\mathbf{C}^2$, represents the total number of two stage links from C_i to C_j. In our example

$$\mathbf{C}^2 = \begin{bmatrix} 1 & 0 & 0 & 2 \\ 1 & 0 & 0 & 0 \\ 1 & 1 & 1 & 0 \\ 0 & 1 & 1 & 0 \end{bmatrix}.$$

This shows that C_1 and C_3 can communicate with themselves via the two-stage links $C_1 \to C_3 \to C_1$ and $C_3 \to C_1 \to C_3$. Also, there are two two-stage links from C_1 to C_4, namely $C_1 \to C_2 \to C_4$ and $C_1 \to C_3 \to C_4$.

EXERCISES 1.4

1. Consider a two-sector economy with technology matrix

$$\mathbf{A} = \begin{bmatrix} .2 & .3 \\ .5 & .6 \end{bmatrix}.$$

(a) Suppose that the outside demands are

$$\mathbf{b} = \begin{bmatrix} 10 \\ 20 \end{bmatrix}.$$

Find the production levels $\mathbf{x}$.

(b) If the values added are

$$\mathbf{v} = \begin{bmatrix} 1.5 \\ .75 \end{bmatrix},$$

find the prices $\mathbf{p}$.

2. What is the economic meaning of $a_{ii} = 1$ in the technology matrix?

3. **(a)** Find the transfer matrix for the following network.

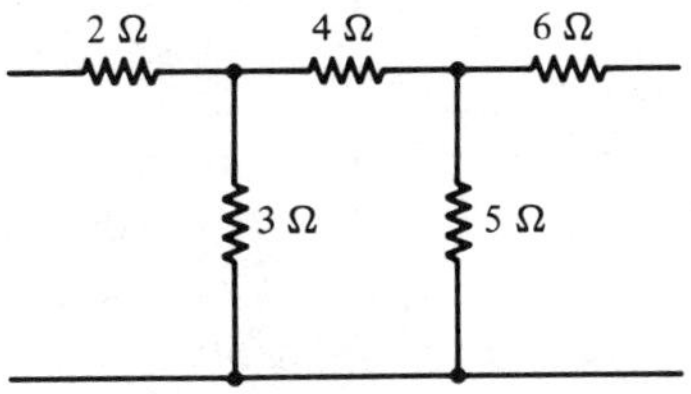

(b) Suppose that the input voltage is 8.4 V and the input current is 2 A. What is the output voltage and current?

4. **(a)** Let

$$\mathbf{S} = \begin{bmatrix} 1 & 1 & 1 \\ 0 & 1 & 0 \\ 2 & 1 & 0 \end{bmatrix}.$$

Verify that

$$\mathbf{S}^{-1} = \begin{bmatrix} 0 & -\frac{1}{2} & \frac{1}{2} \\ 0 & 1 & 0 \\ 1 & -\frac{1}{2} & -\frac{1}{2} \end{bmatrix}.$$

(b) Show that

$$\mathbf{H} = \mathbf{S}^{-1}\mathbf{diag}\,(1, \tfrac{1}{2}, 1)\mathbf{S},$$

where **H** is the matrix on page 36.

(c) Conclude by Exercise 22 of Section 1.3 that

$$\begin{aligned} \mathbf{H}^n &= \mathbf{S}^{-1}\mathbf{diag}\,(1, 2^{-n}, 1)\mathbf{S} \\ &= \begin{bmatrix} 1 & 2^{-1} - 2^{-(n+1)} & 0 \\ 0 & 2^{-n} & 0 \\ 0 & 2^{-1} - 2^{-(n+1)} & 1 \end{bmatrix}. \end{aligned}$$

5. Give the communication matrices for the following networks.

(a)

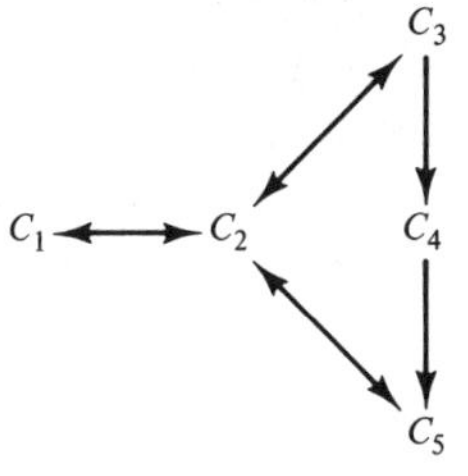

(b)

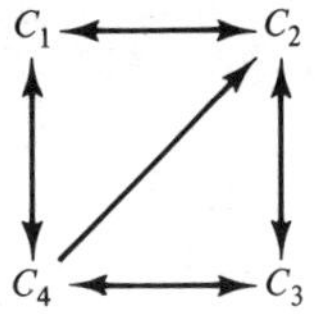

6. **(a)** Draw the network diagram for the following communication matrix.

$$\mathbf{C} = \begin{bmatrix} 0 & 1 & 0 & 0 \\ 0 & 0 & 1 & 1 \\ 1 & 1 & 0 & 0 \\ 1 & 1 & 1 & 0 \end{bmatrix}.$$

(b) Compute the two-stage communication matrix $\mathbf{C}^2$.

(c) What interpretation can be given to the matrix $\mathbf{C} + \mathbf{C}^2$?

GLOSSARY

column vector An $n \times 1$ matrix.
component An entry in a row or column vector.
consistent system A linear system that has a solution.
diagonal matrix A square matrix $\mathbf{A}$ that satisfies $a_{ij} = 0$ for $i \neq j$.
dimension The number of components in a row or column vector.
entry The number associated with a given row and column of a matrix.
identity matrix A diagonal matrix $\mathbf{I}$ each of whose diagonal entries is 1.
inverse matrix The inverse of a matrix $\mathbf{A}$ is the matrix $\mathbf{A}^{-1}$ satisfying $\mathbf{A}\mathbf{A}^{-1} = \mathbf{A}^{-1}\mathbf{A} = \mathbf{I}$.
lower triangular matrix A matrix $\mathbf{A}$ satisfying $a_{ij} = 0$ for $j > i$.
matrix A rectangular array of numbers.
overdetermined system A linear system with more equations than unknowns.
row vector A $1 \times n$ matrix.
size The number of rows and columns in a matrix.
square matrix A matrix with the same number of rows and columns.
symmetric matrix A matrix $\mathbf{A}$ satisfying $\mathbf{A} = \mathbf{A}^T$.
transpose matrix The transpose of an $m \times n$ matrix $\mathbf{A}$ is the $n \times m$ matrix $\mathbf{A}^T$ satisfying $a_{ij}^T = a_{ji}$.
underdetermined system A linear system with fewer equations than unknowns.
upper triangular matrix A matrix $\mathbf{A}$ satisfying $a_{ij} = 0$ if $i > j$.
zero matrix A matrix all of whose entries are zero.

NOTES AND COMMENTS

W. W. Leontief published the first input–output economic analysis, a 46-sector analysis of the U.S. economy, in 1936. A very readable introduction to input–output analysis is C. S. Yan, *Introduction to Input–Output Economics* (New York: Holt, Rinehart and Winston, 1969). More information on matrices in genetics and networks can be found in J. G. Kemeny, J. L. Snell, and G. L. Thompson, *Introduction to Finite Mathematics* (Englewood Cliffs, N.J.: Prentice-Hall, 1956). For more on matrices in electric circuit analysis, see E. A. Guillemin, *Mathematics of Circuit Analysis* (Cambridge, Mass.: The MIT Press, 1949).

2

Gaussian Elimination

In this chapter we introduce Gaussian elimination, one of the oldest, simplest, yet most effective methods for solving linear systems. We show that this method is essentially a way of factoring the coefficient matrix into two triangular matrices and solving two very simple linear systems. A naive computational complexity analysis is presented and it is shown how the algorithm can be adapted to compute inverses and to solve linear programming problems. A general discussion of linear systems is the subject of Chapter 4.

2.1 Simple Examples

The reader has probably already discovered that the solution of an $n \times n$ system of linear equations with a triangular coefficient matrix is a fairly simple matter (see Exercises 5–7 and especially 8, of Section 1.1). We illustrate with an example.

Example 2.1

To solve the system

$$\begin{aligned} x_1 + 2x_2 + x_3 &= 1 \\ 3x_2 - x_3 &= 3 \\ 2x_3 &= 6 \end{aligned}$$

note that from the third equation, $x_3 = 3$. Substituting this into the second equation we obtain

$$3x_2 - 3 = 3 \quad \text{and hence} \quad x_2 = 2.$$

Finally, substituting both of these results into the first equation gives

$$x_1 + 2 \cdot 2 + 3 = 1 \quad \text{and hence} \quad x_1 = -6.$$

■

This "back substitution" method of solution will work on any upper triangular system, provided that all of the diagonal coefficients are nonzero. It would therefore be very handy if there were a procedure for converting any $n \times n$ linear system into upper triangular form. In fact, there is such a procedure and you may already have discovered it yourself (see Exercises 5–8 of Section 1.1). The procedure is illustrated with a simple example.

Example 2.2

We will solve the 3×3 linear system

$$\begin{aligned} x_1 + 2x_2 + x_3 &= 1 \\ 2x_1 + 7x_2 + x_3 &= 5 \\ -x_1 + 10x_2 - 3x_3 &= 17 \end{aligned}$$

by first reducing it to upper triangular form. To do this we must produce zero coefficients in the second and third rows of the first column. Notice that if we subtract twice the first equation from the second, we obtain

$$\begin{aligned} x_1 + 2x_2 + x_3 &= 1 \\ 3x_2 - x_3 &= 3 \\ -x_1 + 10x_2 - 3x_3 &= 17. \end{aligned}$$

Adding the first equation to the third gives

$$\begin{aligned} x_1 + 2x_2 + x_3 &= 1 \\ 3x_2 - x_3 &= 3 \\ 12x_2 - 2x_3 &= 18. \end{aligned}$$

Finally, subtracting four times the second equation from the third equation, we obtain the upper triangular system

$$\begin{aligned} x_1 + 2x_2 + x_3 &= 1 \\ 3x_2 - x_3 &= 3 \\ 2x_3 &= 6 \end{aligned}$$

whose solution

$$\begin{bmatrix} -6 \\ 2 \\ 3 \end{bmatrix}$$

was found in Example 2.1. We will show in Section 2.2 that the upper triangular system that results from this procedure has the same solutions as the original system.

■

The method used in Examples 2.1 and 2.2 is a simple but powerful tool for solving linear systems called **Gaussian elimination** (after C. F. Gauss, 1777–1855). It consists of two stages, an **elimination phase**, in which multiples of rows are subtracted from other rows to produce an upper triangular system as in Example 2.2, and a **back substitution phase** in which the upper triangular system is solved, starting with the last equation, as in Example 2.1. In Section 2.2 we examine Gaussian elimination in more detail. But first we point out that all of the work which is done in Gaussian elimination is performed on the coefficients and the right-hand sides. Therefore, the computations can be streamlined by ignoring the names of the unknowns and dealing with the **augmented matrix** obtained by appending to the coefficient matrix the column vector of right-hand sides. We use this approach in the following example.

Example 2.3

To solve

$$\begin{aligned} 2x_1 - x_2 + 2x_3 &= 11 \\ 4x_1 \qquad - x_3 &= 5 \\ -x_1 + x_2 + 3x_3 &= 6 \end{aligned}$$

we first form the augmented matrix

$$\left[\begin{array}{rrr|r} 2 & -1 & 2 & 11 \\ 4 & 0 & -1 & 5 \\ -1 & 1 & 3 & 6 \end{array}\right].$$

In this matrix each row represents an equation of the system. The first step in the reduction to triangular form consists in subtracting twice the first row from the second:

$$\left[\begin{array}{rrr|r} 2 & -1 & 2 & 11 \\ 0 & 2 & -5 & -17 \\ -1 & 1 & 3 & 6 \end{array}\right].$$

Now we want to subtract some multiple of the first row from the third row in order to produce a 0 in the 3,1-entry. Call this multiple m. Then we require that $-1 - 2m = 0$, that is, $m = -\frac{1}{2}$. Thus we multiply the first row by $-\frac{1}{2}$ and subtract from the third row:

$$\left[\begin{array}{rrr|r} 2 & -1 & 2 & 11 \\ 0 & 2 & -5 & -17 \\ 0 & \frac{1}{2} & 4 & \frac{23}{2} \end{array}\right].$$

Finally, subtracting one-fourth of the second row from the third, we obtain the upper triangular form

$$\left[\begin{array}{rrr|r} 2 & -1 & 2 & 11 \\ 0 & 2 & -5 & -17 \\ 0 & 0 & \frac{21}{4} & \frac{63}{4} \end{array}\right].$$

Note that, throughout the computation, each row represents an equation.

We can solve this upper triangular system by back substitution. That is, we solve the system

$$\begin{aligned} 2x_1 - x_2 + 2x_3 &= 11 \\ 2x_2 - 5x_3 &= -17 \\ \tfrac{21}{4}x_3 &= \tfrac{63}{4} \end{aligned}$$

in reverse order. This gives

$$x_3 = \tfrac{63}{4} \div \tfrac{21}{4} = 3.$$

Substituting this back into the second equation gives

$$2x_2 - 5(3) = -17$$

and hence $x_2 = -1$. Finally, substituting these values for x_3 and x_2 into the first equation, we have

$$2x_1 - 1(-1) + 2(3) = 11$$

and hence $x_1 = 2$.

■

The method of back substitution illustrated in Example 2.3 can be used to solve an upper triangular system

$$\begin{aligned} a_{11}x_1 + a_{12}x_2 + \cdots + a_{1n}x_n &= b_1 \\ a_{22}x_2 + \cdots + a_{2n}x_n &= b_2 \\ \vdots \quad & \quad \vdots \\ a_{nn}x_n &= b_n \end{aligned} \tag{1}$$

provided that $a_{ii} \neq 0$ for all i. The process is quite simple and can easily be implemented on a computer (see Exercise 9).

EXERCISES 2.1

Use Gaussian elimination to solve the systems in Exercises 1–5. Use the augmented matrix as in Example 2.3.

1. $\begin{aligned} 2x_1 - x_2 &= 1 \\ -x_1 + 3x_2 &= 7 \end{aligned}$

2. $\begin{aligned} -3x_1 + 2x_2 &= 2 \\ x_1 + x_2 &= 6 \end{aligned}$

3. $\begin{aligned} x_1 + x_2 &= 1 \\ x_1 + x_3 &= 1 \\ x_2 + x_3 &= 1 \end{aligned}$

4. $\begin{aligned} x_1 + 2x_2 + 3x_3 &= 1 \\ 2x_1 + x_2 + x_3 &= 0 \\ 2x_2 - x_3 &= 1 \end{aligned}$

5. $\begin{aligned} 2x_1 + x_2 - 4x_3 + x_4 &= 3 \\ -x_1 + x_3 + 7x_4 &= 7 \\ x_1 + 2x_2 - x_3 + x_4 &= 1 \\ 7x_1 - 6x_2 + 4x_4 &= -3 \end{aligned}$

6. Let **M** be given by

$$\mathbf{M} = \begin{bmatrix} 1 & 0 & 0 & 0 \\ 0 & 1 & 0 & 0 \\ m & 0 & 1 & 0 \\ 0 & 0 & 0 & 1 \end{bmatrix}.$$

(a) Compute the matrix product **MA**, where **A** is a 4×4 matrix.
(b) Describe the effect on **A** of premultiplication by **M**.
(c) If

$$\mathbf{A} = \begin{bmatrix} 4 & -1 & 5 & 1 \\ 1 & 2 & 8 & 0 \\ 2 & 1 & 6 & -1 \\ 0 & 1 & 2 & 1 \end{bmatrix},$$

choose m in **M** such that the 3,2-entry of **MA** is zero.

7. Let

$$\mathbf{N} = \begin{bmatrix} 1 & 0 & 0 & 0 \\ 0 & 1 & 0 & 0 \\ -m & 0 & 1 & 0 \\ 0 & 0 & 0 & 1 \end{bmatrix}$$

and perform the multiplication **MN** and **NM**, where **M** is given in Exercise 6.

8. Let **P** be the 4×4 matrix which is obtained by interchanging the second and fourth rows of $\mathbf{I}_4$.

(a) Compute the matrix product **PA**, where **A** is a 4×4 matrix.

(b) Describe the effect on **A** of premultiplication by **P**.

(c) From part (b) conclude that $\mathbf{P}^2 = \mathbf{PP} = \mathbf{I}_4$ and hence $\mathbf{P} = \mathbf{P}^{-1}$.

9. (Computer Exercise) The back substitution method for solving (1) is easily implemented on a digital computer. The algorithm is essentially a single DO loop:

$$x_n \leftarrow b_n/a_{nn}$$

$$\text{DO} \quad i = n - 1, \ldots, 1;$$

$$x_i \leftarrow \left(b_i - \sum_{j=i+1}^{n} a_{ij}x_j\right) / a_{ii}.$$

Note that the unknowns are solved for in reverse order within the DO loop ($\leftarrow$ indicates value assignment).

(a) Write a program to perform back substitution.

(b) Use your program to solve the system

$$\begin{aligned} -2.10x_1 + 3.20x_2 + 5.79x_3 - 4.11x_4 &= 3.78 \\ 4.70x_2 - 3.21x_3 + 2.14x_4 &= 9.75 \\ 9.72x_3 - 9.69x_4 &= 5.11 \\ 8.78x_4 &= 4.53. \end{aligned}$$

2.2 Elementary Row Operations and Equivalent Systems

In Section 2.1 we saw that a basic row operation in the elimination process is the subtraction of a multiple of one row from another row. In this section we examine the elementary row operations which are used in Gaussian elimination and their connection with matrices.

The **elementary row operations** used in Gaussian elimination are:

1. Subtract a multiple, say l, of one row from another.
2. Interchange two rows.

We have not had occasion to use the latter operation; however, we shall see in Section 2.3 that this operation is crucial when applying elimination to some linear systems. Associated with each elementary row operation is a matrix called an **elementary matrix**. The elementary matrix produces the required elementary row operation when used as a premultiplier. To be more precise, suppose that $\mathbf{A}$ is a matrix on which we want to perform an elementary row operation. The corresponding elementary matrix, call it $\mathbf{E}$, accomplishes the required row operation by premultiplying $\mathbf{A}$, that is, the matrix product $\mathbf{EA}$ results in the desired row operation.

Let us consider the first type of elementary row operation and corresponding elementary matrix. Consider the linear system

$$\begin{aligned} 2x_1 + 4x_2 &= -1 \\ 3x_1 + 6x_2 + x_3 &= 2 \\ -x_1 + x_2 + 2x_3 &= 4, \end{aligned}$$

or equivalently,

$$\mathbf{Ax} = \begin{bmatrix} 2 & 4 & 0 \\ 3 & 6 & 1 \\ -1 & 1 & 2 \end{bmatrix} \begin{bmatrix} x_1 \\ x_2 \\ x_3 \end{bmatrix} = \begin{bmatrix} -1 \\ 2 \\ 4 \end{bmatrix} = \mathbf{b}.$$

For now our primary interest is the interpretation of elementary row operations on the coefficient matrix $\mathbf{A}$. Of course, in solving a linear system the same operations could be performed on $\mathbf{b}$ by dealing with the augmented matrix.

The first step in the elimination process is to subtract a multiple, call it l_{21}, of the first row of $\mathbf{A}$ from the second row of $\mathbf{A}$ to produce a 0 in the 2,1-entry. Thus l_{21} is chosen such that

$$a_{21} - l_{21}a_{11} = 0.$$

Clearly, $l_{21} = \frac{3}{2}$ and the corresponding elementary matrix $\mathbf{E}_{21}$ which accomplishes the first step is (see Exercise 6 of Section 2.1)

$$\mathbf{E}_{21} = \begin{bmatrix} 1 & 0 & 0 \\ -l_{21} & 1 & 0 \\ 0 & 0 & 1 \end{bmatrix} = \begin{bmatrix} 1 & 0 & 0 \\ -\frac{3}{2} & 1 & 0 \\ 0 & 0 & 1 \end{bmatrix},$$

which differs from the 3×3 identity matrix only in the 2,1-entry. Indeed, we note that

$$\mathbf{E}_{21}\mathbf{A} = \begin{bmatrix} 1 & 0 & 0 \\ -l_{21} & 1 & 0 \\ 0 & 0 & 1 \end{bmatrix} \begin{bmatrix} a_{11} & a_{12} & a_{13} \\ a_{21} & a_{22} & a_{23} \\ a_{31} & a_{32} & a_{33} \end{bmatrix}$$

$$= \begin{bmatrix} a_{11} & a_{12} & a_{13} \\ a_{21} - l_{21}a_{11} & a_{22} - l_{21}a_{12} & a_{23} - l_{21}a_{13} \\ a_{31} & a_{32} & a_{33} \end{bmatrix}$$

and hence

$$\mathbf{E}_{21}\mathbf{A} = \begin{bmatrix} 2 & 4 & 0 \\ 0 & 0 & 1 \\ -1 & 1 & 2 \end{bmatrix} \qquad \text{since } l_{21} = \tfrac{3}{2}.$$

To illustrate further, suppose that we consider the second step of the elimination method. To produce a zero in the 3,1-entry we find the multiplier $l_{31} = -\frac{1}{2}$ and corresponding elementary matrix

$$\mathbf{E}_{31} = \begin{bmatrix} 1 & 0 & 0 \\ 0 & 1 & 0 \\ -l_{31} & 0 & 1 \end{bmatrix} = \begin{bmatrix} 1 & 0 & 0 \\ 0 & 1 & 0 \\ \frac{1}{2} & 0 & 1 \end{bmatrix}.$$

Then we have

$$\mathbf{E}_{31}\,\mathbf{E}_{21}\,\mathbf{A} = \mathbf{E}_{31}\begin{bmatrix} 2 & 4 & 0 \\ 0 & 0 & 1 \\ -1 & 1 & 2 \end{bmatrix} = \begin{bmatrix} 2 & 4 & 0 \\ 0 & 0 & 1 \\ 0 & 3 & 2 \end{bmatrix}.$$

By now it should be clear how these elementary matrices are formed. To subtract a multiple l_{ij} of row j from row i, form $\mathbf{E}_{ij}$ as follows: Replace the zero in the i,j-entry of the identity matrix $\mathbf{I}$ by $-l_{ij}$.

Before considering row interchanges and the corresponding matrices, we make two observations about the elementary matrices $\mathbf{E}_{ij}$. First, each of the elementary matrices is lower triangular, and hence by Exercise 18 of Section 1.3 it follows that the product of two such matrices is lower triangular. For example,

$$\mathbf{E}_{31}\,\mathbf{E}_{21} = \begin{bmatrix} 1 & 0 & 0 \\ 0 & 1 & 0 \\ -l_{31} & 0 & 1 \end{bmatrix}\begin{bmatrix} 1 & 0 & 0 \\ -l_{21} & 1 & 0 \\ 0 & 0 & 1 \end{bmatrix}$$
$$= \begin{bmatrix} 1 & 0 & 0 \\ -l_{21} & 1 & 0 \\ -l_{31} & 0 & 1 \end{bmatrix}.$$

Another important property of these elementary matrices is that they are invertible, and moreover their inverses are easy to determine. Consider, for example, the first step in the elimination process on the previous linear system. We subtract $l_{21} = \frac{3}{2}$ times the first row from the second row. To undo this operation we *add* $\frac{3}{2}$ times the first row to the second row. Thus the inverse of

$$\mathbf{E}_{21} = \begin{bmatrix} 1 & 0 & 0 \\ -\frac{3}{2} & 1 & 0 \\ 0 & 0 & 1 \end{bmatrix}$$

is simply given by

$$\mathbf{E}_{21}^{-1} = \begin{bmatrix} 1 & 0 & 0 \\ \frac{3}{2} & 1 & 0 \\ 0 & 0 & 1 \end{bmatrix}.$$

The reader can check that

$$\begin{bmatrix} 1 & 0 & 0 \\ \frac{3}{2} & 1 & 0 \\ 0 & 0 & 1 \end{bmatrix}\begin{bmatrix} 1 & 0 & 0 \\ -\frac{3}{2} & 1 & 0 \\ 0 & 0 & 1 \end{bmatrix} = \mathbf{I}$$

and

$$\begin{bmatrix} 1 & 0 & 0 \\ -\frac{3}{2} & 1 & 0 \\ 0 & 0 & 1 \end{bmatrix} \begin{bmatrix} 1 & 0 & 0 \\ \frac{3}{2} & 1 & 0 \\ 0 & 0 & 1 \end{bmatrix} = \mathbf{I}.$$

Therefore, the inverse of $\mathbf{E}_{ij}$ is found by changing the sign of its i,j-entry.

The interchange of two rows can be accomplished in matrix terms by premultiplication by a different type of elementary matrix called an **elementary permutation matrix**. To interchange rows i and j of $\mathbf{A}$ we form the elementary permutation matrix $\mathbf{P}_{ij}$ as follows: Interchange rows i and j of the identity matrix. (Note that $\mathbf{P}_{ij} = \mathbf{P}_{ji}$.) For example, to interchange rows 2 and 3 of the matrix

$$\begin{bmatrix} 2 & 4 & 0 \\ 0 & 0 & 1 \\ 0 & 3 & 2 \end{bmatrix},$$

we form

$$\mathbf{P}_{23} = \begin{bmatrix} 1 & 0 & 0 \\ 0 & 0 & 1 \\ 0 & 1 & 0 \end{bmatrix}$$

and calculate the matrix product

$$\mathbf{P}_{23} \begin{bmatrix} 2 & 4 & 0 \\ 0 & 0 & 1 \\ 0 & 3 & 2 \end{bmatrix} = \begin{bmatrix} 1 & 0 & 0 \\ 0 & 0 & 1 \\ 0 & 1 & 0 \end{bmatrix} \begin{bmatrix} 2 & 4 & 0 \\ 0 & 0 & 1 \\ 0 & 3 & 2 \end{bmatrix}$$

$$= \begin{bmatrix} 2 & 4 & 0 \\ 0 & 3 & 2 \\ 0 & 0 & 1 \end{bmatrix}.$$

Thus the elementary permutation matrix that is formed by interchanging rows i and j of the identity has the effect of interchanging rows i and j of the matrix that it premultiplies (see Exercise 8 of Section 2.1).

To undo a row interchange, we perform the same row interchange again, that is,

$$\mathbf{P}_{ij}(\mathbf{P}_{ij}\mathbf{A}) = \mathbf{A}.$$

It follows that

$$\mathbf{P}_{ij}\mathbf{P}_{ij} = \mathbf{P}_{ij}^2 = \mathbf{I}$$

and hence $\mathbf{P}_{ij}^{-1} = \mathbf{P}_{ij}$ for any elementary permutation matrix.

The product of any number of elementary permutation matrices is simply called a **permutation matrix**. For example,

$$\mathbf{P} = \begin{bmatrix} 0 & 1 & 0 \\ 0 & 0 & 1 \\ 1 & 0 & 0 \end{bmatrix}$$

is a permutation matrix since $\mathbf{P} = \mathbf{P}_{12}\mathbf{P}_{13}$. A permutation matrix has a single 1 in each row and column—all other entries are zero.

Next we introduce the idea of equivalent linear systems and its connection with Gaussian elimination and elementary row operations. We say that two linear systems

$$\mathbf{Ax} = \mathbf{b} \quad \text{and} \quad \mathbf{Ux} = \mathbf{c}$$

are **equivalent** if every solution of one system is a solution of the other. Thus $\mathbf{Ax} = \mathbf{b}$ and $\mathbf{Ux} = \mathbf{c}$ are equivalent if they have the same set of solutions. Now suppose that $\mathbf{Ax} = \mathbf{b}$ is a given system to which an elementary row operation is applied, say $\mathbf{E}_{ij}$. Let $\mathbf{U} = \mathbf{E}_{ij}\mathbf{A}$ and $\mathbf{c} = \mathbf{E}_{ij}\mathbf{b}$; then $\mathbf{Ux} = \mathbf{c}$ is equivalent to $\mathbf{Ax} = \mathbf{b}$. Clearly, if $\mathbf{x}$ satisfies $\mathbf{Ax} = \mathbf{b}$, then $\mathbf{Ux} = \mathbf{E}_{ij}\mathbf{Ax} = \mathbf{E}_{ij}\mathbf{b} = \mathbf{c}$ and hence $\mathbf{x}$ is a solution of $\mathbf{Ux} = \mathbf{c}$. Conversely, if $\mathbf{x}$ is a solution of $\mathbf{Ux} = \mathbf{c}$, then $\mathbf{Ax} = \mathbf{E}_{ij}^{-1}\mathbf{Ux} = \mathbf{E}_{ij}^{-1}\mathbf{c} = \mathbf{b}$ and $\mathbf{x}$ satisfies $\mathbf{Ax} = \mathbf{b}$. Similarly, it follows that premultiplication by an elementary permutation matrix produces an equivalent system. We have the following theorem.

Theorem 2.1. Suppose that $\mathbf{Ax} = \mathbf{b}$ is subjected to a sequence of elementary row operations to produce another system $\mathbf{Ux} = \mathbf{c}$. Then $\mathbf{Ax} = \mathbf{b}$ and $\mathbf{Ux} = \mathbf{c}$ are equivalent linear systems.

■

This theorem has important consequence for Gaussian elimination. It guarantees that the upper triangular system which results from the elimination process has the same solutions as the original problem.

EXERCISES 2.2

1. Let

$$\mathbf{A} = \begin{bmatrix} 2 & 1 & 1 & 1 \\ 1 & -1 & 2 & 0 \\ 4 & 3 & 6 & -3 \end{bmatrix}.$$

Find the matrices obtained by performing the following elementary row operations on $\mathbf{A}$.

(a) Interchange the second and third rows.

(b) Subtract -2 times the second row from the third row.

(c) Perform part (a) *and* then part (b).

(d) Perform part (b) *and* then part (a).

(e) Compare the result of parts(c) and (d).

2. Find the elementary matrices corresponding to parts (a) and (b) of Exercise 1. Verify that premultiplication by the appropriate elementary matrix agrees with the results of parts (a) and (b).

3. Find the following 4×4 permutation matrices.

(a) $\mathbf{P}_{24}$ **(b)** $\mathbf{P}_{13}$ **(c)** $\mathbf{P}_{42}^{-1}$ **(d)** $\mathbf{P} = \mathbf{P}_{13}\mathbf{P}_{24}$

4. Find the inverse of

$$\mathbf{B} = \begin{bmatrix} 1 & 0 & 0 & 0 \\ 0 & 1 & 0 & 0 \\ 0 & -2 & 1 & 0 \\ 0 & 3 & 1 & 1 \end{bmatrix}.$$

(*Hint:* Write $\mathbf{B}$ as the product of elementary matrices.)

5. Let **B** be given as in Exercise 4. Find elementary permutation matrices **P** and **Q** such that

(a)
$$\mathbf{PB} = \begin{bmatrix} 0 & -2 & 1 & 0 \\ 0 & 1 & 0 & 0 \\ 1 & 0 & 0 & 0 \\ 0 & 3 & 1 & 1 \end{bmatrix}$$

(b)
$$\mathbf{Q}\begin{bmatrix} 0 & 3 & 1 & 1 \\ 0 & 1 & 0 & 0 \\ 0 & -2 & 1 & 0 \\ 1 & 0 & 0 & 0 \end{bmatrix} = \mathbf{B}$$

6. What is the inverse of the following product of 3×3 elementary matrices: $\mathbf{P}_{23}\mathbf{E}_{31}\mathbf{E}_{21}$?

7. What is the effect of postmultiplying an $n \times n$ matrix **A** by the $n \times n$ permutation matrix $\mathbf{P}_{ij}$? That is, how is $\mathbf{AP}_{ij}$ related to **A**?

8. (a) Compute $\mathbf{P}_{12}\mathbf{P}_{12}^{\mathrm{T}}$ for the 3×3 permutation matrix $\mathbf{P}_{12}$.

(b) Show that if $\mathbf{P}_{ij}$ is the $n \times n$ permutation matrix obtained from **I** by interchanging the *ith* and *jth* rows, then $\mathbf{P}_{ij}\mathbf{P}_{ij}^{\mathrm{T}} = \mathbf{I}$ and $\mathbf{P}_{ij}^{\mathrm{T}}\mathbf{P}_{ij} = \mathbf{I}$.

9. (a) Suppose that **P** is an $n \times n$ permutation matrix. Verify that $\mathbf{PP}^{\mathrm{T}} = \mathbf{I} = \mathbf{P}^{\mathrm{T}}\mathbf{P}$. [*Hint:* See part (b) of Exercise 8.]

(b) Conclude that $\mathbf{P}^{-1} = \mathbf{P}^{\mathrm{T}}$ if **P** is an $n \times n$ permutation matrix.

2.3 LU Factorization

Previously, it was demonstrated how elimination can be implemented using an augmented matrix and elementary row operations. These elementary row operations can be accomplished by premultiplication with appropriate elementary matrices. Our aim in this section is to consider Gaussian elimination from a different point of view—one in which the elimination phase is interpreted as a factorization of the coefficient matrix. This viewpoint is primarily of a theoretical nature but has important practical consequences. We also show, by example, that row interchanges are required in the elimination process for certain linear systems.

We begin by reconsidering Example 2.2 in terms of premultiplication by elementary matrices. The system of Example 2.2 in matrix form is

$$\mathbf{Ax} = \begin{bmatrix} 1 & 2 & 1 \\ 2 & 7 & 1 \\ -1 & 10 & -3 \end{bmatrix}\begin{bmatrix} x_1 \\ x_2 \\ x_3 \end{bmatrix} = \mathbf{b} = \begin{bmatrix} 1 \\ 5 \\ 17 \end{bmatrix}.$$

Our first step in the elimination process for this system was to subract $l_{21} = 2$ times the first row from the second. We form $\mathbf{E}_{21}$ and premultiply **A** and **b** to find

$$\mathbf{E}_{21}\mathbf{A} = \begin{bmatrix} 1 & 0 & 0 \\ -2 & 1 & 0 \\ 0 & 0 & 1 \end{bmatrix}\begin{bmatrix} 1 & 2 & 1 \\ 2 & 7 & 1 \\ -1 & 10 & -3 \end{bmatrix} = \begin{bmatrix} 1 & 2 & 1 \\ 0 & 3 & -1 \\ -1 & 10 & -3 \end{bmatrix}$$

$$\mathbf{E}_{21}\mathbf{b} = \begin{bmatrix} 1 & 0 & 0 \\ -2 & 1 & 0 \\ 0 & 0 & 1 \end{bmatrix}\begin{bmatrix} 1 \\ 5 \\ 17 \end{bmatrix} = \begin{bmatrix} 1 \\ 3 \\ 17 \end{bmatrix}.$$

The next step in the elimination process is subtracting $l_{31} = -1$ times the first equation from the third. This is accomplished by premultiplying by the elementary matrix

$$\mathbf{E}_{31} = \begin{bmatrix} 1 & 0 & 0 \\ 0 & 1 & 0 \\ 1 & 0 & 1 \end{bmatrix}.$$

The coefficient matrix and right-hand side are then transformed as follows:

$$\mathbf{E}_{31}\,\mathbf{E}_{21}\,\mathbf{A} = \begin{bmatrix} 1 & 0 & 0 \\ 0 & 1 & 0 \\ 1 & 0 & 1 \end{bmatrix} \begin{bmatrix} 1 & 2 & 1 \\ 0 & 3 & -1 \\ -1 & 10 & -3 \end{bmatrix} = \begin{bmatrix} 1 & 2 & 1 \\ 0 & 3 & -1 \\ 0 & 12 & -2 \end{bmatrix}$$

$$\mathbf{E}_{31}\,\mathbf{E}_{21}\,\mathbf{b} = \begin{bmatrix} 1 & 0 & 0 \\ 0 & 1 & 0 \\ 1 & 0 & 1 \end{bmatrix} \begin{bmatrix} 1 \\ 3 \\ 17 \end{bmatrix} = \begin{bmatrix} 1 \\ 3 \\ 18 \end{bmatrix}.$$

The last step, subtracting $4 = l_{32}$ times the second equation from the third equation, is accomplished by premultiplying by the elementary matrix

$$\mathbf{E}_{32} = \begin{bmatrix} 1 & 0 & 0 \\ 0 & 1 & 0 \\ 0 & -4 & 1 \end{bmatrix}.$$

Indeed, we have

$$\begin{bmatrix} 1 & 0 & 0 \\ 0 & 1 & 0 \\ 0 & -4 & 1 \end{bmatrix} \begin{bmatrix} 1 & 2 & 1 \\ 0 & 3 & -1 \\ 0 & 12 & -2 \end{bmatrix} = \begin{bmatrix} 1 & 2 & 1 \\ 0 & 3 & -1 \\ 0 & 0 & 2 \end{bmatrix}$$

$$\begin{bmatrix} 1 & 0 & 0 \\ 0 & 1 & 0 \\ 0 & -4 & 1 \end{bmatrix} \begin{bmatrix} 1 \\ 3 \\ 18 \end{bmatrix} = \begin{bmatrix} 1 \\ 3 \\ 6 \end{bmatrix}.$$

Therefore, we see that the original coefficient matrix

$$\mathbf{A} = \begin{bmatrix} 1 & 2 & 1 \\ 2 & 7 & 1 \\ -1 & 10 & -3 \end{bmatrix}$$

is transformed, in the elimination phase, to the final upper triangular matrix

$$\mathbf{U} = \begin{bmatrix} 1 & 2 & 1 \\ 0 & 3 & -1 \\ 0 & 0 & 2 \end{bmatrix}$$

in the following way:

$$\mathbf{U} = \mathbf{E}_{32}\,\mathbf{E}_{31}\,\mathbf{E}_{21}\,\mathbf{A}. \tag{2}$$

Recall that these elementary matrices are invertible and in fact their inverses are very easy to find: Simply change the sign of the off-diagonal entry. Therefore, we find

$$\mathbf{E}_{32}^{-1} = \begin{bmatrix} 1 & 0 & 0 \\ 0 & 1 & 0 \\ 0 & 4 & 1 \end{bmatrix}$$

and similarly

$$\mathbf{E}_{31}^{-1} = \begin{bmatrix} 1 & 0 & 0 \\ 0 & 1 & 0 \\ -1 & 0 & 1 \end{bmatrix}, \qquad \mathbf{E}_{21}^{-1} = \begin{bmatrix} 1 & 0 & 0 \\ 2 & 1 & 0 \\ 0 & 0 & 1 \end{bmatrix}.$$

From (2) we then have

$$\mathbf{A} = \mathbf{E}_{21}^{-1}\, \mathbf{E}_{31}^{-1}\, \mathbf{E}_{32}^{-1}\, \mathbf{U}. \tag{3}$$

Since each of the matrices $\mathbf{E}_{21}^{-1}$, $\mathbf{E}_{31}^{-1}$, $\mathbf{E}_{32}^{-1}$ is lower triangular, we know (see Exercise 18 of Section 1.3) that their product is lower triangular. But even more is true. Note that

$$\mathbf{E}_{21}^{-1}\, \mathbf{E}_{31}^{-1} = \begin{bmatrix} 1 & 0 & 0 \\ 2 & 1 & 0 \\ 0 & 0 & 1 \end{bmatrix}\begin{bmatrix} 1 & 0 & 0 \\ 0 & 1 & 0 \\ -1 & 0 & 1 \end{bmatrix}$$

$$= \begin{bmatrix} 1 & 0 & 0 \\ 2 & 1 & 0 \\ -1 & 0 & 1 \end{bmatrix}$$

and hence

$$\mathbf{E}_{21}^{-1}\, \mathbf{E}_{31}^{-1}\, \mathbf{E}_{32}^{-1} = \begin{bmatrix} 1 & 0 & 0 \\ 2 & 1 & 0 \\ -1 & 0 & 1 \end{bmatrix}\begin{bmatrix} 1 & 0 & 0 \\ 0 & 1 & 0 \\ 0 & 4 & 1 \end{bmatrix}$$

$$= \begin{bmatrix} 1 & 0 & 0 \\ 2 & 1 & 0 \\ -1 & 4 & 1 \end{bmatrix} = \begin{bmatrix} 1 & 0 & 0 \\ l_{21} & 1 & 0 \\ l_{31} & l_{32} & 1 \end{bmatrix}.$$

That is, the product of such elementary matrices is obtained simply by "filling in" the multipliers l_{ij} in the corresponding below diagonal entries. Let $\mathbf{L} = \mathbf{E}_{21}^{-1}\, \mathbf{E}_{31}^{-1}\, \mathbf{E}_{32}^{-1}$ (note that the order of the product is important); then by (3) we have

$$\mathbf{A} = \mathbf{LU}. \tag{4}$$

To put it another way, we may think of the elimination phase of the Gauss algorithm as a method of *factoring* the matrix **A** into a product of a lower triangular matrix **L** times an upper triangular matrix **U**. The factorization (4) is therefore called the **LU factorization** of **A**.

Let's take a closer look at the matrix **L**. We have seen that it is lower triangular and all of its diagonal entries are 1. But the important thing to notice is that the subdiagonal entries are just the multipliers used in the elimination process: $l_{21} = 2$ (the first row was multiplied by 2 and subtracted from the second), $l_{31} = -1$ (the first row was multiplied by -1 and subtracted from the third), and $l_{32} = 4$ (the second row was multiplied by 4 and subtracted from the third). In general, l_{ij} is the number that is multiplied times the jth row in order to produce a 0 in the i, jth subdiagonal entry on subtraction from the jth row. Therefore, by keeping track of the multipliers in the elimination phase we are able to construct **L** directly.

Example 2.4

We will find the LU factorization of

$$\begin{bmatrix} 2 & -1 & 2 \\ -4 & 1 & -1 \\ 6 & -5 & 16 \end{bmatrix}.$$

The first multiplier in the elimination process is $l_{21} = -2$ and the second multiplier is $l_{31} = 3$. Performing the corresponding row operations, we obtain

$$\begin{bmatrix} 2 & -1 & 2 \\ 0 & -1 & 3 \\ 0 & -2 & 10 \end{bmatrix}.$$

The next multiplier in the elimination is $l_{32} = 2$ and the elimination step gives

$$\mathbf{U} = \begin{bmatrix} 2 & -1 & 2 \\ 0 & -1 & 3 \\ 0 & 0 & 4 \end{bmatrix}.$$

Taking account of the multipliers, we find

$$\mathbf{L} = \begin{bmatrix} 1 & 0 & 0 \\ -2 & 1 & 0 \\ 3 & 2 & 1 \end{bmatrix}.$$

■

Once a factorization $\mathbf{A} = \mathbf{LU}$ has been obtained, solving the system $\mathbf{Ax} = \mathbf{b}$ is equivalent to solving

$$\mathbf{LUx} = \mathbf{b},$$

that is, to solving

$$\mathbf{Ly} = \mathbf{b} \quad \text{and} \quad \mathbf{Ux} = \mathbf{y}.$$

Each of these systems is triangular and hence very easy to solve. In fact, solving $\mathbf{Ly} = \mathbf{b}$ is equivalent to performing the elimination phase on the original right-hand side, and solving $\mathbf{Ux} = \mathbf{y}$ is just the back-substitution phase. It is important to note that the LU factorization is independent of the right-hand-side vector **b**. Thus, given **L** and **U**, it is easy to solve $\mathbf{Ax} = \mathbf{b}$ for *any* right-hand side **b**.

Example 2.5

In order to solve the system

$$\begin{aligned} 2x_1 - x_2 + 2x_3 &= 11 \\ -4x_1 + x_2 - x_3 &= -12 \\ 6x_1 - 5x_2 + 16x_3 &= 65, \end{aligned}$$

note that from Example 2.4 we have the *LU* factorization

$$\begin{bmatrix} 2 & -1 & 2 \\ -4 & 1 & -1 \\ 6 & -5 & 16 \end{bmatrix} = \begin{bmatrix} 1 & 0 & 0 \\ -2 & 1 & 0 \\ 3 & 2 & 1 \end{bmatrix} \begin{bmatrix} 2 & -1 & 2 \\ 0 & -1 & 3 \\ 0 & 0 & 4 \end{bmatrix}.$$

Solving $\mathbf{Ly} = \mathbf{b}$, that is, solving

$$\begin{aligned} y_1 \qquad\qquad &= 11 \\ -2y_1 + y_2 \qquad &= -12 \\ 3y_1 + 2y_2 + y_3 &= 65 \end{aligned}$$

gives $y_1 = 11$, $y_2 = 10$, $y_3 = 12$ (in this order). Solving $\mathbf{Ux} = \mathbf{y}$,

$$\begin{aligned} 2x_1 - x_2 + 2x_3 &= 11 \\ - x_2 + 3x_3 &= 10 \\ 4x_3 &= 12, \end{aligned}$$

we find $x_3 = 3$, $x_2 = -1$, $x_1 = 2$.

■

In Examples 2.1 through 2.5 we found that the *i*th diagonal entry which occurs during the *i*th stage of the elimination process is nonzero; that is, the diagonal entries of the matrix **U** are nonzero. Such an entry is called a **pivot** for the elimination process. Clearly, if some pivot is zero, then in general no multiplier will serve to produce zeros in the column under the pivot. This difficulty is usually solved quite easily by interchanging two equations (or equivalently, two rows in the augmented matrix).

Example 2.6

In solving the system

$$\mathbf{Ax} = \begin{bmatrix} 0 & 1 \\ 2 & 3 \end{bmatrix} \begin{bmatrix} x_1 \\ x_2 \end{bmatrix} = \begin{bmatrix} 4 \\ 5 \end{bmatrix} = \mathbf{b},$$

the very first pivot is zero. No multiple of the first row of **A** when subtracted from the second row will produce a zero in the 2,1-entry. However, if we simply exchange the first and second equations, we obtain the equivalent system

$$\begin{bmatrix} 2 & 3 \\ 0 & 1 \end{bmatrix} \begin{bmatrix} x_1 \\ x_2 \end{bmatrix} = \begin{bmatrix} 5 \\ 4 \end{bmatrix},$$

which is already in upper triangular form.

■

This interchange of rows can be described in matrix terms by premultiplication by the 2×2 elementary permutation matrix $\mathbf{P}_{12}$. In Example 2.6 note that

$$\mathbf{P}_{12}\mathbf{A} = \begin{bmatrix} 0 & 1 \\ 1 & 0 \end{bmatrix} \begin{bmatrix} 0 & 1 \\ 2 & 3 \end{bmatrix} = \begin{bmatrix} 2 & 3 \\ 0 & 1 \end{bmatrix}.$$

The elementary permutation matrix

$$\mathbf{P}_{12} = \begin{bmatrix} 0 & 1 \\ 1 & 0 \end{bmatrix}$$

has the effect of interchanging the first and second rows of a matrix which it premultiplies.

Recall that the product of elementary permutation matrices is a permutation matrix. Also, if we keep track of the row interchanges needed to produce an *LU* factorization of a matrix $\mathbf{A}$ and record them in a permutation matrix $\mathbf{P}$, then we can imagine that these interchanges are performed ahead of time and in the end obtain the factorization in the form

$$\mathbf{PA} = \mathbf{LU}.$$

Example 2.7

Consider the matrix

$$\mathbf{A} = \begin{bmatrix} 1 & 3 & 1 \\ 2 & 6 & 1 \\ 1 & 1 & 2 \end{bmatrix}.$$

The first stage of the elimination process leads to

$$\begin{bmatrix} 1 & 3 & 1 \\ 0 & 0 & -1 \\ 0 & -2 & 1 \end{bmatrix}.$$

In view of the zero pivot in the 2,2-entry, it is necessary to exchange the second and third rows to continue. Therefore, to produce an *LU* factorization it is necessary to premultiply by the permutation matrix

$$\mathbf{P} = \mathbf{P}_{23} = \begin{bmatrix} 1 & 0 & 0 \\ 0 & 0 & 1 \\ 0 & 1 & 0 \end{bmatrix}.$$

Then elimination on $\mathbf{PA}$ gives the result

$$\mathbf{PA} = \mathbf{LU}.$$

Indeed, we have

$$\mathbf{PA} = \begin{bmatrix} 1 & 0 & 0 \\ 0 & 0 & 1 \\ 0 & 1 & 0 \end{bmatrix} \begin{bmatrix} 1 & 3 & 1 \\ 2 & 6 & 1 \\ 1 & 1 & 2 \end{bmatrix} = \begin{bmatrix} 1 & 3 & 1 \\ 1 & 1 & 2 \\ 2 & 6 & 1 \end{bmatrix}$$

and the first stage of elimination applied to **PA** gives

$$\begin{bmatrix} 1 & 3 & 1 \\ 0 & -2 & 1 \\ 0 & 0 & -1 \end{bmatrix}$$

with corresponding multipliers $l_{21} = 1$ and $l_{31} = 2$. It follows that

$$\mathbf{PA} = \mathbf{LU} = \begin{bmatrix} 1 & 0 & 0 \\ 1 & 1 & 0 \\ 2 & 0 & 1 \end{bmatrix} \begin{bmatrix} 1 & 3 & 1 \\ 0 & -2 & 1 \\ 0 & 0 & -1 \end{bmatrix}.$$

■

Thus the net effect of the elimination process, which requires row interchanges, is first to perform the row interchanges on **A** and then to perform Gaussian elimination (without row interchanges) on **PA** to produce **LU**.

In the next example we show how to obtain the *LU* factorization of **PA** as a by-product of the elimination process without reapplying elimination to **PA** as we did in Example 2.7.

Example 2.8

Consider the system $\mathbf{Ax} = \mathbf{b}$ with

$$\mathbf{A} = \begin{bmatrix} -1 & 1 & 2 & -2 \\ 3 & -3 & -6 & 12 \\ -2 & 4 & 3 & 2 \\ -5 & 1 & 17 & -24 \end{bmatrix} \qquad \mathbf{b} = \begin{bmatrix} -1 \\ 3 \\ 3 \\ -20 \end{bmatrix}.$$

We apply elimination to the corresponding augmented matrix and keep track of the process by recording two matrices and a "pointer" vector **p**. The vector **p** will record the row interchanges and initially will be $\mathbf{p} = [1, 2, 3, 4]^T$ since no interchanges have yet been made. The two matrices that we record are the augmented matrix and a 4×4 lower triangular matrix whose below-diagonal entries will be used to store the multipliers. These below-diagonal entries are specified as elimination proceeds.

We start with

$$\begin{array}{ccc} \text{Augmented Matrix} & \text{Lower Triangular Matrix} & \mathbf{p} \\ \left[\begin{array}{cccc|c} -1 & 1 & 2 & -2 & -1 \\ 3 & -3 & -6 & 12 & 3 \\ -2 & 4 & 3 & 2 & 3 \\ -5 & 1 & 17 & -24 & -20 \end{array}\right], & \begin{bmatrix} 1 & 0 & 0 & 0 \\ - & 1 & 0 & 0 \\ - & - & 1 & 0 \\ - & - & - & 1 \end{bmatrix}, & \begin{bmatrix} 1 \\ 2 \\ 3 \\ 4 \end{bmatrix}. \end{array}$$

For the first stage of elimination the pivot is -1 (the 1,1-entry), with multipliers -3, 2, and 5 corresponding to rows 2, 3, and 4. We record these multipliers in the corresponding rows of the first column of the lower triangular matrix. After the first stage we obtain

$$\left[\begin{array}{rrrr|r} -1 & 1 & 2 & -2 & -1 \\ 0 & 0 & 0 & 6 & 0 \\ 0 & 2 & -1 & 6 & 5 \\ 0 & -4 & 7 & -14 & -15 \end{array}\right], \quad \begin{bmatrix} 1 & 0 & 0 & 0 \\ -3 & 1 & 0 & 0 \\ 2 & - & 1 & 0 \\ 5 & - & - & 1 \end{bmatrix}, \quad \begin{bmatrix} 1 \\ 2 \\ 3 \\ 4 \end{bmatrix}.$$

For the next stage of elimination we encounter a zero pivot and hence interchange rows 2 and 3 of the augmented matrix. Thus we have

$$\left[\begin{array}{rrrr|r} -1 & 1 & 2 & -2 & -1 \\ 0 & 2 & -1 & 6 & 5 \\ 0 & 0 & 0 & 6 & 0 \\ 0 & -4 & 7 & -14 & -15 \end{array}\right], \quad \begin{bmatrix} 1 & 0 & 0 & 0 \\ 2 & 1 & 0 & 0 \\ -3 & - & 1 & 0 \\ 5 & - & - & 1 \end{bmatrix}, \quad \begin{bmatrix} 1 \\ 3 \\ 2 \\ 4 \end{bmatrix}.$$

The vector **p** has been updated to reflect the interchange of rows 2 and 3. Note also that rows 2 and 3 of the first column of the lower triangular matrix have been interchanged accordingly since they correspond to multipliers for the rows of the augmented matrix. For the second stage of elimination the multipliers are 0 and -2 (corresponding to rows 3 and 4) and are recorded in the corresponding rows of the second column of the lower triangular matrix. After the second stage we have

$$\left[\begin{array}{rrrr|r} -1 & 1 & 2 & -2 & -1 \\ 0 & 2 & -1 & 6 & 5 \\ 0 & 0 & 0 & 6 & 0 \\ 0 & 0 & 5 & -2 & -5 \end{array}\right], \quad \begin{bmatrix} 1 & 0 & 0 & 0 \\ 2 & 1 & 0 & 0 \\ -3 & 0 & 1 & 0 \\ 5 & -2 & - & 1 \end{bmatrix}, \quad \begin{bmatrix} 1 \\ 3 \\ 2 \\ 4 \end{bmatrix}.$$

Again we encounter a zero pivot, so we must interchange rows 3 and 4. Accordingly, we interchange these rows in **p** and in the first two columns of the lower triangular matrix. This yields

$$\left[\begin{array}{rrrr|r} -1 & 1 & 2 & -2 & -1 \\ 0 & 2 & -1 & 6 & 5 \\ 0 & 0 & 5 & -2 & -5 \\ 0 & 0 & 0 & 6 & 0 \end{array}\right], \quad \begin{bmatrix} 1 & 0 & 0 & 0 \\ 2 & 1 & 0 & 0 \\ 5 & -2 & 1 & 0 \\ -3 & 0 & - & 1 \end{bmatrix}, \quad \begin{bmatrix} 1 \\ 3 \\ 4 \\ 2 \end{bmatrix}.$$

Since the first four columns of the augmented matrix constitute an upper triangular matrix, the elimination phase is complete. Consequently, the multiplier for row 4 is 0, which we record in the 4,3-entry of the lower triangular matrix. Our final matrices are

$$\left[\begin{array}{rrrr|r} -1 & 1 & 2 & -2 & -1 \\ 0 & 2 & -1 & 6 & 5 \\ 0 & 0 & 5 & -2 & -5 \\ 0 & 0 & 0 & 6 & 0 \end{array}\right], \quad \begin{bmatrix} 1 & 0 & 0 & 0 \\ 2 & 1 & 0 & 0 \\ 5 & -2 & 1 & 0 \\ -3 & 0 & 0 & 1 \end{bmatrix}, \quad \begin{bmatrix} 1 \\ 3 \\ 4 \\ 2 \end{bmatrix}.$$

From these matrices we can identify **L**, **U**, and a permutation matrix **P** such that **PA** = **LU**. To find **P** we rearrange the rows of **I** in the order given by the final vector **p**, that is

$$\mathbf{P} = \begin{bmatrix} 1 & 0 & 0 & 0 \\ 0 & 0 & 1 & 0 \\ 0 & 0 & 0 & 1 \\ 0 & 1 & 0 & 0 \end{bmatrix}.$$

The final lower triangular matrix is **L**, and **U** is given by the first four columns of the final augmented matrix. We leave it to the reader to verify that

$$\mathbf{PA} = \begin{bmatrix} 1 & 0 & 0 & 0 \\ 0 & 0 & 1 & 0 \\ 0 & 0 & 0 & 1 \\ 0 & 1 & 0 & 0 \end{bmatrix} \begin{bmatrix} -1 & 1 & 2 & -2 \\ 3 & -3 & -6 & 12 \\ -2 & 4 & 3 & 2 \\ -5 & 1 & 17 & -24 \end{bmatrix}$$

$$= \mathbf{LU} = \begin{bmatrix} 1 & 0 & 0 & 0 \\ 2 & 1 & 0 & 0 \\ 5 & -2 & 1 & 0 \\ -3 & 0 & 0 & 1 \end{bmatrix} \begin{bmatrix} -1 & 1 & 2 & -2 \\ 0 & 2 & -1 & 6 \\ 0 & 0 & 5 & -2 \\ 0 & 0 & 0 & 6 \end{bmatrix}.$$

Finally, the solution of $\mathbf{Ax} = \mathbf{b}$ is found by back-substitution on the final augmented matrix and results in $x_4 = 0$, $x_3 = -1$, $x_2 = 2$, $x_1 = 1$.

■

An important point to appreciate about the *LU* factorization is that once it is obtained for a given square matrix **A**, many systems of the form

$$\mathbf{Ax} = \mathbf{b},$$

for many different right-hand sides **b**, can be solved with relative ease using the same triangular factors **L** and **U**. That is, if the right-hand side changes but the coefficient matrix does not, it is not necessary to repeat the entire Gaussian algorithm to compute the solution of the new system. We shall make use of this observation in Section 2.4 when discussing a means of computing $\mathbf{A}^{-1}$.

In the next example we consider a system with the same coefficient matrix as in Example 2.8 but a different right-hand side.

Example 2.9

For the matrix **A** of Example 2.8 consider the problem $\mathbf{Ax} = \mathbf{c}$, where

$$\mathbf{c} = \begin{bmatrix} 1 \\ 3 \\ 7 \\ -2 \end{bmatrix}.$$

From Example 2.8 we know **P**, **L**, and **U** such that $\mathbf{PA} = \mathbf{LU}$. Consequently, we have

$$\mathbf{PAx} = \mathbf{LUx} = \mathbf{Pc} = [1, 7, -2, 3]^T,$$

and to find $\mathbf{x}$ we solve the two triangular systems

$$\mathbf{Ly} = \mathbf{Pc} \quad \text{and} \quad \mathbf{Ux} = \mathbf{y}.$$

Therefore, we solve

$$\begin{aligned} y_1 \qquad\qquad\qquad &= 1 \\ 2y_1 + y_2 \qquad\qquad &= 7 \\ 5y_1 - 2y_2 + y_3 \qquad &= -2 \\ -3y_1 \qquad\qquad + y_4 &= 3 \end{aligned}$$

to yield $y_1 = 1$, $y_2 = 5$, $y_3 = 3$, $y_4 = 6$. Using this, we solve $\mathbf{Ux} = \mathbf{y}$:

$$\begin{aligned} -x_1 + x_2 + 2x_3 - 2x_4 &= 1 \\ 2x_2 - x_3 + 6x_4 &= 5 \\ 5x_3 - 2x_4 &= 3 \\ 6x_4 &= 6 \end{aligned}$$

to give $x_4 = 1$, $x_3 = 1$, $x_2 = 0$, $x_1 = -1$.

■

To summarize, the previous discussion shows that the appearance of a zero pivot in the elimination process is not catastrophic as long as there is a nonzero entry in the column below the pivot. In this case elimination can proceed after a row interchange is performed to produce a nonzero pivot. The net result of the elimination process with row interchanges is that a row rearrangement of $\mathbf{A}$, that is, $\mathbf{PA}$, has an *LU* factorization.

Unfortunately, worse things can happen. To see this, consider the following.

Example 2.10

In applying elimination to the augmented matrix of the system

$$\begin{aligned} x_1 + 2x_2 &= 2 \\ 2x_1 + 4x_2 &= 3, \end{aligned}$$

we obtain at the first elimination step,

$$\left[\begin{array}{cc|c} 1 & 2 & 2 \\ 0 & 0 & 1 \end{array}\right].$$

Here a zero pivot is unavoidable and we see from the second equation,

$$0 \cdot x_1 + 0 \cdot x_2 = -1,$$

that, in fact, the system has no solution.

■

In case the elimination process leads, as it does in Example 2.10, to an unavoidable zero pivot, we say that the coefficient matrix is **singular**. It is important to realize that *a singular matrix is not invertible*. To see this, suppose that we continue with the elimination process to arrive at a factorization $\mathbf{PA} = \mathbf{LU}$, where $\mathbf{L}$ is lower triangular with 1s on the diagonal. But the upper triangular matrix $\mathbf{U}$ must have at least one 0 (the unavoidable zero pivot) on the diagonal and hence is not invertible (see Exercise 25 of Section 1.3). Then we have $\mathbf{U} = \mathbf{L}^{-1}\mathbf{PA}$ and hence if $\mathbf{A}$ is invertible, we find $\mathbf{U}^{-1} = \mathbf{A}^{-1}\mathbf{P}^{-1}\mathbf{L}$, which contradicts the fact that $\mathbf{U}$ has no inverse.

EXERCISES 2.3

1. Find the LU factorization of

$$\mathbf{A} = \begin{bmatrix} 3 & 2 \\ 7 & 5 \end{bmatrix}.$$

2. Find the LU factorization of

$$\mathbf{A} = \begin{bmatrix} 1 & 0 \\ 3 & 0 \end{bmatrix}.$$

3. Find the LU factorization of

$$\mathbf{A} = \begin{bmatrix} 2 & 1 & 0 \\ 4 & 3 & 2 \\ 6 & 7 & 8 \end{bmatrix}.$$

4. Find the LU factorization of

$$\mathbf{A} = \begin{bmatrix} 1 & 0 & 0 & 0 \\ -1 & -2 & 0 & 1 \\ 2 & 4 & 1 & 0 \\ -1 & -6 & 0 & 2 \end{bmatrix}.$$

5. Verify that if an $n \times n$ matrix has the form

$$\mathbf{A} = \begin{bmatrix} 1 & 0 & 0 & \cdots & 0 & \cdots & \cdots & 0 \\ 0 & 1 & 0 & \cdots & 0 & \cdots & \cdots & 0 \\ 0 & 0 & 1 & & & & & \\ \vdots & \vdots & & & 1 & & & \vdots \\ & & & & -m_{j+1,j} & 1 & & \\ & & & & -m_{j+2,j} & & & \\ \vdots & \vdots & & & \vdots & & & \\ 0 & 0 & & & -m_{n,j} & \cdots & \cdots & 1 \end{bmatrix},$$

then

$$
\mathbf{A}^{-1} = \begin{bmatrix}
1 & 0 & 0 & \cdots & 0 & & \cdots & \cdots & 0 \\
0 & 1 & 0 & \cdots & 0 & & \cdots & \cdots & 0 \\
0 & 0 & 1 & & & & & & \\
& & & & & & & & \\
\vdots & \vdots & & & 1 & & & & \\
& & & & & & & & \\
& & & & m_{j+1,j} & 1 & & & \vdots \\
& & & & m_{j+2,j} & & & & \\
\vdots & \vdots & & & \vdots & & & & \\
0 & 0 & & & m_{n,j} & & \cdots & \cdots & 1
\end{bmatrix}.
$$

6. Find the inverse of

$$
\begin{bmatrix}
1 & 0 & 0 & 0 & 0 \\
0 & 1 & 0 & 0 & 0 \\
0 & 2 & 1 & 0 & 0 \\
0 & -2 & 0 & 1 & 0 \\
0 & -1 & 0 & 0 & 1
\end{bmatrix}.
$$

7. Let

$$
\mathbf{A} = \begin{bmatrix} 0 & 7 \\ 0 & 1 \end{bmatrix}.
$$

(a) Show that $\mathbf{A}$ is not invertible.

(b) Verify that $\mathbf{A} = \mathbf{LU}$, where

$$
\mathbf{L} = \begin{bmatrix} 1 & 0 \\ 2 & 1 \end{bmatrix} \quad \text{and} \quad \mathbf{U} = \begin{bmatrix} 0 & 7 \\ 0 & -13 \end{bmatrix}.
$$

(c) Verify that $\mathbf{A} = \mathbf{LU}$, where

$$
\mathbf{L} = \begin{bmatrix} 1 & 0 \\ -3 & 1 \end{bmatrix} \quad \text{and} \quad \mathbf{U} = \begin{bmatrix} 0 & 7 \\ 0 & 22 \end{bmatrix}.
$$

8. Suppose that $\mathbf{A}$ is invertible and has LU factorization $\mathbf{A} = \mathbf{LU}$. Show that $\mathbf{A}^{-1}$ can be factored as $\mathbf{A}^{-1} = \mathbf{BC}$, where $\mathbf{B}$ is upper triangular and $\mathbf{C}$ is lower triangular.

9. (a) Find the LU factorization of

$$
\mathbf{A} = \begin{bmatrix} 2 & 1 & -1 \\ 4 & 3 & 1 \\ -2 & 2 & 12 \end{bmatrix}.
$$

(b) Use the LU factorization to solve $\mathbf{Ax} = \mathbf{b}$ if

$$
\mathbf{b} = \begin{bmatrix} -3 \\ -3 \\ 14 \end{bmatrix}.
$$

(c) Use the LU factorization to solve $\mathbf{Ax} = \mathbf{b}$ if

$$\mathbf{b} = \begin{bmatrix} 4 \\ 10 \\ 2 \end{bmatrix}.$$

10. The technology matrix for a three-sector economy is

$$\mathbf{A} = \begin{bmatrix} .8 & .1 & .1 \\ .1 & .2 & .2 \\ .2 & .1 & .2 \end{bmatrix}.$$

(See Section 1.4.)
(a) Find the LU factorization for $\mathbf{I} - \mathbf{A}$.
(b) The projected quarterly demands for the products are

$$\begin{bmatrix} 0 \\ 60 \\ 60 \end{bmatrix}, \quad \begin{bmatrix} 20 \\ 40 \\ 30 \end{bmatrix}, \quad \begin{bmatrix} 50 \\ 0 \\ 110 \end{bmatrix}, \quad \text{and} \quad \begin{bmatrix} 50 \\ 200 \\ 80 \end{bmatrix},$$

respectively. How should the outputs be set in each quarter?

11. Let

$$\mathbf{A} = \begin{bmatrix} 1 & 2 & 3 \\ 3 & 6 & 1 \\ -1 & 2 & 1 \end{bmatrix}.$$

Find a $\mathbf{PA} = \mathbf{LU}$ factorization for $\mathbf{A}$.

12. Find a $\mathbf{PA} = \mathbf{LU}$ factorization for

$$\mathbf{A} = \begin{bmatrix} 2 & 1 & 6 & 3 \\ 2 & 1 & 6 & 0 \\ -2 & 4 & 0 & 1 \\ 1 & 4 & 2 & 1 \end{bmatrix}.$$

13. Show that the matrix

$$\begin{bmatrix} 1 & 12 & 6 & 1 \\ -2 & -4 & 0 & 2 \\ 4 & 5 & 1 & 1 \\ -1 & 8 & 6 & 3 \end{bmatrix}$$

is singular.

14. **(a)** Show that the matrix

$$\mathbf{A} = \begin{bmatrix} 3 & 2 & 1 \\ -1 & -1 & 0 \\ 3 & 1 & 2 \end{bmatrix}$$

is singular.
(b) Find a nonzero vector $\mathbf{x}$ such that $\mathbf{Ax} = \boldsymbol{\theta}$.

2.4 Computation of Inverses

The Gaussian elimination algorithm can be easily adapted to compute the inverse of a matrix. Suppose that $\mathbf{A}$ is a given $n \times n$ invertible matrix. Then

$$\mathbf{A}\mathbf{A}^{-1} = \mathbf{I}$$

and recall that (Exercise 27 of Section 1.2)

$$j\text{th column of } (\mathbf{A}\mathbf{A}^{-1}) = \mathbf{A}(j\text{th column of } \mathbf{A}^{-1}).$$

The jth column of $\mathbf{A}^{-1}$ is denoted by $\mathbf{A}_j^{-1}$ and that of $\mathbf{I}$ is denoted by $\mathbf{e}^{(j)}$, that is,

$$\mathbf{e}^{(j)} = \begin{bmatrix} 0 \\ \vdots \\ 0 \\ 1 \\ 0 \\ \vdots \\ 0 \end{bmatrix} \begin{matrix} \\ \\ \\ \leftarrow j\text{th entry.} \\ \\ \\ \\ \end{matrix}$$

It follows that

$$\mathbf{A}\mathbf{A}_j^{-1} = \mathbf{e}^{(j)}, \qquad 1 \le j \le n. \tag{5}$$

Therefore, we can find the columns of $\mathbf{A}^{-1}$ by solving the n systems (5) for the unknown column vectors $\mathbf{A}_1^{-1}, \ldots, \mathbf{A}_n^{-1}$.

Example 2.11

We find the inverse of

$$\mathbf{A} = \begin{bmatrix} 1 & 2 \\ -1 & 1 \end{bmatrix}$$

by solving the systems

$$\mathbf{A}\mathbf{A}_1^{-1} = \begin{bmatrix} 1 \\ 0 \end{bmatrix} = \mathbf{e}^{(1)} \quad \text{and} \quad \mathbf{A}\mathbf{A}_2^{-1} = \begin{bmatrix} 0 \\ 1 \end{bmatrix} = \mathbf{e}^{(2)}.$$

By elimination we find

$$\left[\begin{array}{rr|r} 1 & 2 & 1 \\ -1 & 1 & 0 \end{array}\right] \longrightarrow \left[\begin{array}{rr|r} 1 & 2 & 1 \\ 0 & 3 & 1 \end{array}\right]$$

and hence, by back substitution, the first column of $\mathbf{A}^{-1}$ is

$$\mathbf{A}_1^{-1} = \begin{bmatrix} \frac{1}{3} \\ \frac{1}{3} \end{bmatrix}.$$

Similarly, we have

$$\left[\begin{array}{rr|r} 1 & 2 & 0 \\ -1 & 1 & 1 \end{array}\right] \longrightarrow \left[\begin{array}{rr|r} 1 & 2 & 0 \\ 0 & 3 & 1 \end{array}\right],$$

and hence back substitution gives

$$\mathbf{A}_2^{-1} = \begin{bmatrix} -\frac{2}{3} \\ \frac{1}{3} \end{bmatrix}.$$

Thus the inverse of **A** is

$$\mathbf{A}^{-1} = \begin{bmatrix} \frac{1}{3} & -\frac{2}{3} \\ \frac{1}{3} & \frac{1}{3} \end{bmatrix}.$$

■

In solving (5) with the various right-hand sides $\mathbf{e}^{(1)}, \ldots, \mathbf{e}^{(n)}$ it is convenient to augment the matrix **A** with all of the columns of the identity in performing the elimination. Back substitution on the jth augmented column will then result in the jth column of $\mathbf{A}^{-1}$. In this way the elimination phase on **A** is not repeated for each system.

Example 2.12

To compute $\mathbf{A}^{-1}$, where

$$\begin{bmatrix} 2 & 2 & 3 \\ 1 & 0 & 1 \\ 1 & 1 & 1 \end{bmatrix},$$

we first augment **A** with the identity matrix

$$\left[\begin{array}{ccc|ccc} 2 & 2 & 3 & 1 & 0 & 0 \\ 1 & 0 & 1 & 0 & 1 & 0 \\ 1 & 1 & 1 & 0 & 0 & 1 \end{array}\right].$$

Elimination then reduces **A** to upper triangular form:

$$\left[\begin{array}{ccc|ccc} 2 & 2 & 3 & 1 & 0 & 0 \\ 1 & 0 & 1 & 0 & 1 & 0 \\ 1 & 1 & 1 & 0 & 0 & 1 \end{array}\right] \longrightarrow \left[\begin{array}{ccc|ccc} 2 & 2 & 3 & 1 & 0 & 0 \\ 0 & -1 & -\frac{1}{2} & -\frac{1}{2} & 1 & 0 \\ 0 & 0 & -\frac{1}{2} & -\frac{1}{2} & 0 & 1 \end{array}\right].$$

Back substitution on columns 4, 5, and 6 gives

$$\mathbf{A}_1^{-1} = \begin{bmatrix} -1 \\ 0 \\ 1 \end{bmatrix}, \quad \mathbf{A}_2^{-1} = \begin{bmatrix} 1 \\ -1 \\ 0 \end{bmatrix}, \quad \text{and} \quad \mathbf{A}_3^{-1} = \begin{bmatrix} 2 \\ 1 \\ -2 \end{bmatrix},$$

respectively. It follows that

$$\mathbf{A}^{-1} = \begin{bmatrix} -1 & 1 & 2 \\ 0 & -1 & 1 \\ 1 & 0 & -2 \end{bmatrix}.$$

■

For hand computation of small matrices there is a method, called **Gauss–Jordan elimination**, which is somewhat more convenient in its layout. For computing $\mathbf{A}^{-1}$ this method is computationally equivalent to the elimination method as presented in

Example 2.12. In the Gauss–Jordan method we use elimination as before to reduce **A** to upper triangular form. Then, instead of performing back substitution, elimination is carried out above the pivots to produce a diagonal matrix. The inverse is found simply by dividing by the diagonal entries (pivots). The procedure should be clear with the aid of the following example.

Example 2.13

We apply the Gauss–Jordan method to the same matrix as in Example 2.12. We augment **A** by the identity **I** and reduce **A** to upper triangular form as before:

$$\left[\begin{array}{ccc|ccc} 2 & 2 & 3 & 1 & 0 & 0 \\ 1 & 0 & 1 & 0 & 1 & 0 \\ 1 & 1 & 1 & 0 & 0 & 1 \end{array}\right] \longrightarrow \left[\begin{array}{ccc|ccc} 2 & 2 & 3 & 1 & 0 & 0 \\ 0 & -1 & -\frac{1}{2} & -\frac{1}{2} & 1 & 0 \\ 0 & 0 & -\frac{1}{2} & -\frac{1}{2} & 0 & 1 \end{array}\right]$$

Next we use elementary row operations to reduce the first three columns to a diagonal matrix. The first stage is to produce zeros in the column above the pivot in the 3,3-entry. We subtract 1 times row 3 from row 2 and subtract -6 times row 3 from row 1, to yield

$$\left[\begin{array}{ccc|ccc} 2 & 2 & 3 & 1 & 0 & 0 \\ 0 & -1 & -\frac{1}{2} & -\frac{1}{2} & 1 & 0 \\ 0 & 0 & -\frac{1}{2} & -\frac{1}{2} & 0 & 1 \end{array}\right] \longrightarrow \left[\begin{array}{ccc|ccc} 2 & 2 & 0 & -2 & 0 & 6 \\ 0 & -1 & 0 & 0 & 1 & -1 \\ 0 & 0 & -\frac{1}{2} & -\frac{1}{2} & 0 & 1 \end{array}\right]$$

Finally, we subtract -2 times row 2 from row 1:

$$\left[\begin{array}{ccc|ccc} 2 & 2 & 0 & -2 & 0 & -6 \\ 0 & -1 & 0 & 0 & 1 & -1 \\ 0 & 0 & -\frac{1}{2} & -\frac{1}{2} & 0 & 1 \end{array}\right] \longrightarrow \left[\begin{array}{ccc|ccc} 2 & 0 & 0 & -2 & 2 & 4 \\ 0 & -1 & 0 & 1 & 1 & -1 \\ 0 & 0 & -\frac{1}{2} & -\frac{1}{2} & 0 & 0 \end{array}\right]$$

To find $\mathbf{A}^{-1}$ we divide the first row by 2, the second row by -1, and the last row by $-\frac{1}{2}$. That is, we form

$$\left[\begin{array}{ccc|ccc} 1 & 0 & 0 & -1 & 1 & 2 \\ 0 & 1 & 0 & 0 & -1 & 1 \\ 0 & 0 & 1 & 1 & 0 & -2 \end{array}\right].$$

It follows that the last three columns give $\mathbf{A}^{-1}$.

■

Thus in the Gauss–Jordan method we reduce $[\mathbf{A} \mid \mathbf{I}]$ to $[\mathbf{I} \mid \mathbf{A}^{-1}]$.

EXERCISES 2.4

1. Compute the inverse of

$$\mathbf{A} = \begin{bmatrix} -2 & 1 \\ \frac{3}{2} & -\frac{1}{2} \end{bmatrix}$$

by solving each of the systems $\mathbf{A}\mathbf{A}_1^{-1} = \mathbf{e}^{(1)}$ and $\mathbf{A}\mathbf{A}_2^{-1} = \mathbf{e}^{(2)}$ using Gaussian elimination.

2. Compute the inverse of

$$\begin{bmatrix} 1 & -1 & 0 \\ 0 & 1 & -1 \\ 1 & 0 & 1 \end{bmatrix}$$

by augmenting the 3×3 identity matrix and using Gaussian elimination.

In Exercises 3–6, use Gauss–Jordan elimination to compute the inverse of the given matrix **A**.

3. $\mathbf{A} = \begin{bmatrix} -\frac{1}{2} & -\frac{1}{2} & \frac{1}{2} \\ \frac{3}{2} & \frac{1}{2} & -\frac{1}{2} \\ 1 & 1 & 0 \end{bmatrix}$

4. $\mathbf{A} = \begin{bmatrix} 1 & 0 & 0 & 0 \\ 0 & -2 & 0 & 0 \\ 0 & 0 & 2 & 0 \\ 0 & 0 & 0 & 1 \end{bmatrix}$

5. $\mathbf{A} = \begin{bmatrix} 1 & 0 & 0 & 0 \\ -1 & 2 & 0 & 0 \\ 3 & 1 & -1 & 0 \\ 2 & 0 & 4 & 3 \end{bmatrix}$

6. $\mathbf{A} = \begin{bmatrix} 1 & 0 & 2 & 0 \\ 0 & 3 & 0 & 4 \\ 5 & 0 & 6 & 0 \\ 0 & 7 & 0 & 8 \end{bmatrix}$

7. **(a)** What effect does a row interchange have on the Gauss–Jordan method?
(b) Use the Gauss–Jordan method to compute the inverse of the following matrix.

$$\begin{bmatrix} 1 & 1 & 1 & 1 \\ 1 & 1 & 1 & 2 \\ 0 & 1 & 1 & 1 \\ -1 & 2 & 1 & 2 \end{bmatrix}$$

8. The matrix

$$\begin{bmatrix} 1 & -2 & -1 \\ 2 & -1 & -1 \\ 1 & 7 & 2 \end{bmatrix}$$

is singular and therefore has no inverse. At what point does the Gauss–Jordan method "break down"?

9. Use Gauss–Jordan elimination to show that if $abc \neq 0$, then

$$\begin{bmatrix} 0 & 0 & a \\ 0 & b & 0 \\ c & 0 & 0 \end{bmatrix}^{-1} = \begin{bmatrix} 0 & 0 & \frac{1}{c} \\ 0 & \frac{1}{b} & 0 \\ \frac{1}{a} & 0 & 0 \end{bmatrix}$$

2.5 Computational Considerations

With the advent of digital computers it has become important to assess the computational complexity of algorithms so that the relative efficiency (in terms of computer time) of various algorithms can be compared. A rough measure of the computational complexity of a method is its **operation count.** Since additions and subtractions generally require less computer time than multiplications and divisions, it is traditional to count only the latter operations. By the "operation count" for an algorithm we mean the total number of multiplications and divisions required to perform the algorithm.

Suppose that we are solving the $n \times n$ nonsingular linear system $\mathbf{Ax} = \mathbf{b}$ by Gaussian elimination. Suppose further that the LU factorization has already been obtained and we are solving $\mathbf{LUx} = \mathbf{b}$ by first solving $\mathbf{Ly} = \mathbf{b}$ and then $\mathbf{Ux} = \mathbf{y}$. Now $\mathbf{Ly} = \mathbf{b}$ has the form

$$\begin{array}{lll} (0) & y_1 & = b_1 \\ (1) & l_{21} y_1 + y_2 & = b_2 \\ & \vdots & \vdots \\ (n-1) & l_{n1} y_1 + \cdots + l_{n,n-1} y_{n-1} + y_n & = b_n. \end{array}$$

To solve for $\mathbf{y}$ we find $y_1 = b_1$ (0 multiplications required), then substitute y_1 in the second equation to find $y_2 = b_2 - l_{21} y_1$ (one multiplication required), then substitute y_1 and y_2 into the third equation to find $y_3 = b_3 - l_{31} y_1 - l_{32} y_2$ (two multiplications required), and so on. The number of multiplicative operations is indicated in parentheses to the left of each equation in $\mathbf{Ly} = \mathbf{b}$. Therefore, the total number of operations in this phase is

$$0 + 1 + 2 + \cdots + (n-1) = \frac{(n-1)n}{2}.$$

(This can be proved by a process called mathematical induction, but see Exercise 1.)

In solving $\mathbf{Ux} = \mathbf{y}$ one extra operation (i.e., a division) is required for each equation since the diagonal entries of $\mathbf{U}$ are not necessarily equal to 1. Therefore, the total number of operations in solving $\mathbf{Ux} = \mathbf{y}$ is

$$1 + 2 + \cdots + n = \frac{n(n+1)}{2}.$$

It follows that the operation count for the solution phase of the Gaussian elimination algorithm is

$$\frac{(n-1)n}{2} + \frac{n(n+1)}{2} = n^2.$$

Consider now the elimination phase of the Gaussian algorithm in which the entries of $\mathbf{L}$ and $\mathbf{U}$ are determined. The first step is to form the multipler a_{21}/a_{11}, which requires one division, and then to multiply this times the numbers $a_{12}, a_{13}, \ldots, a_{1n}$ (in row 1) and subtract (note that since we *know* that a_{21} will be transformed to zero, there is no need to take this operation into account) from the corresponding numbers in row 2. Therefore, to produce 0 in the 2,1-entry requires $1 + (n-1) = n$ operations. A similar procedure is applied to produce the rest of the zeros in the first column. Since there will be $n - 1$ zeros in the first column, the total operation count for the first column is $n(n-1)$. In a similar way, taking into account the rest of the

$n - 1$ columns on which elimination is performed, we find a total operation count for the elimination phase of

$$n(n-1) + (n-1)(n-2) + \cdots + 2 \cdot 1 = \frac{n(n-1)(n+1)}{3}.$$

Therefore, the total operation count for both phases of the algorithm is

$$n^2 + \frac{n(n-1)(n+1)}{3} = \frac{n^3}{3} + n^2 - \frac{n}{3}.$$

Note that for large n the dominant term is $n^3/3$, and in this case we say that the operation count is on the order of $n^3/3$.

In the preceding section it was demonstrated how to compute $\mathbf{A}^{-1}$ using Gaussian elimination or the Gauss–Jordan method. Having shown this we now wish to point out that it is rarely necessary to compute $\mathbf{A}^{-1}$. The usual motivation for computing $\mathbf{A}^{-1}$ is that it may be necessary to compute solutions of many systems

$$\mathbf{Ax} = \mathbf{b}^{(i)}, \tag{6}$$

for many right-hand sides $\mathbf{b}^{(1)}, \mathbf{b}^{(2)}, \ldots$. Conventional wisdom then has it that $\mathbf{A}^{-1}$ should be computed and the solutions of (6) obtained as

$$\mathbf{A}^{-1}\mathbf{b}^{(1)}, \mathbf{A}^{-1}\mathbf{b}^{(2)}, \ldots.$$

Note that each of these matrix products requires n^2 operations, which is the same number of operations as in the solution phase of the Gauss elimination algorithm (see the earlier discussion). Therefore, computing the inverse will be computationally advantageous only if it requires fewer operations than forming the LU factorization of $\mathbf{A}$. However, computing $\mathbf{A}^{-1}$ *first* requires the computation of the LU factorization and then the *additional* work of back substitution on the columns, $\mathbf{e}^{(1)}, \ldots, \mathbf{e}^{(n)}$.

Therefore, to solve systems of the type (6), particularly for large values of n, it makes much more sense computationally to compute the LU factorization and perform the solution phase on the various right-hand sides $\mathbf{b}^{(1)}, \mathbf{b}^{(2)}, \ldots$, than to compute $\mathbf{A}^{-1}$ and form the products $\mathbf{A}^{-1}\mathbf{b}^{(1)}, \mathbf{A}^{-1}\mathbf{b}^{(2)}, \ldots$. Therefore, to solve $\mathbf{Ax} = \mathbf{b}^{(i)}$ for several right-hand sides $\mathbf{b}^{(i)}$, one should first determine $\mathbf{P}$, $\mathbf{L}$, and $\mathbf{U}$ such that $\mathbf{PA} = \mathbf{LU}$. Then solve each system $\mathbf{Ax} = \mathbf{b}^{(i)}$ by solving

$$\mathbf{Ly} = \mathbf{Pb}^{(i)} \quad \text{and} \quad \mathbf{Ux} = \mathbf{y}.$$

In fact, most routines in computer libraries for solving linear systems are coded in exactly this way.

EXERCISES 2.5

1. Let $S = \sum_{k=1}^{n} k$. Then

$$S = 1 + 2 \cdots + (n-1) + n \quad \text{and} \quad S = n + (n-1) + \cdots + 2 + 1.$$

Therefore, $2S = n(n+1)$ (add the two previous equations and simplify). Conclude that $S = n(n+1)/2$.

2. Suppose that $\mathbf{x}$ and $\mathbf{y}$ are n-vectors (i.e., $n \times 1$ matrices). How many multiplications are required to compute $\mathbf{x}^T\mathbf{y}$? $\mathbf{x}\mathbf{y}^T$?

3. Suppose that Gauss–Jordan elimination is used on a general $n \times n$ linear system $\mathbf{Ax} = \mathbf{b}$. How many multiplicative operations are performed? Compare with the operation count for Gauss elimination.

4. **(a)** Show by a careful operation count that the inverse of an $n \times n$ matrix can be computed in n^3 multiplicative operations by Gauss–Jordan elimination.
(b) Show that Gauss elimination can be performed in n^3 multiplicative operations to compute the inverse of an $n \times n$ matrix.

5. Show that if $\mathbf{A}$ is an $n \times n$ matrix and $\mathbf{x}$ is an n-vector, then the computation of $\mathbf{Ax}$ requires n^2 multiplicative operations.

6. An $n \times n$ matrix $\mathbf{A}$ is called *tridiagonal* if $a_{ij} = 0$ for $|i - j| > 1$. Show that if $\mathbf{A}$ is tridiagonal and $\mathbf{x}$ is an n-vector, then the computation of $\mathbf{Ax}$ requires $3n - 2$ multiplicative operations.

7. Let

$$\mathbf{L} = \begin{bmatrix} 1 & 0 & 0 & \cdots & & 0 \\ l_{21} & 1 & 0 & \cdots & & 0 \\ \cdot & l_{32} & 1 & & & \cdot \\ \cdot & & \ddots & & & \cdot \\ \cdot & & & \ddots & & \cdot \\ \cdot & & & & \ddots & 0 \\ 0 & \cdot & \cdot & & l_{n,n-1} & 1 \end{bmatrix}, \quad \mathbf{U} = \begin{bmatrix} u_{11} & u_{12} & 0 & \cdot & \cdots & 0 \\ 0 & u_{22} & u_{23} & 0 & \cdots & 0 \\ \cdot & \cdot & \cdot & \cdot & & \cdot \\ \cdot & & & \ddots & \ddots & \cdot \\ \cdot & & & & \ddots & \cdot \\ \cdot & & & & \cdot & u_{n-1,n} \\ 0 & \cdot & \cdot & \cdot & 0 & u_{nn} \end{bmatrix}.$$

(a) Show that $\mathbf{LU}$ is tridiagonal.
(b) If

$$\mathbf{A} = \begin{bmatrix} 2 & 1 & 0 & 0 & 0 \\ 1 & 2 & 1 & 0 & 0 \\ 0 & 1 & 2 & 1 & 0 \\ 0 & 0 & 1 & 2 & 1 \\ 0 & 0 & 0 & 1 & 2 \end{bmatrix},$$

determine $\mathbf{L}$, $\mathbf{U}$, in the form above such that $\mathbf{A} = \mathbf{LU}$. (*Hint:* Equate entries of $\mathbf{LU}$ and $\mathbf{A}$.)

2.6 Linear Programming

The Gauss–Jordan elimination scheme is a basic step in the simplex method for linear programming. Linear programming is an important mathematical tool in resource management and economics. The basic problem of linear programming is to minimize (or maximize) a linear function of certain variables subject to some linear inequality constraints on the variables.

Example 2.14

A physician recommends to a patient that he should take at least 30 units of vitamin B_1 and at most 80 units of vitamin B_2 daily. The vitamins are not available in pure form, but health food stores sell Megavit tablets, each of which contains 3 units of B_1 and 10 units of B_2, and Vitatrim pills, each of which contains 2 units of B_1 and 4 units of B_2. The tablets sell for 15 cents each and the pills for 11 cents each. The problem is to find a tablet/pill mix which will satisfy the daily vitamin requirements at a minimum cost. If we let x_1 and x_2 stand for the daily dose of tablets and pills, respectively, then the problem may be stated:

$$\begin{aligned} \text{Minimize} \quad & 15x_1 + 11x_2 \\ \text{subject to:} \quad & 3x_1 + 2x_2 \geq 30 \\ & 10x_1 + 4x_2 \leq 80 \\ & x_1 \geq 0, \quad x_2 \geq 0. \end{aligned}$$

■

The inequality constraints can easily be converted into equalities by the introduction of additional variables. For example, if we subtract a nonnegative *surplus variable* x_3 which is exactly enough to make the first inequality an equality, we can convert the first constraint to

$$3x_1 + 2x_2 - x_3 = 30, \qquad x_3 \geq 0.$$

In a similar way, by adding a *slack variable* x_4, the second constraint becomes

$$10x_1 + 4x_2 + x_4 = 80, \qquad x_4 \geq 0.$$

Therefore, the original problem may be reformulated as

$$\begin{aligned} \text{Minimize} \quad & 15x_1 + 11x_2 \\ \text{subject to:} \quad & 3x_1 + 2x_2 - x_3 = 30 \\ & 10x_1 + 4x_2 + x_4 = 80 \\ & x_1 \geq 0, \quad x_2 \geq 0, \quad x_3 \geq 0, \quad x_4 \geq 0. \end{aligned}$$

It should be clear that any inequality constraint can, as in Example 2.14, be converted into an equality constraint by the introduction of a slack or surplus variable. Therefore, the basic linear programming problem can be converted to the form

$$\begin{aligned} \text{Minimize (or maximize)} \quad & c_1x_1 + c_2x_2 + \cdots + c_nx_n \\ \text{subject to:} \quad & \mathbf{Ax} = \mathbf{b} \\ & x_i \geq 0, \quad i = 1, 2, \ldots, n, \end{aligned}$$

where $\mathbf{A}$ is an $m \times n$ matrix (the *constraint matrix*) and $\mathbf{b}$ is a given m-vector. An n-vector $\mathbf{x}$ is called a *solution* of the linear program if $\mathbf{Ax} = \mathbf{b}$. If, in addition, $x_i \geq 0$, $i = 1, \ldots, n$, we call $\mathbf{x}$ a *feasible solution*. Finally, if $\mathbf{x}$ is a feasible solution which satisfies the minimization (or maximization) criterion, then we call $\mathbf{x}$ an *optimal feasible solution*. Of course, our object is to find an optimal feasible solution.

Suppose that we choose m columns of the matrix $\mathbf{A}$ to form an $m \times m$ nonsingular matrix $\mathbf{A}^{(m)}$. By a *basic solution* of $\mathbf{Ax} = \mathbf{b}$ we mean a solution each of whose components *not* corresponding to these m columns is zero. Therefore, there may be many basic solutions corresponding to different choices of the m columns. Basic solutions may be found by Gauss–Jordan elimination as illustrated in the following example.

Example 2.15

We will find the basic solutions of the system

$$3x_1 + 2x_2 - x_3 = 30$$
$$10x_1 + 4x_2 + x_4 = 80$$

which arises from the constraints in Example 2.14. First we form the augmented matrix

$$\left[\begin{array}{cccc|c} 3 & 2 & -1 & 0 & 30 \\ 10 & 4 & 0 & 1 & 80 \end{array}\right]$$

and applying Gauss–Jordan elimination to the third and fourth columns gives

$$\left[\begin{array}{cccc|c} -3 & -2 & 1 & 0 & -30 \\ 10 & 4 & 0 & 1 & 80 \end{array}\right]$$

and hence we obtain the basic solution

$$\mathbf{x} = \begin{bmatrix} 0 \\ 0 \\ -30 \\ 80 \end{bmatrix}.$$

Choosing -2 as a pivot and applying Gauss–Jordan elimination to the second and fourth columns gives

$$\left[\begin{array}{cccc|c} \frac{3}{2} & 1 & -\frac{1}{2} & 0 & 15 \\ 4 & 0 & 2 & 1 & 20 \end{array}\right]$$

and therefore

$$\mathbf{x} = \begin{bmatrix} 0 \\ 15 \\ 0 \\ 20 \end{bmatrix}$$

is another basic solution. Continuing to apply Gauss–Jordan elimination in this way, we obtain the following augmented matrices and corresponding basic solutions:

$$\left[\begin{array}{cccc|c} 1 & \frac{2}{3} & \frac{1}{3} & 0 & 10 \\ 0 & -\frac{8}{3} & \frac{10}{3} & 1 & -20 \end{array}\right], \qquad \mathbf{x} = \begin{bmatrix} 10 \\ 0 \\ 0 \\ -20 \end{bmatrix}$$

$$\left[\begin{array}{cccc|c} 1 & 0 & \frac{1}{2} & \frac{1}{4} & 5 \\ 0 & 1 & -\frac{5}{4} & -\frac{3}{8} & \frac{15}{2} \end{array}\right], \qquad \mathbf{x} = \begin{bmatrix} 5 \\ \frac{15}{2} \\ 0 \\ 0 \end{bmatrix}$$

$$\left[\begin{array}{cccc|c} 1 & \frac{2}{5} & 0 & \frac{1}{10} & 8 \\ 0 & -\frac{4}{5} & 1 & \frac{3}{10} & -6 \end{array}\right], \qquad \mathbf{x} = \begin{bmatrix} 8 \\ 0 \\ -6 \\ 0 \end{bmatrix}$$

$$\left[\begin{array}{cccc|c} \frac{5}{2} & 1 & 0 & \frac{1}{4} & 20 \\ 2 & 0 & 1 & \frac{1}{2} & 10 \end{array}\right], \qquad \mathbf{x} = \begin{bmatrix} 0 \\ 20 \\ 10 \\ 0 \end{bmatrix}.$$

Note that the zero components of each basic solution correspond to those columns that were *not* used in the Gauss–Jordan process.

■

The importance of basic solutions derives from the *fundamental theorem of linear programming*, which states that *if a linear programming problem has an optimal feasible solution, then it has a basic optimal feasible solution*. Therefore, we may find an optimal solution by finding all basic solutions, eliminating those which are not feasible, and checking the value of the objective function for the rest.

Example 2.16

To find an optimal solution of the linear programming problem in Example 2.14, we check the objective function $15x_1 + 11x_2$ for each of the basic feasible solutions found in Example 2.15:

solution: $\begin{bmatrix} 0 \\ 15 \\ 0 \\ 20 \end{bmatrix}$ objective value: $11 \cdot 15 = 165$

solution: $\begin{bmatrix} 5 \\ \frac{15}{2} \\ 0 \\ 0 \end{bmatrix}$ objective value: $15 \cdot 5 + 11 \cdot \frac{15}{2} = 157.5$

solution: $\begin{bmatrix} 0 \\ 20 \\ 10 \\ 0 \end{bmatrix}$ objective value: $11 \cdot 20 = 220$

Therefore, an optimal solution is $x_1 = 5$ and $x_2 = \frac{15}{2}$ and the patient, if buying a 30-day supply, should purchase 150 Megavits and 225 Vitatrims. ■

It should be clear that solving linear programming problems by finding and checking all basic solutions can involve quite a lot of work. For example, in a linear programming problem with 10 variables and 10 inequality constraints, introduction of slack and surplus variables gives rise to a 10×20 linear system. Such a system has possibly

$$\binom{20}{10} = \frac{20!}{10!10!} = 184{,}756$$

basic solutions.

The **simplex method** for linear programming is a clever algorithm for reducing the number of basic solutions that must be checked. This is done by arranging the computations so that (1) only *feasible* basic solutions are found, and (2) the algorithm, in moving from one basic feasible solution to the next, strictly improves the objective function. Moreover, the simplex method contains an automatic optimality check and terminates in a finite number of steps. We refer the reader to the ample literature on the simplex method for further information.

EXERCISES 2.6

1. Find all basic solutions of the system

$$\begin{aligned} x_1 + 3x_2 - x_3 &= 1 \\ 3x_1 + 2x_2 + x_3 &= 7. \end{aligned}$$

2. Consider the linear programming problem

$$\begin{array}{ll} \text{Maximize} & x_1 + x_2 \\ \text{Subject to:} & x_1 + 2x_2 \le 4 \\ & 3x_1 + x_2 \le 3 \\ & x_1 \ge 0, \quad x_2 \ge 0. \end{array}$$

(a) Introduce slack variables to convert the constraints into equalities.
(b) Solve the linear programming problem by checking all basic feasible solutions.

3. A coffee company sells two brands of coffee: Mocha Delight and Hava Java. Mocha Delight is packaged in $2\frac{1}{2}$-lb bags which contain 1 lb of Brazilian coffee and $1\frac{1}{2}$ lb of Colombian coffee. Hava Java comes in $3\frac{1}{2}$-lb bags containing 2 lb of Brazilian coffee and $1\frac{1}{2}$ lb of Colombian coffee. The company has 10,000 lb of Brazilian coffee and 9000 lb of Colombian coffee on hand. The packaging equipment can handle only one brand at a time and requires 12 seconds to fill and label a bag of Hava Java and 30 seconds to process of bag of Mocha Delight. The company makes a profit of \$3 on a bag of Mocha Delight and \$2.50 on a bag of Hava Java. The management is not on good terms with labor, which has called a strike to begin in 37.5 working hours. Management would like to plan production so as to maximize profit before the strike. Set this up as a linear programming problem.

4. Solve the linear programming problem:

$$\begin{array}{ll} \text{Minimize} & -2x_1 - x_2 \\ \text{subject to:} & x_1 + x_2 \leq 2 \\ & x_1 + 6x_2 \leq 3 \\ & x_1 \geq 0, \quad x_2 \geq 0. \end{array}$$

5. Set up and solve the following linear programming problem. A refinery blends two grades of crude oil to make heating oil and gasoline. Each barrel of heating oil requires .3 barrel of grade 1 crude and .7 barrel of grade 2 crude. Each barrel of gasoline requires .7 barrel of grade 1 crude and .3 gallon of grade 2 crude. The company has 1000 barrels of each type of crude on hand. It makes a $12 profit on a barrel of gasoline and a $10 profit on a barrel of heating oil. How should its crude be allocated so as to maximize profit?

GLOSSARY

augmented matrix An array used in the elimination method.

back substitution A method of solution for an upper triangular system.

elementary matrix A matrix that affects an elementary row operation when used as a premultiplier.

elementary permutation matrix A matrix obtained by interchanging two rows of an identity matrix.

elementary row operations The operations used to reduce a matrix to upper triangular form.

equivalent systems Systems of linear equations which have the same solutions.

Gaussian elimination An efficient method for solving linear systems.

Gauss–Jordan elimination A convenient elimination method for computing inverses by reduction of $[\mathbf{A} \mid \mathbf{I}]$ to $[\mathbf{I} \mid \mathbf{A}^{-1}]$.

linear programming A mathematical method for optimization of linear functions with inequality constraints.

LU factorization In the absence of row interchanges during elimination on **A**, the elimination phase is equivalent to factoring **A** into the product of lower triangular matrix **L** times upper triangular matrix **U**.

operation count The total number of multiplications and divisions required in a given algorithm.

permutation matrix A matrix that is the product of elementary permutation matrices.

pivot The k,k-entry of the array on which the kth stage of elimination is applied.

simplex method An algorithm for solving linear programming problems by examination of basic feasible solutions.

singular matrix A matrix in which the elimination process leads to an unavoidable zero pivot.

NOTES AND COMMENTS

More complete information on Gaussian elimination, including computer codes and suggestions for minimizing round-off error, may be found in G. Forsythe and C. Moler, *Computer Solution of Linear Algebraic Systems* (Englewood Cliffs, N.J.: Prentice-Hall, 1967). We have shown that Gaussian elimination solves an $n \times n$ system with the order of n^3 multiplicative operations. In recent years a number of algorithms have been developed which do the job with the order of n^α operations with $\alpha < 3$. The competition to find algorithms with ever smaller α continues.

For a discussion of linear programming, the simplex method and the ellipsoidal method for linear programming, which made lots of headlines a few years ago, see W. McLewin, *Linear Programming and Applications* (London, Input-Output Press, 1980). Because of the tremendous economic importance of linear programming, it continues to be an area of active mathematical research. For stories on two recent entries in the linear programming algorithm competition, see *The New York Times* articles by James Gleick (November 19, 1984) and Calvin Sims (December 10, 1986).

3

Determinants

Determinants were quite the mathematical rage at the turn of the century. Nowadays they are rarely used in practical matrix computations. Nevertheless, determinants have an appealing theory and there are important connections between determinants and eigenvalues (Chapter 6) as well as between determinants and multiple integrals.

In this brief chapter we point out some basic properties of determinants, but our main aim is to establish an important theorem. The theorem states that a square matrix is invertible if and only if its determinant is nonzero. This result will play a fundamental role in our discussion of eigenvalues in Chapter 6.

3.1 Basic Properties

With each square matrix we associate a number called its *determinant*. The determinant of a matrix **A** will be denoted det **A**. If **A** is written out in long form,

$$\mathbf{A} = \begin{bmatrix} a_{11} & a_{12} & \cdots & a_{1n} \\ a_{21} & a_{22} & \cdots & a_{2n} \\ \vdots & \vdots & & \vdots \\ a_{n1} & a_{n2} & \cdots & a_{nn} \end{bmatrix},$$

we will denote its determinant by enclosing the array between vertical bars:

$$\det \mathbf{A} = \begin{vmatrix} a_{11} & a_{12} & \cdots & a_{1n} \\ a_{21} & a_{22} & \cdots & a_{2n} \\ \vdots & \vdots & & \vdots \\ a_{n1} & a_{n2} & \cdots & a_{nn} \end{vmatrix}.$$

We give a case-by-case definition of the determinant. If **A** is a 1×1 matrix, that is, a scalar, then the determinant of **A** is **A** itself. For 2×2 matrices the determinant is defined by

$$\begin{vmatrix} a_{11} & a_{12} \\ a_{21} & a_{22} \end{vmatrix} = a_{11} \cdot a_{22} - a_{12} \cdot a_{21}.$$

Example 3.1

For

$$\mathbf{A} = \begin{bmatrix} 3 & 4 \\ 6 & 2 \end{bmatrix}$$

we find

$$\det \mathbf{A} = \begin{vmatrix} 3 & 4 \\ 6 & 2 \end{vmatrix} = 3 \cdot 2 - 4 \cdot 6 = -18.$$

■

For 3×3 matrices we define the determinant in terms of the 2×2 case:

$$\begin{vmatrix} a_{11} & a_{12} & a_{13} \\ a_{21} & a_{22} & a_{23} \\ a_{31} & a_{32} & a_{33} \end{vmatrix} = a_{11} \begin{vmatrix} a_{22} & a_{23} \\ a_{32} & a_{33} \end{vmatrix} - a_{12} \begin{vmatrix} a_{21} & a_{23} \\ a_{31} & a_{33} \end{vmatrix} + a_{13} \begin{vmatrix} a_{21} & a_{22} \\ a_{31} & a_{32} \end{vmatrix}.$$

Example 3.2

If

$$\mathbf{A} = \begin{bmatrix} 1 & -2 & -1 \\ 1 & 0 & 3 \\ 4 & 2 & 5 \end{bmatrix},$$

then

$$\begin{aligned} \det \mathbf{A} &= \begin{vmatrix} 1 & -2 & -1 \\ 1 & 0 & 3 \\ 4 & 2 & 5 \end{vmatrix} = (1) \begin{vmatrix} 0 & 3 \\ 2 & 5 \end{vmatrix} - (-2) \begin{vmatrix} 1 & 3 \\ 4 & 5 \end{vmatrix} + (-1) \begin{vmatrix} 1 & 0 \\ 4 & 2 \end{vmatrix} \\ &= 0 \cdot 5 - 3 \cdot 2 + 2(1 \cdot 5 - 3 \cdot 4) + (-1)(1 \cdot 2 - 0 \cdot 4) \\ &= -22. \end{aligned}$$

■

In order to define the determinant for an $n \times n$ matrix we must first introduce some notation and terminology. The determinant of an $n \times n$ matrix will be defined in terms of determinants of certain $(n - 1) \times (n - 1)$ matrices. We define the *i,j*th minor of an $n \times n$ matrix $\mathbf{A}$ to be the determinant M_{ij} of the $(n - 1) \times (n - 1)$ matrix obtained by deleting the *i*th row and *j*th column of $\mathbf{A}$. Hence there are n^2 such minors associated with an $n \times n$ matrix, one associated with each entry position. The *i,j*th **cofactor** of $\mathbf{A}$ is the number A_{ij} defined by

$$A_{ij} = (-1)^{i+j} M_{ij}.$$

Notice that the cofactor differs from the corresponding minor by at most a sign. A convenient way to remember the sign which is associated with a given minor is to notice that they form the following "checkerboard" pattern:

$$\begin{bmatrix} + & - & + & - & \cdots \\ - & + & - & + & \cdots \\ + & - & + & - & \cdots \\ \cdot & \cdot & \cdot & & \\ \cdot & \cdot & \cdot & & \\ \cdot & \cdot & \cdot & & \end{bmatrix}.$$

Example 3.3

Let

$$\mathbf{A} = \begin{bmatrix} 1 & 2 & 3 \\ 4 & 5 & 6 \\ 7 & 8 & 9 \end{bmatrix}.$$

Then

$$M_{13} = \begin{vmatrix} 4 & 5 \\ 7 & 8 \end{vmatrix} = -3, \qquad M_{23} = \begin{vmatrix} 1 & 2 \\ 7 & 8 \end{vmatrix} = -6$$

$$A_{13} = (-1)^{1+3}(-3) = -3, \qquad A_{23} = (-1)^{2+3}(-6) = 6.$$

■

We may now give a general definition of the determinant of an $n \times n$ matrix $\mathbf{A}$:

$$\det \mathbf{A} = a_{11}A_{11} + a_{12}A_{12} + \cdots + a_{1n}A_{1n}$$
$$= \sum_{j=1}^{n} a_{1j}A_{1j}.$$

Before discussing computation of determinants, we point out two simple but important properties of determinants which follow immediately from the definition:

(1) If the first row of a matrix $\mathbf{A}$ is multiplied by a scalar r, then the determinant changes by a factor of r. That is,

$$\begin{vmatrix} ra_{11} & ra_{12} & \cdots & ra_{1n} \\ a_{21} & a_{22} & \cdots & a_{2n} \\ \vdots & \vdots & \cdots & \vdots \\ a_{n1} & a_{n2} & \cdots & a_{nn} \end{vmatrix} = r \begin{vmatrix} a_{11} & a_{12} & \cdots & a_{1n} \\ a_{21} & a_{22} & \cdots & a_{2n} \\ \vdots & \vdots & \cdots & \vdots \\ a_{n1} & a_{n2} & \cdots & a_{nn} \end{vmatrix}.$$

(2) If the first row is composed of the sum of two rows, then the determinant is the sum of the corresponding determinants:

$$\begin{vmatrix} a_{11}+b_{11} & a_{12}+b_{12} & \cdots & a_{1n}+b_{1n} \\ a_{21} & a_{22} & \cdots & a_{2n} \\ \vdots & \vdots & \cdots & \vdots \\ a_{n1} & a_{n2} & & a_{nn} \end{vmatrix}$$

$$= \begin{vmatrix} a_{11} & a_{12} & \cdots & a_{1n} \\ a_{21} & a_{22} & \cdots & a_{2n} \\ \vdots & \vdots & \cdots & \vdots \\ a_{n1} & a_{n2} & \cdots & a_{nn} \end{vmatrix} + \begin{vmatrix} b_{11} & b_{12} & \cdots & b_{1n} \\ a_{21} & a_{22} & \cdots & a_{2n} \\ \vdots & \vdots & \cdots & \vdots \\ a_{n1} & a_{n2} & \cdots & a_{nn} \end{vmatrix}.$$

The following property of determinants does not follow easily from the definition:

(3) Interchanging two distinct rows of a matrix changes the sign of its determinant.

We will not prove property (3), but we give an example to illustrate its meaning.

Example 3.4

We illustrate properties (1)–(3) for 3×3 matrices in (a)–(c) below.

(a) To illustrate (1), consider the determinant

$$\begin{vmatrix} 3 & 6 & 9 \\ 1 & 0 & -1 \\ 4 & 2 & 1 \end{vmatrix} = 3\begin{vmatrix} 0 & -1 \\ 2 & 1 \end{vmatrix} - 6\begin{vmatrix} 1 & -1 \\ 4 & 1 \end{vmatrix} + 9\begin{vmatrix} 1 & 0 \\ 4 & 2 \end{vmatrix}$$

$$= 3 \cdot 2 - 6 \cdot 5 + 9 \cdot 2 = -6.$$

Factoring 3 from the first row, we have

$$\begin{vmatrix} 3 & 6 & 9 \\ 1 & 0 & -1 \\ 4 & 2 & 1 \end{vmatrix} = \begin{vmatrix} 3 \cdot 1 & 3 \cdot 2 & 3 \cdot 3 \\ 1 & 0 & -1 \\ 4 & 2 & 1 \end{vmatrix} = 3\begin{vmatrix} 1 & 2 & 3 \\ 1 & 0 & -1 \\ 4 & 2 & 1 \end{vmatrix}$$

$$= 3\left\{\begin{vmatrix} 0 & -1 \\ 2 & 1 \end{vmatrix} - 2\begin{vmatrix} 1 & -1 \\ 4 & 1 \end{vmatrix} + 3\begin{vmatrix} 1 & 0 \\ 4 & 2 \end{vmatrix}\right\}$$

$$= 3(1 \cdot 2 - 2 \cdot 5 + 3 \cdot 2) = 3(-2) = -6.$$

(b) Consider the determinant

$$\begin{vmatrix} 0 & 2 & 1 \\ -1 & 3 & 4 \\ 2 & 1 & 1 \end{vmatrix} = 0\begin{vmatrix} 3 & 4 \\ 1 & 1 \end{vmatrix} - 2\begin{vmatrix} -1 & 4 \\ 2 & 1 \end{vmatrix} + 1\begin{vmatrix} -1 & 3 \\ 2 & 1 \end{vmatrix}$$

$$= -2(-9) + 1(-7) = 11.$$

Writing the first row as a sum, we find

$$\begin{vmatrix} 0 & 2 & 1 \\ -1 & 3 & 4 \\ 2 & 1 & 1 \end{vmatrix} = \begin{vmatrix} 1 + (-1) & 2 + 0 & -1 + 2 \\ -1 & 3 & 4 \\ 2 & 1 & 1 \end{vmatrix},$$

but

$$\begin{vmatrix} 1 & 2 & -1 \\ -1 & 3 & 4 \\ 2 & 1 & 1 \end{vmatrix} = 1\begin{vmatrix} 3 & 4 \\ 1 & 1 \end{vmatrix} - 2\begin{vmatrix} -1 & 4 \\ 2 & 1 \end{vmatrix} - 1\begin{vmatrix} -1 & 3 \\ 2 & 1 \end{vmatrix}$$

$$= 1(-1) - 2(-9) - (-7) = 24$$

and

$$\begin{vmatrix} 1 & 0 & 2 \\ -1 & 3 & 4 \\ 2 & 1 & 1 \end{vmatrix} = -1\begin{vmatrix} 3 & 4 \\ 1 & 1 \end{vmatrix} - 0\begin{vmatrix} -1 & 4 \\ 2 & 1 \end{vmatrix} + 2\begin{vmatrix} -1 & 3 \\ 2 & 1 \end{vmatrix}$$

$$= -1(-1) - 0 + 2(-7) = -13.$$

finally,

$$11 = 24 + (-13).$$

(c) We observe the result of interchanging the first and third rows of the matrix in part (a). In part (a) we found that

$$\begin{vmatrix} 3 & 6 & 9 \\ 1 & 0 & -1 \\ 4 & 2 & 1 \end{vmatrix} = -6.$$

However,

$$\begin{vmatrix} 4 & 2 & 1 \\ 1 & 0 & -1 \\ 3 & 6 & 9 \end{vmatrix} = 4\begin{vmatrix} 0 & -1 \\ 6 & 9 \end{vmatrix} - 2\begin{vmatrix} 1 & -1 \\ 3 & 9 \end{vmatrix} + 1\begin{vmatrix} 1 & 0 \\ 3 & 6 \end{vmatrix}$$

$$= 4(6) - 2(12) + 6 = 6 = -(-6).$$

■

It turns out that there are many other ways to compute a determinant. The definition we gave above is called the *expansion by cofactors across the first row*. Although we will not prove it, the determinant can also be computed by a similar expansion across any row or down any column. The following formulas, called *Laplace's formulas*, can also be used to compute the determinant:

$$\det \mathbf{A} = \sum_{k=1}^{n} a_{ik}A_{ik} \qquad \text{for any } i = 1, 2, \ldots, n \tag{1}$$

$$\det \mathbf{A} = \sum_{k=1}^{n} a_{kj}A_{kj} \qquad \text{for any } j = 1, 2, \ldots, n. \tag{2}$$

The formulas in (1) are called *expansions by cofactors across the* ith *row* and those in (2) are called *expansions by cofactors down the* jth *column*. Although these $2n$ formulas look quite different, they all result in the same number, namely det **A**. This is illustrated in the following example.

Example 3.5

We compute

$$\begin{vmatrix} 1 & 3 & 2 \\ 0 & 1 & -1 \\ 2 & -1 & 4 \end{vmatrix}$$

in several ways using (1) and (2)

Expanding across the first row yields

$$1\begin{vmatrix} 1 & -1 \\ -1 & 4 \end{vmatrix} - 3\begin{vmatrix} 0 & -1 \\ 2 & 4 \end{vmatrix} + 2\begin{vmatrix} 0 & 1 \\ 2 & -1 \end{vmatrix} = 1\cdot 3 - 3\cdot 2 + 2(-2) = -7.$$

Expanding across the last row yields

$$2\begin{vmatrix} 3 & 2 \\ 1 & -1 \end{vmatrix} + 1\begin{vmatrix} 1 & 2 \\ 0 & -1 \end{vmatrix} + 4\begin{vmatrix} 1 & 3 \\ 0 & 1 \end{vmatrix} = 2(-5) + 1\cdot(-1) + 4\cdot 1 = -7.$$

Expanding down the middle column yields

$$-3\begin{vmatrix} 0 & -1 \\ 2 & 4 \end{vmatrix} + 1\begin{vmatrix} 1 & 2 \\ 2 & 4 \end{vmatrix} + 1\begin{vmatrix} 1 & 2 \\ 0 & -1 \end{vmatrix} = -3\cdot 2 + 1\cdot 0 + 1(-1) = -7.$$

■

The extra flexibility of Laplace's formulas allows one to take advantage of the zero structure of a matrix (if there is any) in computing a determinant. We illustrate with Example 3.6.

Example 3.6

Suppose that

$$\mathbf{A} = \begin{bmatrix} 7 & 0 & 8 & 0 & 0 \\ 1 & 2 & 5 & 3 & 6 \\ 0 & 0 & 2 & 0 & 0 \\ 1 & 4 & 2 & 0 & 3 \\ 9 & 1 & 6 & 0 & 2 \end{bmatrix}.$$

If we compute det $\mathbf{A}$ by using one of Laplace's formulas, then a natural choice would be to expand down the fourth column or across the third row. Going down the fourth column gives

$$\det \mathbf{A} = 3\begin{vmatrix} 7 & 0 & 8 & 0 \\ 0 & 0 & 2 & 0 \\ 1 & 4 & 2 & 3 \\ 9 & 1 & 6 & 2 \end{vmatrix}.$$

Expanding this determinant across the second row then gives

$$\det \mathbf{A} = 3(-2)\begin{vmatrix} 7 & 0 & 0 \\ 1 & 4 & 3 \\ 9 & 1 & 2 \end{vmatrix}.$$

Finally, expanding across the first row results in

$$\det \mathbf{A} = 3(-2)(7)\begin{vmatrix} 4 & 3 \\ 1 & 2 \end{vmatrix}$$
$$= 3(-2)(7)(5) = -210.$$

■

In Chapter 2 we saw that the triangular matrices were very important. We now show that the determinant of a triangular matrix is very easy to compute. In fact, *the determinant of a triangular matrix is the product of its diagonal entries*. For example, by repeatedly expanding down the first column, we find

$$\begin{vmatrix} a_{11} & a_{12} & \cdots & a_{1n} \\ 0 & a_{22} & \cdots & a_{2n} \\ 0 & 0 & \cdots & a_{3n} \\ \vdots & \vdots & & \vdots \\ 0 & 0 & \cdots & a_{nn} \end{vmatrix} = a_{11} \begin{vmatrix} a_{22} & \cdots & a_{2n} \\ \vdots & \ddots & \vdots \\ 0 & \cdots & a_{nn} \end{vmatrix}$$

$$= a_{11}a_{22} \begin{vmatrix} a_{33} & \cdots & a_{3n} \\ \vdots & \ddots & \vdots \\ 0 & \cdots & a_{nn} \end{vmatrix}$$

$$= a_{11}a_{22} \quad \cdots \quad a_{nn},$$

and in a similar way we can show that the determinant of a lower triangular matrix is the product of its diagonal entries.

In Chapter 2 we developed an efficient algorithm, Gaussian elimination, for reducing a matrix to upper triangular form. Since the determinant of an upper triangular matrix is so easy to compute, it would be useful to know the effect of the basic Gauss elimination operation, that is, an elementary row operation, on the determinant of a matrix. We could then determine the effect of Gaussian elimination on the determinant of a matrix.

To determine this effect, we make two observations. The first is that *if* $\mathbf{A}$ *is a square matrix with two identical rows, then* $\det \mathbf{A} = 0$. This follows from property (3): If we interchange the two identical rows, then the matrix is unchanged and hence

$$\det \mathbf{A} = -\det \mathbf{A},$$

that is,

$$\det \mathbf{A} = 0.$$

The other observation is that formula (1) allows expansion across any row and therefore properties (1) and (2) hold for any row, not just the first row.

The basic operation in Gaussian elimination is the subtraction of a multiple of one row from another row. We can now see what effect this operation has on the determinant. For example, if t times the second row is subtracted from the first row, we have by properties (1) and (2),

$$\begin{vmatrix} a_{11} - ta_{21} & a_{12} - ta_{22} & \cdots & a_{1n} - ta_{2n} \\ a_{21} & a_{22} & \cdots & a_{2n} \\ \vdots & \vdots & \vdots & \vdots \\ a_{n1} & a_{n2} & \cdots & a_{nn} \end{vmatrix}$$

$$= \begin{vmatrix} a_{11} & a_{12} & \cdots & a_{1n} \\ a_{21} & a_{22} & \cdots & a_{2n} \\ \vdots & \vdots & \vdots & \vdots \\ a_{n1} & a_{n2} & \cdots & a_{nn} \end{vmatrix} - t \begin{vmatrix} a_{21} & a_{22} & \cdots & a_{2n} \\ a_{21} & a_{22} & \cdots & a_{2n} \\ \vdots & \vdots & \vdots & \vdots \\ a_{n1} & a_{n2} & \cdots & a_{nn} \end{vmatrix}.$$

The last determinant is zero because its first two rows are identical. Therefore,

$$\det \mathbf{A} = \begin{vmatrix} a_{11} & a_{12} & \cdots & a_{1n} \\ a_{21} & a_{22} & \cdots & a_{2n} \\ \vdots & \vdots & \vdots & \vdots \\ a_{n1} & a_{n2} & \cdots & a_{nn} \end{vmatrix} = \begin{vmatrix} a_{11} - ta_{21} & a_{12} - ta_{22} & \cdots & a_{1n} - ta_{2n} \\ a_{21} & a_{22} & \cdots & a_{2n} \\ \vdots & \vdots & \vdots & \vdots \\ a_{n1} & a_{n2} & \cdots & a_{nn} \end{vmatrix}.$$

Of course, the same result holds no matter what two rows we choose, and hence subtracting a multiple of one row of a matrix from another row does not change its determinant. Therefore, Gaussian elimination, combined with the fact that the determinant of a triangular matrix is a product of its diagonal entries, can be used as an efficient method of computing the determinant. Should it actually be necessary to compute the numerical value of a determinant, then this reduction to triangular form is the method of choice. We illustrate with an example.

Example 3.7

We calculate the determinant of the matrix

$$\begin{bmatrix} 2 & 1 & 0 \\ 4 & 5 & 1 \\ -2 & 2 & 6 \end{bmatrix}$$

by first using elementary row operation to reduce it to upper triangular form:

$$\begin{vmatrix} 2 & 1 & 0 \\ 4 & 5 & 1 \\ -2 & 2 & 6 \end{vmatrix} = \begin{vmatrix} 2 & 1 & 0 \\ 0 & 3 & 1 \\ -2 & 2 & 6 \end{vmatrix} = \begin{vmatrix} 2 & 1 & 0 \\ 0 & 3 & 1 \\ 0 & 3 & 6 \end{vmatrix} = \begin{vmatrix} 2 & 1 & 0 \\ 0 & 3 & 1 \\ 0 & 0 & 5 \end{vmatrix}$$

$$= 2 \cdot 3 \cdot 5 = 30.$$

■

In this section we have seen that there are many equivalent definitions of the determinant. Some definitions may be better than others in developing the basic properties of the concept that is being defined, but it is important for the reader to realize that a good definition is not necessarily a good computational tool. We defined the determinant in terms of cofactor expansions and this definition is useful in establishing some basic properties of determinants. However, if a determinant must actually be computed numerically, then cofactor expansions are of little use (unless the matrix is very small). For example, it would take many *centuries* to compute a 25×25 determinant using cofactor expansions on a modern computer (see Exercise 11). On the other hand, Gaussian elimination is a very effective computational tool for determinants (see Exercise 11) even though a definition of the determinant in terms of Gaussian elimination is not very useful from the point of view of developing theoretical concepts.

EXERCISES 3.1

1. Compute

$$\begin{vmatrix} 3 & -7 \\ -1 & -2 \end{vmatrix}.$$

2. Compute

$$\begin{vmatrix} 1 & -2 & -3 \\ 0 & 1 & 6 \\ 0 & 4 & 5 \end{vmatrix}$$

by
(a) expanding down the first column
(b) expanding across the second row

3. Compute

$$\begin{vmatrix} 16 & 76 & -4 & 1 & -20 \\ 0 & 144 & 0 & 0 & 0 \\ 100 & 0 & 0 & 0 & -2 \\ 1 & 0 & 1 & 0 & 0 \\ 100 & 4 & 0 & 0 & -1 \end{vmatrix}.$$

4. Show that if $\mathbf{A}$ has an entire row or an entire column of zeros, then $\det \mathbf{A} = 0$.

5. Find A_{23} and A_{22} if

$$\mathbf{A} = \begin{bmatrix} -1 & 1 & 3 \\ 2 & 4 & 1 \\ -5 & 6 & 2 \end{bmatrix}.$$

6. Verify that

$$\begin{vmatrix} -1 & 1 & 2 \\ 3 & 0 & 1 \\ 2 & 1 & 1 \end{vmatrix} - \begin{vmatrix} 3 & 4 & -6 \\ 3 & 0 & 1 \\ 2 & 1 & 1 \end{vmatrix} = \begin{vmatrix} -4 & -3 & 8 \\ 3 & 0 & 1 \\ 2 & 1 & 1 \end{vmatrix}.$$

7. Verify that

$$\begin{vmatrix} 3 & 1 \\ 4 & -2 \end{vmatrix} = \begin{vmatrix} 3 & 1 \\ 0 & -\frac{10}{3} \end{vmatrix}.$$

What elementary row operation produced this result?

8. Find all numbers t such that

$$\begin{vmatrix} 6 & t & 0 \\ 1-t & 0 & 1 \\ 2 & 1 & 2 \end{vmatrix} = 2.$$

9. Verify that the determinant of a lower triangular matrix is the product of its diagonal entries.

10. Compute the determinant of an $n \times n$ matrix $\mathbf{A}$ of the form

$$\mathbf{A} = \begin{bmatrix} a & a & a & \cdots & a & a \\ b & a & a & \cdots & a & a \\ 0 & b & a & \cdots & a & a \\ \vdots & \vdots & \vdots & & \vdots & \vdots \\ 0 & 0 & 0 & \cdots & b & a \end{bmatrix}.$$

11. **(a)** Show that at least $n!$ multiplicative operations are required to compute the determinant of an $n \times n$ matrix using Laplace's formulas.

(b) Suppose that a 20×20 determinant is computed using Laplace's method on a computer which is capable of performing a million multiplications per second. Suppose further that time on the machine costs \$500 per hour. About how much would it cost to compute the determinant?

(c) How many multiplicative operations are required to compute the determinant of an $n \times n$ matrix by the Gauss elimination method as outlined in Example 3.7?

(d) How much would it cost to compute the determinant in part (b) using the Gauss elimination method?

12. Show that if $\mathbf{A}$ is an $n \times n$ matrix and λ is a scalar, then $\det(\lambda\mathbf{A}) = \lambda^n \det \mathbf{A}$.

13. Construct an example, using 2×2 matrices, to show that generally $\det(\mathbf{A} + \mathbf{B}) \neq \det \mathbf{A} + \det \mathbf{B}$.

14. **(a)** Use equations (1) and (2) to show that $\det(\mathbf{A}^T) = \det \mathbf{A}$. (*Hint:* Expanding down the ith column of $\mathbf{A}^T$ is equivalent to expanding across the ith row of $\mathbf{A}$.)

(b) From part (a) conclude that properties analogous to (1)–(3) hold for columns (e.g., interchanging two columns changes the sign of the determinant).

15. Suppose that λ is a scalar and

$$\mathbf{A} = \begin{bmatrix} 2 & 3 & -1 \\ 3 & 1 & 2 \\ -1 & 2 & 0 \end{bmatrix}.$$

Compute $\det(\mathbf{A} - \lambda\mathbf{I})$.

16. Compute $\det \mathbf{A}$, where

$$\mathbf{A} = \begin{bmatrix} 2 & -6 & 1 \\ -1 & 3 & -2 \\ -4 & 12 & 3 \end{bmatrix},$$

using Gauss elimination.

17. Suppose that an unavoidable zero pivot occurs when the Gauss elimination method is applied to $\mathbf{A}$. What is $\det \mathbf{A}$?

3.2 Inverses and Determinants

We just saw, in Example 3.7, that if no row interchanges occur during the elimination process, the determinant is equal to the product of the pivots. But what if row interchanges are necessary? In this case property (3) tells us what happens: Each time a row interchange occurs, the determinant changes sign. Therefore, the determinant will be $(-1)^k$ times the product of the pivots, where k is the number of row interchanges that have occurred. In other words, the determinant is the product of the pivots or minus the product of the pivots depending on whether the number of row interchanges is even or odd, respectively. In particular, we have the following result.

Theorem 3.1. **A** is singular if and only if det **A** = 0.

■

The theorem is true since the product of the pivots is zero if and only if some pivot is zero.

Example 3.8

Using elementary row operations, we find

$$\begin{vmatrix} 1 & 2 & 3 \\ 1 & 2 & 1 \\ 3 & 0 & -1 \end{vmatrix} = \begin{vmatrix} 1 & 2 & 3 \\ 0 & 0 & -2 \\ 0 & -6 & -10 \end{vmatrix} = -\begin{vmatrix} 1 & 2 & 3 \\ 0 & -6 & -10 \\ 0 & 0 & -2 \end{vmatrix}$$

$$= -(1)\cdot(-6)\cdot(-2) = -12,$$

and hence the matrix is nonsingular.

■

Example 3.9

We have $\begin{vmatrix} 1 & 2 \\ 2 & 4 \end{vmatrix} = 4 - 4 = 0$, and therefore the matrix

$$\mathbf{A} = \begin{bmatrix} 1 & 2 \\ 2 & 4 \end{bmatrix}$$

is singular. Note that the elimination process gives

$$\begin{bmatrix} 1 & 2 \\ 2 & 4 \end{bmatrix} \longrightarrow \begin{bmatrix} 1 & 2 \\ 0 & 0 \end{bmatrix},$$

and hence the second pivot is unavoidably zero, verifying that **A** is indeed singular.

■

We now come to one of the more curious properties of the determinant. It involves something called the adjugate matrix (many books call this matrix the adjoint, but we will use this term for another matrix). By the **adjugate** of an $n \times n$ matrix **A** we mean the $n \times n$ matrix, denoted **adj A,** which is the transpose of the matrix of cofactors of **A**, that is,

$$\textbf{adj A} = \begin{bmatrix} A_{11} & A_{21} & \cdots & A_{n1} \\ A_{12} & A_{22} & \cdots & A_{n2} \\ \vdots & \vdots & & \vdots \\ A_{1n} & A_{2n} & \cdots & A_{nn} \end{bmatrix}.$$

Example 3.10

If

$$\mathbf{A} = \begin{bmatrix} 1 & 2 \\ -1 & 3 \end{bmatrix},$$

then $A_{11} = 3$, $A_{12} = 1$, $A_{21} = -2$, $A_{22} = 1$; therefore,

$$\textbf{adj A} = \begin{bmatrix} 3 & -2 \\ 1 & 1 \end{bmatrix}.$$

■

Suppose we premultiply the adjugate of **A** by **A** to obtain a matrix **C**, that is,

$$\mathbf{A}\ \textbf{adj}\ \mathbf{A} = \mathbf{C}.$$

Since the ith column of **adj A** is the ith row of the cofactor matrix, the diagonal entries of **C** are given by

$$c_{ii} = \sum_{k=1}^{n} a_{ik}A_{ik} = \det \mathbf{A}, \qquad 1 \le i \le n \tag{3}$$

[the last equality follows from (1)]. Moreover, if $i \neq j$, then

$$c_{ij} = \sum_{k=1}^{n} a_{ik}A_{jk}. \tag{4}$$

This expression looks suspiciously like a determinant expansion [see (1)]. In fact,

$$\sum_{k=1}^{n} a_{ik}A_{jk} = \det \mathbf{B},$$

where **B** is the matrix obtained from **A** by replacing the jth row of **A** with the ith row of **A** and det **B** is computed by expanding across the jth row of **B**. Since **B** has two identical rows (the ith and jth), det **B** = 0 and hence $c_{ij} = 0$ for $i \neq j$. From (3) and (4) we see that **C** = (det **A**)**I**, that is,

$$\mathbf{A}\ \textbf{adj}\ \mathbf{A} = (\det \mathbf{A})\mathbf{I}.$$

In a similar way, using column expansions, one can show that

$$(\textbf{adj}\ \mathbf{A})\mathbf{A} = (\det \mathbf{A})\mathbf{I}.$$

From these equations it follows that if det **A** ≠ 0, then **A** is invertible and

$$\mathbf{A}^{-1} = \frac{1}{\det \mathbf{A}}\textbf{adj}\ \mathbf{A}. \tag{5}$$

We previously encountered this formula for 2 × 2 matrices in Exercise 1 of Section 1.3.

Although this formula is of some independent interest, it should not be considered a good method for computing $\mathbf{A}^{-1}$. If an inverse really must be computed, then the Gauss–Jordan method introduced in Chapter 2 is a much more efficient technique and should be used instead of (5).

Example 3.11

We compute the inverse of

$$\mathbf{A} = \begin{bmatrix} 2 & 2 & 3 \\ 1 & 0 & 1 \\ 2 & 2 & 2 \end{bmatrix}$$

using formula (5). First,

$$\begin{aligned} \det \mathbf{A} &= 2\begin{vmatrix} 0 & 1 \\ 2 & 2 \end{vmatrix} - 2\begin{vmatrix} 1 & 1 \\ 2 & 2 \end{vmatrix} + 3\begin{vmatrix} 1 & 0 \\ 2 & 2 \end{vmatrix} \\ &= 2(-2) - 2\cdot 0 + 3\cdot 2 = 2. \end{aligned}$$

Also,

$$A_{11} = \begin{vmatrix} 0 & 1 \\ 2 & 2 \end{vmatrix} = -2, \quad A_{12} = -\begin{vmatrix} 1 & 1 \\ 2 & 2 \end{vmatrix} = 0, \quad A_{13} = \begin{vmatrix} 1 & 0 \\ 2 & 2 \end{vmatrix} = 2,$$

$$A_{21} = -\begin{vmatrix} 2 & 3 \\ 2 & 2 \end{vmatrix} = 2, \quad A_{22} = \begin{vmatrix} 2 & 3 \\ 2 & 2 \end{vmatrix} = -2, \quad A_{23} = -\begin{vmatrix} 2 & 2 \\ 2 & 2 \end{vmatrix} = 0,$$

$$A_{31} = \begin{vmatrix} 2 & 3 \\ 0 & 1 \end{vmatrix} = 2, \quad A_{32} = -\begin{vmatrix} 2 & 3 \\ 1 & 1 \end{vmatrix} = 1, \quad A_{33} = \begin{vmatrix} 2 & 2 \\ 1 & 0 \end{vmatrix} = -2,$$

and hence

$$\textbf{adj A} = \begin{bmatrix} -2 & 2 & 2 \\ 0 & -2 & 1 \\ 2 & 0 & -2 \end{bmatrix}.$$

Therefore,

$$\mathbf{A}^{-1} = \frac{\textbf{adj A}}{\det \mathbf{A}} = \begin{bmatrix} -1 & 1 & 1 \\ 0 & -1 & \frac{1}{2} \\ 1 & 0 & -1 \end{bmatrix}.$$

■

It should be clear from this example that Gauss–Jordan elimination is a much more efficient method for computing the inverse. Indeed, as a computational tool (5) is quite impractical; however, it does establish the important fact that $\mathbf{A}^{-1}$ exists when $\det \mathbf{A} \neq 0$. It also happens that $\det \mathbf{A} \neq 0$ when $\mathbf{A}^{-1}$ exists. This follows easily from the fact that determinants are multiplicative, that is,

$$\det \mathbf{AB} = \det \mathbf{A} \det \mathbf{B}. \tag{6}$$

Accepting this for now (a proof is indicated in the exercises) we see that if $\mathbf{A}^{-1}$ exists, then

$$1 = \det \mathbf{I} = \det (\mathbf{A}^{-1}\mathbf{A}) = \det \mathbf{A}^{-1} \det \mathbf{A}$$

and hence $\det \mathbf{A} \neq 0$. Therefore we have established the following fact.

Theorem 3.2. $\mathbf{A}^{-1}$ exists if and only if $\det \mathbf{A} \neq 0$.

■

This theorem is the most important result of this section and we will put it to use in our discussion of eigenvalues in Chapter 6. Combining Theorems 3.1 and 3.2 gives

Corollary 3.3. $\mathbf{A}^{-1}$ exists if and only if $\mathbf{A}$ is nonsingular.

Example 3.12

Let

$$\mathbf{A} = \begin{bmatrix} 3 & 2 \\ -1 & 1 \end{bmatrix}, \quad \mathbf{B} = \begin{bmatrix} -2 & 1 \\ 4 & 5 \end{bmatrix}, \quad \text{and} \quad \mathbf{C} = \mathbf{AB} = \begin{bmatrix} 2 & 13 \\ 6 & 4 \end{bmatrix}.$$

Then $\det \mathbf{A} = 5$, $\det \mathbf{B} = -14$, and $\det \mathbf{C} = -70 = \det \mathbf{A} \det \mathbf{B}$.

■

Finally, we point out another consequence of (5), an algebraic formula known as *Cramer's rule:* If $\mathbf{A}$ is an invertible matrix and $\mathbf{Ax} = \mathbf{b}$, then $x_j = \det \mathbf{A}^{(j)}/\det \mathbf{A}$, where $\mathbf{A}^{(j)}$ is the matrix obtained from $\mathbf{A}$ by replacing its jth column by $\mathbf{b}$. To see this, note that by (5), $\mathbf{x} = \mathbf{A}^{-1}\mathbf{b} = (\mathbf{adj\ A})\mathbf{b}/\det \mathbf{A}$. However, expanding $\mathbf{A}^{(j)}$ down its jth column we find

$$\det \mathbf{A}^{(j)} = b_1 A_{1j} + b_2 A_{2j} + \cdots + b_n A_{nj},$$

which is the jth component of $(\mathbf{adj\ A})\mathbf{b}$, i.e., $(\det \mathbf{A})x_j$. Therefore, $x_j = \det \mathbf{A}^{(j)}/\det \mathbf{A}$.

We illustrate Cramer's rule with a simple example.

Example 3.13

Consider the system

$$\begin{aligned} 2x_1 - x_2 &= 3 \\ x_1 + 3x_2 &= 5 \end{aligned}$$

Then

$$\mathbf{A}^{(1)} = \begin{bmatrix} 3 & -1 \\ 5 & 3 \end{bmatrix}, \quad \mathbf{A}^{(2)} = \begin{bmatrix} 2 & 3 \\ 1 & 5 \end{bmatrix},$$

$\det \mathbf{A} = 7$, and the solution by Cramer's rule is

$$x_1 = \frac{\begin{vmatrix} 3 & -1 \\ 5 & 3 \end{vmatrix}}{\begin{vmatrix} 2 & -1 \\ 1 & 3 \end{vmatrix}} = \frac{14}{7} = 2, \qquad x_2 = \frac{\begin{vmatrix} 2 & 3 \\ 1 & 5 \end{vmatrix}}{\begin{vmatrix} 2 & -1 \\ 1 & 3 \end{vmatrix}} = \frac{7}{7} = 1.$$

■

For computational purposes the utility of Cramer's rule is limited to very small systems as in Example 3.13, and it should not be taken seriously as a computational device. We include it primarily for the sake of tradition and because it appears in explanations of formulas in many mathematics, science, and engineering textbooks.

EXERCISES 3.2

1. Explain why the coefficient matrix of the system

$$\begin{aligned} 3x_1 - x_2 + x_3 + 5x_4 &= 7 \\ x_1 + 2x_2 - x_3 + 7x_4 &= 16 \\ 10x_1 + 3x_2 - 4x_3 + x_4 &= -2 \\ 6x_1 - 2x_2 + 2x_3 + 10x_4 &= 12 \end{aligned}$$

is singular.

2. (a) Let

$$\mathbf{A} = \begin{bmatrix} 1 & 3 & 0 \\ -1 & 2 & 4 \\ 1 & -2 & 0 \end{bmatrix} \quad \text{and} \quad \mathbf{B} = \begin{bmatrix} 2 & 0 & 1 \\ -1 & 3 & 2 \\ 4 & 0 & 0 \end{bmatrix}.$$

Verify that det $\mathbf{AB}$ = det $\mathbf{A}$ det $\mathbf{B}$.

(b*) In more advanced books it is shown that properties (1)–(3) plus one more property, namely (4), det $\mathbf{I} = 1$, are *characteristic* properties of the determinant. This means that $f(\mathbf{A}) = \det \mathbf{A}$ is the *only* real-valued function f defined on $n \times n$ matrices which satisfies (1)–(4). Suppose that det $\mathbf{B} \neq 0$ and let $f(\mathbf{A}) = \det(\mathbf{AB})/\det \mathbf{B}$. Show that f satisfies conditions (1)–(4) and hence $f(\mathbf{A}) = \det \mathbf{A}$, that is, $\det(\mathbf{AB}) = \det \mathbf{A} \det \mathbf{B}$.

3. Suppose that

$$\begin{bmatrix} 1 & 1 \\ 3 & 2 \end{bmatrix} \mathbf{A} \begin{bmatrix} 3 & 1 \\ 1 & 3 \end{bmatrix} = 4\mathbf{I}.$$

What is det $\mathbf{A}$?

4. Suppose that

$$\mathbf{A} = \begin{bmatrix} 1 & 2 \\ -1 & a \end{bmatrix}$$

and det $\mathbf{A}^{-1} = 9$. What is a?

5. Find **adj A** if

$$\mathbf{A} = \begin{bmatrix} 1 & 0 & 1 \\ 2 & -1 & 1 \\ 3 & 0 & 4 \end{bmatrix}.$$

6. Suppose that

$$\mathbf{A} = \begin{bmatrix} 2 & 2 \\ 1 & 1 \end{bmatrix}.$$

Find all numbers λ such that $(\mathbf{A} - \lambda\mathbf{I})$ is not invertible.

7. Let

$$\mathbf{A} = \begin{bmatrix} 1 & 2 & 1 \\ -1 & -3 & 5 \\ 2 & 1 & 21 \end{bmatrix}.$$

Use a watch to time yourself in solving parts (a) and (b).
(a) Find $\mathbf{A}^{-1}$ using (5).
(b) Find $\mathbf{A}^{-1}$ using Gauss–Jordan elimination.

8. Consider the system

$$\begin{aligned} 2x_1 - x_2 + 2x_3 + x_4 &= 4 \\ x_1 + 7x_2 - x_4 &= -5 \\ -3x_1 + x_2 + 3x_3 &= -1 \\ 14x_2 + 3x_3 - 2x_4 &= -9. \end{aligned}$$

Use a watch to time yourself in parts (a) and (b).
(a) Solve the system using Cramer's rule.
(b) Solve the system using Gaussian elimination.

9. Suppose that **A**, **B**, and **C** are $n \times n$ matrices.
(a) Show that if $\mathbf{BA} = \mathbf{I}$, then $\mathbf{B} = \mathbf{A}^{-1}$. (*Hint:* First show that det $\mathbf{A} \neq 0$ and hence $\mathbf{A}^{-1}$ exists. Then postmultiply the first equation by $\mathbf{A}^{-1}$.)
(b) Show that if $\mathbf{AC} = \mathbf{I}$, then $\mathbf{C} = \mathbf{A}^{-1}$.

3.3 Age Distributions

Consider a population of animals that is divided into various age groups. For example, we might have x_1 animals less than 1 year old, x_2 animals which are 1 year old, x_3 which are 2 years old, and so on through x_N which are $N - 1$ years old, where this is the oldest possible age. We can envision the age distribution of the population as a vector

$$\mathbf{x} = \begin{bmatrix} x_1 \\ x_2 \\ \vdots \\ x_N \end{bmatrix}.$$

Suppose that b_i and d_i are the birth rate and death rate for the ith age group. If we denote the age distribution 1 year hence by the vector **y**, then we have

$$y_1 = b_1x_1 + b_2x_2 + \cdots + b_Nx_N,$$

reflecting the fact that new borns have been produced by each age group in proportion to the birth rate for the group. Moreover,

$$\begin{aligned} y_2 &= (1 - d_1)x_1 \\ y_3 &= (1 - d_2)x_2 \\ &\vdots \\ y_N &= (1 - d_{N-1})x_{N-1}, \end{aligned}$$

as only the survivors age another year. In matrix terms the relationship between the age distribution **x** and the later age distribution **y** is $\mathbf{y} = \mathbf{A}\mathbf{x}$, where

$$\mathbf{A} = \begin{bmatrix} b_1 & b_2 & \cdots & b_{N-1} & b_N \\ 1 - d_1 & 0 & \cdots & 0 & 0 \\ 0 & 1 - d_2 & \cdots & 0 & 0 \\ \vdots & \vdots & & \vdots & \vdots \\ 0 & 0 & \cdots & 1 - d_{N-1} & 0 \end{bmatrix}. \tag{7}$$

Example 3.14

Consider a population of insects that is divided into two age groups. Suppose that the birth rates and death rates for the groups are $b_1 = .4$, $d_1 = .3$, $b_2 = .6$ $(d_2 = 1)$, respectively. If the initial age distribution is

$$\begin{bmatrix} 1000 \\ 2000 \end{bmatrix},$$

then in two successive time intervals we obtain age distributions

$$\begin{bmatrix} 1600 \\ 700 \end{bmatrix} = \begin{bmatrix} .4 & .6 \\ .7 & 0 \end{bmatrix} \begin{bmatrix} 1000 \\ 2000 \end{bmatrix}$$

and

$$\begin{bmatrix} 1060 \\ 1120 \end{bmatrix} = \begin{bmatrix} .4 & .6 \\ .7 & 0 \end{bmatrix} \begin{bmatrix} 1600 \\ 700 \end{bmatrix}.$$

■

It is interesting to speculate on the existence of a *static* age distribution, that is, an age distribution **x** which does not change over time. According to (7), such a distribution must satisfy

$$\mathbf{x} = \mathbf{A}\mathbf{x},$$

or equivalently,

$$(\mathbf{I} - \mathbf{A})\mathbf{x} = \boldsymbol{\theta}. \tag{8}$$

If $(\mathbf{I} - \mathbf{A})$ is invertible, then we have

$$\mathbf{x} = (\mathbf{I} - \mathbf{A})^{-1}(\mathbf{I} - \mathbf{A})\mathbf{x} = (\mathbf{I} - \mathbf{A})^{-1}\boldsymbol{\theta} = \boldsymbol{\theta},$$

and hence if there is a nonzero static age distribution, then necessarily $\mathbf{I} - \mathbf{A}$ is not invertible and hence

$$\det(\mathbf{I} - \mathbf{A}) = 0 \tag{9}$$

by Theorem 3.2. However, we shall show in Chapter 4 that (9) also *implies* the existence of an $\mathbf{x} \neq \boldsymbol{\theta}$ which satisfies (8). Therefore, the existence of a static age distribution is equivalent to (9).

Example 3.15

Consider a two-age-group population with $b_1 = .6$, $b_2 = .8$, $d_1 = .5$. Then

$$\det(\mathbf{I} - \mathbf{A}) = \begin{vmatrix} .4 & -.8 \\ -.5 & 1 \end{vmatrix} = 0,$$

and hence a static age distribution exists. To find such a distribution we solve (8), forming the augmented matrix

$$\left[\begin{array}{cc:c} .4 & -.8 & 0 \\ -.5 & 1 & 0 \end{array}\right]$$

and performing Gauss elimination:

$$\left[\begin{array}{cc:c} .4 & -.8 & 0 \\ 0 & 0 & 0 \end{array}\right].$$

From this we see that x_2 is arbitrary and $x_1 = 2x_2$. Therefore, any age distribution of the form

$$t\begin{bmatrix} 2 \\ 1 \end{bmatrix}$$

with $t > 0$ is static for this population. This is quite easy to verify directly:

$$\mathbf{A}\left(t\begin{bmatrix} 2 \\ 1 \end{bmatrix}\right) = t\mathbf{A}\begin{bmatrix} 1 \\ 2 \end{bmatrix}$$

$$= t\begin{bmatrix} .6 & .8 \\ .5 & 0 \end{bmatrix}\begin{bmatrix} 2 \\ 1 \end{bmatrix} = t\begin{bmatrix} 2 \\ 1 \end{bmatrix}.$$

■

EXERCISES 3.3

1. Consider a two-age-group model with

$$\mathbf{A} = \begin{bmatrix} .1 & .4 \\ .6 & 0 \end{bmatrix}.$$

Take some age distribution and trace it through five time periods. What do you predict about this population?

2. **(a)** Show that a two-age-group population with $b_1 = .9$, $b_2 = .4$, $d_1 = .75$ has a static age distribution.

(b) Find such a static age distribution.

3. Consider a population of insects that is divided into two age groups. The birth rates of the groups are found to be $b_1 = .2$ and $b_2 = .9$, respectively, but the death rate d_1 cannot be estimated because the insects immediately and unceremoniously devour the departed. In tracing a certain age distribution it is found to be static, however. What is d_1?

4. In a two-age-group population the initial age distribution

$$\begin{bmatrix} 1000 \\ 2000 \end{bmatrix} \text{ is transformed to } \begin{bmatrix} 700 \\ 800 \end{bmatrix} \text{ and then to } \begin{bmatrix} 310 \\ 560 \end{bmatrix}$$

in two successive time intervals. Find the birth and death rates.

GLOSSARY

adjugate The transpose of the matrix of cofactors of a given square matrix.

cofactor The i,j-th cofactor is $(-1)^{i+j}M_{ij}$, where M_{ij} is the i,j-th minor.

determinant A number associated with a square matrix. May be defined in terms of cofactor expansions.

minor The i, j-th minor of an $n \times n$ matrix $\mathbf{A}$ is the determinant of the $(n-1) \times (n-1)$ matrix obtained by deleting the ith row and jth column of $\mathbf{A}$.

NOTES AND COMMENTS

Determinants have a long history; in fact, they predate matrices. Gottfried Leibniz (1646–1716), one of the cofounders of calculus, formulated a notion of determinants in 1693. Augustin Cauchy (1789–1857) coined the term "determinant" about a century and a half later and gave a formula in terms of determinants for the volume of the parallelopiped formed by three vectors. It was Karl Weierstrass (1815–1897) who recognized the characteristic nature of properties (1)–(4) given in Exercise 2 of Section 3.2. An historical account of the arcane and curious lore of determinants can be found in Sir Thomas Muir's monumental four-volume work *The Theory of Determinants* (1906).

4

Theory of Solutions

This chapter is devoted to the study of existence and uniqueness of solutions for linear systems of equations. To address these questions we introduce the concepts of linear independence and row echelon form. We shall find an intimate connection between these concepts and the method of Gaussian elimination. Finally, we present a fundamental result, called the rank-nullity theorem, which answers many of the important theoretical questions regarding linear systems of equations.

4.1 Existence and Uniqueness of Solutions

In Chapter 2 we discussed how one solves a linear system with square coefficient matrix by the method of Gaussian elimination. This procedure provides a conceptually simple and effective method for solving such systems. In fact, computer programs based on this procedure are a standard part of the libraries of "canned" routines at computing centers throughout the world. In our presentation of the elimination method we have concentrated, in a mechanical way, on the individual entries of the coefficient matrix to produce an upper triangular matrix on which back substitution can be performed. Our aim in this chapter is to view linear systems from a broader perspective in order to obtain a deeper understanding of the problem. Moreover, we shall study the theory of solutions for rectangular systems, that is, for ones in which the number of equations and number of unknowns are not necessarily the same. The elimination method will again play a key role; however, our aim now is to produce an echelon form (to be discussed later) whose structure will enable us effectively to study the existence and uniqueness of solutions.

We find it convenient and instructive to view multiplication of an n-vector by an $m \times n$ matrix $\mathbf{A}$ a bit more abstractly. We denote by $\mathbf{R}^n$ the set of all n-vectors (i.e., $n \times 1$ matrices) and similarly for $\mathbf{R}^m$. Thus a vector $\mathbf{x}$ is in $\mathbf{R}^n$ if

$$\mathbf{x} = \begin{bmatrix} x_1 \\ x_2 \\ \vdots \\ x_n \end{bmatrix} \qquad \text{with } x_i \in \mathbf{R} \quad \text{for } 1 \le i \le n.$$

The multiplication of $\mathbf{x} \in \mathbf{R}^n$ by an $m \times n$ matrix $\mathbf{A}$ gives a (unique) m-vector $\mathbf{y} = \mathbf{Ax}$. We can think of $\mathbf{A}$ as defining a function $L_{\mathbf{A}}$ given by the rule $\mathbf{x} \to \mathbf{Ax}$. That is, $L_{\mathbf{A}}$ is the function defined by $L_{\mathbf{A}}(\mathbf{x}) = \mathbf{A}\mathbf{x}$. This function is defined on $\mathbf{R}^n$ and has values that are vectors in $\mathbf{R}^m$.

Example 4.1

For

$$\mathbf{A} = \begin{bmatrix} 1 & -2 & 1 \\ 3 & 0 & 4 \end{bmatrix} \quad \text{and} \quad \mathbf{x} = \begin{bmatrix} -1 \\ 0 \\ 1 \end{bmatrix}$$

we have

$$L_{\mathbf{A}}(\mathbf{x}) = \mathbf{A}\mathbf{x} = \mathbf{y} = \begin{bmatrix} 0 \\ 1 \end{bmatrix}.$$

This can be represented pictorially as shown in Figure 4.1.

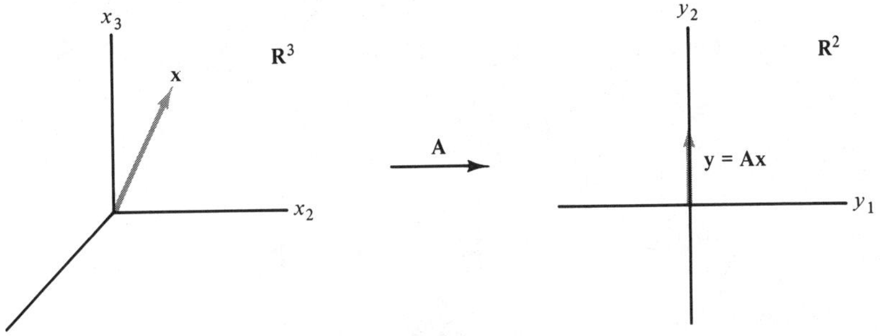

Figure 4.1

■

An important observation to make about $L_{\mathbf{A}}$ is that it is **linear**. By this we mean that for any two n-vectors, $\mathbf{u}$ and $\mathbf{v}$, and scalars, α and β,

$$L_{\mathbf{A}}(\alpha\mathbf{u} + \beta\mathbf{v}) = \alpha L_{\mathbf{A}}(\mathbf{u}) + \beta L_{\mathbf{A}}(\mathbf{v}).$$

The linearity of $L_{\mathbf{A}}$ follows immediately from properties of matrix multiplication:

$$\begin{aligned} L_{\mathbf{A}}(\alpha\mathbf{u} + \beta\mathbf{v}) &= \mathbf{A}(\alpha\mathbf{u} + \beta\mathbf{v}) \\ &= \mathbf{A}(\alpha\mathbf{u}) + \mathbf{A}(\beta\mathbf{v}) = \alpha\mathbf{A}\mathbf{u} + \beta\mathbf{A}\mathbf{v} \\ &= \alpha L_{\mathbf{A}}(\mathbf{u}) + \beta L_{\mathbf{A}}(\mathbf{v}). \end{aligned}$$

In particular, note that for $\alpha = \beta = 1$ we have

$$L_{\mathbf{A}}(\mathbf{u} + \mathbf{v}) = L_{\mathbf{A}}(\mathbf{u}) + L_{\mathbf{A}}(\mathbf{v})$$

and for $\beta = 0$ we get

$$L_{\mathbf{A}}(\alpha\mathbf{u}) = \alpha L_{\mathbf{A}}(\mathbf{u}).$$

$L_{\mathbf{A}}$ is called the **linear transformation** defined by $\mathbf{A}$.

Let us return to the problem $\mathbf{A}\mathbf{x} = \mathbf{b}$ and examine how the linearity of $L_{\mathbf{A}}$ can be used to characterize the type of solutions this problem can have. Suppose that $\mathbf{u}$ is a particular solution and $\mathbf{x}$ is any other solution of $\mathbf{A}\mathbf{x} = \mathbf{b}$. Then $L_{\mathbf{A}}(\mathbf{u}) = \mathbf{b} = L_{\mathbf{A}}(\mathbf{x})$

and hence if $\mathbf{z} = \mathbf{x} - \mathbf{u}$ we find

$$L_{\mathbf{A}}(\mathbf{z}) = L_{\mathbf{A}}(\mathbf{x} - \mathbf{u}) = L_{\mathbf{A}}(\mathbf{x}) - L_{\mathbf{A}}(\mathbf{u}) = \mathbf{b} - \mathbf{b} = \boldsymbol{\theta}.$$

Thus the difference, $\mathbf{z}$, of two solutions is a solution of the homogeneous problem $\mathbf{Az} = \boldsymbol{\theta}$. (Any problem with zero right-hand side is called **homogeneous**.) Note that since $\mathbf{A}\boldsymbol{\theta} = \boldsymbol{\theta}$, $\mathbf{z} = \boldsymbol{\theta}$ is always a solution of the homogeneous problem. If the only solution of $\mathbf{Az} = \boldsymbol{\theta}$ is the trivial solution $\mathbf{z} = \boldsymbol{\theta}$, then $\mathbf{x} = \mathbf{u}$, that is, solutions are unique. Therefore, we have established the following result.

Theorem 4.1. Suppose that $\mathbf{u}$ satisfies $\mathbf{Au} = \mathbf{b}$ and $\mathbf{z}$ satisfies $\mathbf{Az} = \boldsymbol{\theta}$. Then $\mathbf{x} = \mathbf{u} + \mathbf{z}$ is a solution of $\mathbf{Ax} = \mathbf{b}$ and conversely every solution $\mathbf{x}$ of $\mathbf{Ax} = \mathbf{b}$ has the form $\mathbf{x} = \mathbf{u} + \mathbf{z}$ for some $\mathbf{z}$ with $\mathbf{Az} = \boldsymbol{\theta}$. If the only solution of $\mathbf{Az} = \boldsymbol{\theta}$ is $\mathbf{z} = \boldsymbol{\theta}$, then the problem $\mathbf{Ax} = \mathbf{b}$ can have at most one solution.

■

We illustrate the theorem in the next two examples.

Example 4.2

One solution of

$$\mathbf{Ax} = \begin{bmatrix} 1 & 3 & 0 \\ -2 & -4 & 6 \end{bmatrix}\begin{bmatrix} x_1 \\ x_2 \\ x_3 \end{bmatrix} = \begin{bmatrix} 2 \\ -2 \end{bmatrix} = \mathbf{b}$$

is $\mathbf{x} = [-1, 1, 0]^T$. Application of Gaussian elimination to $\mathbf{Az} = \boldsymbol{\theta}$ gives

$$\left[\begin{array}{ccc|c} 1 & 3 & 0 & 0 \\ -2 & -4 & 6 & 0 \end{array}\right] \longrightarrow \left[\begin{array}{ccc|c} 1 & 3 & 0 & 0 \\ 0 & 2 & 6 & 0 \end{array}\right]$$

and hence $z_2 = -3z_3$, $z_1 = -3z_2 = 9z_3$. Therefore, solutions of the homogeneous problem are given by

$$\mathbf{z} = z_3\begin{bmatrix} 9 \\ -3 \\ 1 \end{bmatrix},$$

where z_3 is arbitrary. If we let $\alpha = z_3$, then $\mathbf{Ax} = \mathbf{b}$ has infinitely many solutions given by

$$\mathbf{x} = \begin{bmatrix} -1 \\ 1 \\ 0 \end{bmatrix} + \alpha\begin{bmatrix} 9 \\ -3 \\ 1 \end{bmatrix}.$$

■

Example 4.3

For the problem

$$\mathbf{Ax} = \begin{bmatrix} 1 & -2 \\ 3 & -4 \\ 0 & 6 \end{bmatrix}\begin{bmatrix} x_1 \\ x_2 \end{bmatrix} = \begin{bmatrix} -1 \\ 1 \\ 12 \end{bmatrix},$$

we apply elimination:

$$\left[\begin{array}{cc|c} 1 & -2 & -1 \\ 3 & -4 & 1 \\ 0 & 6 & 12 \end{array}\right] \longrightarrow \left[\begin{array}{cc|c} 1 & -2 & -1 \\ 0 & 2 & 4 \\ 0 & 6 & 12 \end{array}\right] \longrightarrow \left[\begin{array}{cc|c} 1 & -2 & -1 \\ 0 & 2 & 4 \\ 0 & 0 & 0 \end{array}\right].$$

Thus the unique solution is $x_2 = 2$, $x_1 = 3$. The only solution of $\mathbf{Az} = \boldsymbol{\theta}$ is $\mathbf{z} = \boldsymbol{\theta}$. Verify this!

■

Associated with the questions of existence and uniqueness of solutions are two fundamental spaces which we now examine. As before, let $L_{\mathbf{A}}$ be the linear transformation defined by the $m \times n$ matrix $\mathbf{A}$. The **range** of $L_{\mathbf{A}}$ is defined to be the set of all possible values of $L_{\mathbf{A}}$ and is denoted by $R(\mathbf{A})$. That is, $R(\mathbf{A})$ is the subset of $\mathbf{R}^m$ given by

$$R(\mathbf{A}) = \{\mathbf{y} \in \mathbf{R}^m : \mathbf{y} = L_{\mathbf{A}}(\mathbf{x}) = \mathbf{Ax}, \text{ some } \mathbf{x} \in \mathbf{R}^n\}.$$

Thus a vector $\mathbf{b} \in \mathbf{R}^m$ is in $R(\mathbf{A})$ if and only if $\mathbf{Ax} = \mathbf{b}$ for some $\mathbf{x} \in \mathbf{R}^n$. Clearly, $\mathbf{b} \in R(\mathbf{A})$ means that the problem $\mathbf{Ax} = \mathbf{b}$ is consistent. Actually, $R(\mathbf{A})$ is more than merely a subset of $\mathbf{R}^m$; it is a subspace.

We say that a subset S of $\mathbf{R}^n$ is a **subspace** if:

(1) For any two vectors, $\mathbf{w}$ and $\mathbf{y}$ in S, the sum $\mathbf{w} + \mathbf{y}$ is a vector in S.
(2) For $\mathbf{w} \in S$ and any scalar α, the product $\alpha\mathbf{w}$ is a vector in S.

These two conditions say that a subspace is closed under vector addition and scalar multiplication.

We now demonstrate that $R(\mathbf{A})$ is a subspace of $\mathbf{R}^m$. Given two vectors $\mathbf{w}$, $\mathbf{y} \in R(\mathbf{A})$ we have $\mathbf{Ax} = \mathbf{w}$ and $\mathbf{Au} = \mathbf{y}$ for some $\mathbf{x}$, $\mathbf{u} \in \mathbf{R}^n$. It follows that

$$\mathbf{w} + \mathbf{y} = \mathbf{Ax} + \mathbf{Au} = \mathbf{A}(\mathbf{x} + \mathbf{u}) = L_{\mathbf{A}}(\mathbf{x} + \mathbf{u})$$

and hence $\mathbf{w} + \mathbf{y} \in R(\mathbf{A})$. Similarly, since $L_{\mathbf{A}}(\alpha\mathbf{x}) = \alpha L_{\mathbf{A}}(\mathbf{x})$, it follows that $\alpha\mathbf{w} \in R(\mathbf{A})$. Thus $R(\mathbf{A})$ is a subspace of $\mathbf{R}^m$. Knowing how big $R(\mathbf{A})$ is will enable us to say more about the solvability of $\mathbf{Ax} = \mathbf{b}$. In particular, if $R(\mathbf{A}) = \mathbf{R}^m$, then $\mathbf{Ax} = \mathbf{b}$ has a solution for all $\mathbf{b} \in \mathbf{R}^m$.

In the next example we illustrate the range graphically as a plane passing through the origin.

Example 4.4

Consider the 3×2 matrix problem

$$\mathbf{Ax} = \begin{bmatrix} 4 & 1 \\ 1 & 2 \\ 0 & 1 \end{bmatrix} \begin{bmatrix} x_1 \\ x_2 \end{bmatrix} = \begin{bmatrix} b_1 \\ b_2 \\ b_3 \end{bmatrix} = \mathbf{b}.$$

For what type of $\mathbf{b}$ will the problem have a solution? The key in answering this question is to view $\mathbf{Ax} = \mathbf{b}$ as the equivalent vector equation

$$x_1 \begin{bmatrix} 4 \\ 1 \\ 0 \end{bmatrix} + x_2 \begin{bmatrix} 1 \\ 2 \\ 1 \end{bmatrix} = \begin{bmatrix} b_1 \\ b_2 \\ b_3 \end{bmatrix}.$$

Now it is clear that $\mathbf{Ax} = \mathbf{b}$ has a solution if $\mathbf{b}$ can be expressed as above. In this case we say that $\mathbf{b}$ is a linear combination of the columns of $\mathbf{A}$. The range is precisely the set of all such combinations of the columns of $\mathbf{A}$ which, in geometric terms, is the plane P determined by the column vectors (see Figure 4.2).

■

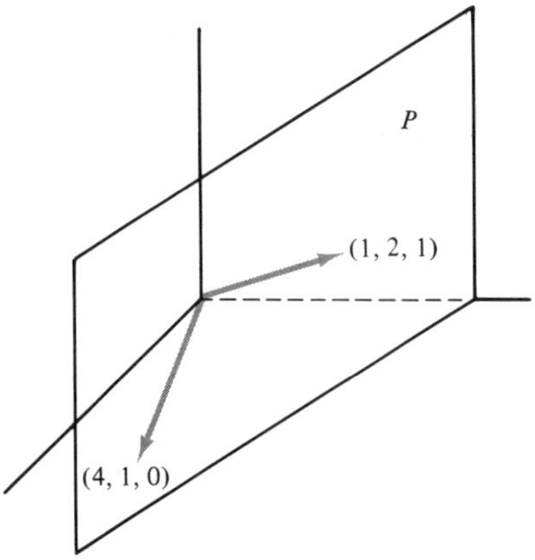

Figure 4.2

We shall expound on the idea of linear combinations of vectors in Section 4.2. For the 3×2 problem of Example 4.4 we are able to visualize the range geometrically, but for m equations in n unknowns it is not so easy to picture the range. What we need is an algebraic way of characterizing the range—this is discussed in more detail later in the chapter.

With regard to uniqueness, Theorem 4.1 tells us that the set

$$N(\mathbf{A}) = \{\mathbf{x} \in \mathbf{R}^n : L_{\mathbf{A}}(\mathbf{x}) = \mathbf{Ax} = \boldsymbol{\theta}\}$$

plays an important role. Indeed, if $N(\mathbf{A})$ consists of simply the zero vector, that is, $N(\mathbf{A}) = \{\boldsymbol{\theta}\}$, then $\mathbf{Ax} = \mathbf{b}$ can have at most one solution. The set $N(\mathbf{A})$ is called the **nullspace** of $L_{\mathbf{A}}$ and is a subspace of $\mathbf{R}^n$. To see this, note that if $\mathbf{x}, \mathbf{y} \in N(\mathbf{A})$ and α and β are scalars, then

$$\begin{aligned} L_{\mathbf{A}}(\alpha\mathbf{x} + \beta\mathbf{y}) &= \alpha L_{\mathbf{A}}(\mathbf{x}) + \beta L_{\mathbf{A}}(\mathbf{y}) \\ &= \alpha\boldsymbol{\theta} + \beta\boldsymbol{\theta} = \boldsymbol{\theta} \end{aligned}$$

and hence $\alpha\mathbf{x} + \beta\mathbf{y} \in N(\mathbf{A})$. Since $N(\mathbf{A})$ is a subspace it follows that $\alpha\mathbf{z}$ is in $N(\mathbf{A})$ for any $\mathbf{z} \in N(\mathbf{A})$ and any scalar α. Thus if $\mathbf{z} \neq \boldsymbol{\theta}$, then $N(\mathbf{A})$ contains infinitely many nonzero vectors. This was the case in Example 4.2. We are now in a position to determine the possible number of solutions of $\mathbf{Ax} = \mathbf{b}$.

Theorem 4.2. The system $\mathbf{Ax} = \mathbf{b}$ has either no solution, exactly one solution, or infinitely many solutions.

Proof. If $\mathbf{b}$ is not in the range, then the problem is inconsistent and no solution exists. Otherwise, a solution takes the form $\mathbf{x} = \mathbf{u} + \mathbf{z}$ with $\mathbf{Au} = \mathbf{b}$ and $\mathbf{Az} = \boldsymbol{\theta}$. If $N(\mathbf{A}) = \{\boldsymbol{\theta}\}$, then $\mathbf{z} = \boldsymbol{\theta}$ and $\mathbf{x} = \mathbf{u}$ is unique. On the other hand, if $\mathbf{z} \neq \boldsymbol{\theta}$, then $\mathbf{u} + \alpha\mathbf{z}$ is a solution for all $\alpha \in \mathbf{R}$.

■

In Examples 4.2 and 4.3 we saw problems in which there were infinitely many solutions and a unique solution, respectively.

Example 4.5

Recall from Example 4.4 that the range consisted of those vectors in the plane P of Figure 4.2. That is, $\mathbf{b}$ is in the range if for some x_1 and x_2,

$$x_1 \begin{bmatrix} 4 \\ 1 \\ 0 \end{bmatrix} + x_2 \begin{bmatrix} 1 \\ 2 \\ 1 \end{bmatrix} = \mathbf{b}.$$

It is a simple matter to check that $\mathbf{b} = [1, 0, 1]^T$ is not a linear combination of the columns of $\mathbf{A}$ and hence $\mathbf{Ax} = \mathbf{b}$ has no solution. If $\mathbf{b} = [5, 3, 1]^T$, the unique solution is $x_1 = x_2 = 1$. Thus existence of a solution, in this example, depends on $\mathbf{b}$.

■

EXERCISES 4.1

1. Suppose that T is a linear function which maps $\mathbf{R}^n$ into $\mathbf{R}^m$. Show that $T(\boldsymbol{\theta}) = \boldsymbol{\theta}$. [*Hint:* Consider $T(\boldsymbol{\theta} + \boldsymbol{\theta})$.]
2. Let

$$\mathbf{A} = \begin{bmatrix} 1 & 2 & 0 \\ -1 & 4 & 1 \end{bmatrix}.$$

Find $L_{\mathbf{A}}(\mathbf{e}^{(j)})$ for $j = 1, 2, 3$.

3. Suppose that $\mathbf{A}$ is an $m \times n$ matrix with columns $\mathbf{A}_1, \ldots, \mathbf{A}_n$. Show that

$$L_{\mathbf{A}}(\mathbf{x}) = x_1 L_{\mathbf{A}}(\mathbf{e}^{(1)}) + x_2 L_{\mathbf{A}}(\mathbf{e}^{(2)}) + \cdots + x_n L_{\mathbf{A}}(\mathbf{e}^{(m)}),$$

with $L_{\mathbf{A}}(\mathbf{e}^{(j)}) = \mathbf{A}_j = j$th column of $\mathbf{A}$.

4. Find all solutions of the homogeneous problem

$$\mathbf{Az} = \begin{bmatrix} 1 & 0 & 1 & 2 \\ 2 & -1 & 1 & 0 \\ -2 & 4 & -1 & -2 \end{bmatrix} \begin{bmatrix} z_1 \\ z_2 \\ z_3 \\ z_4 \end{bmatrix} = \boldsymbol{\theta}.$$

5. Find a particular solution of $\mathbf{Ax} = \mathbf{b}$ where $\mathbf{b} = [1, 0, 3]^T$ and $\mathbf{A}$ is the matrix of Exercise 4. Find the general form of solutions of $\mathbf{Ax} = \mathbf{b}$.
6. If S is a subspace of $\mathbf{R}^n$, show that $\boldsymbol{\theta} \in S$.

7. Consider the set S of all vectors $\mathbf{x}$ in $\mathbf{R}^3$ of the form

$$\mathbf{x} = \begin{bmatrix} x_1 \\ 0 \\ x_3 \end{bmatrix}.$$

Show that S is a subspace of $\mathbf{R}^3$. Can you describe S geometrically?

*8. Consider the set S of all vectors $\mathbf{x}$ in $\mathbf{R}^3$ satisfying $x_1 + x_2 + x_3 = 1$. Is S a subspace of $\mathbf{R}^3$? Justify your answer.

9. In each of the following, find all solutions of the homogeneous problem $\mathbf{Ax} = \mathbf{\theta}$. That is, in each case find the general form of a vector $\mathbf{x} \in N(\mathbf{A})$.

(a) $\mathbf{A} = \begin{bmatrix} 1 & -3 \\ 0 & 1 \end{bmatrix}$ **(b)** $\mathbf{A} = \begin{bmatrix} 0 & 1 & -1 & 3 \\ 0 & 0 & 2 & -1 \\ 0 & 0 & 0 & 0 \end{bmatrix}$

(c) $\mathbf{A} = \begin{bmatrix} 1 & 4 \\ 2 & 6 \\ 1 & -2 \end{bmatrix}$ **(d)** $\mathbf{A} = \begin{bmatrix} 1 & 2 & -1 \\ 4 & 6 & -2 \end{bmatrix}$

10. Suppose that $\mathbf{A}$ is the matrix

$$\begin{bmatrix} 2 & -1 & -3 & 5 \\ 0 & 1 & 1 & -1 \\ -1 & -1 & 0 & -1 \end{bmatrix}.$$

In each case determine if $\mathbf{b} \in R(\mathbf{A})$.

(a) $\mathbf{b} = [0 \quad -2 \quad 1 \quad]^T$ **(b)** $\mathbf{b} = [1 \quad 2 \quad 3]^T$ **(c)** $\mathbf{b} = [2 \quad 0 \quad -1]^T$

11. Suppose that $\mathbf{A}$ is an $m \times n$ matrix whose jth column is $\mathbf{A}_j$. Show that $\mathbf{A}_j \in R(\mathbf{A})$.

12. For the matrix of Example 4.4, determine $N(\mathbf{A})$. How many solutions can $\mathbf{Ax} = \mathbf{b}$ have?

13. Suppose that $\mathbf{A}$ is the 1×2 matrix $[1 \quad 2]$. What is the geometrical relationship between $N(\mathbf{A})$ and $R(\mathbf{A}^T)$?

14. Give examples of $\mathbf{A}$ for which $\mathbf{Ax} = \mathbf{b}$
   (a) has 0 or infinitely many solutions depending on $\mathbf{b}$
   (b) has 0 or 1 solution depending on $\mathbf{b}$
   (c) has exactly one solution for every $\mathbf{b}$.

4.2 Linear Independence

In Example 4.4 we found that the range of $\mathbf{A}$ consists of combinations of the columns of $\mathbf{A}$. In this section we elaborate on this observation for the general $m \times n$ problem. Moreover, we examine the possibility of redundancy in the columns of $\mathbf{A}$. The notion of column redundancy will be made clear (we hope) and will enable us to completely characterize the range algebraically.

We say that $\mathbf{b}$ is a **linear combination** of the m-vectors $\{\mathbf{u}^{(1)}, \ldots, \mathbf{u}^{(n)}\}$ if for some scalars $x_1, \ldots, x_n$,

$$\mathbf{b} = \sum_{j=1}^{n} x_j \mathbf{u}^{(j)}.$$

Let $\mathbf{A}$ be the $m \times n$ matrix whose jth column, $\mathbf{A}_j$, is the vector $\mathbf{u}^{(j)}$. It is then clear that $\mathbf{Ax} = \mathbf{b}$. Conversely, the $m \times n$ matrix problem $\mathbf{Ax} = \mathbf{b}$ can be written as the (column) vector equation

$$\sum_{j=1}^{n} x_j \mathbf{A}_j = \mathbf{b},$$

where $\mathbf{A}_j$ denotes the jth column of $\mathbf{A}$. Thus the question of whether a given vector is a linear combination of $\{\mathbf{u}^{(j)}\}_{j=1}^{n}$ is equivalent to that of existence of solutions of $\mathbf{Ax} = \mathbf{b}$ for the matrix $\mathbf{A}$ whose jth column is $\mathbf{u}^{(j)}$.

Example 4.6

The vector $\mathbf{b} = [1 \quad -2 \quad 1]^T$ is a linear combination of

$$\left\{ \begin{bmatrix} 1 \\ 0 \\ 1 \end{bmatrix}, \begin{bmatrix} 1 \\ 2 \\ 1 \end{bmatrix} \right\}$$

since

$$\mathbf{b} = 2\begin{bmatrix} 1 \\ 0 \\ 1 \end{bmatrix} + (-1)\begin{bmatrix} 1 \\ 2 \\ 1 \end{bmatrix}.$$

■

Example 4.7

The vector $\mathbf{b} = [1 \quad -2 \quad 1]^T$ is not a linear combination of

$$\left\{ \begin{bmatrix} 1 \\ 2 \\ 3 \end{bmatrix}, \begin{bmatrix} 1 \\ 1 \\ 0 \end{bmatrix} \right\}$$

Indeed, if

$$\mathbf{b} = \begin{bmatrix} 1 \\ -2 \\ 1 \end{bmatrix} = x_1\begin{bmatrix} 1 \\ 2 \\ 3 \end{bmatrix} + x_2\begin{bmatrix} 1 \\ 1 \\ 0 \end{bmatrix},$$

then, in matrix form,

$$\begin{bmatrix} 1 & 1 \\ 2 & 1 \\ 3 & 0 \end{bmatrix}\begin{bmatrix} x_1 \\ x_2 \end{bmatrix} = x_1\begin{bmatrix} 1 \\ 2 \\ 3 \end{bmatrix} + x_2\begin{bmatrix} 1 \\ 1 \\ 0 \end{bmatrix} = \begin{bmatrix} 1 \\ -2 \\ 1 \end{bmatrix}$$

and by Gaussian elimination we find

$$\left[\begin{array}{cc|c} 1 & 1 & 1 \\ 2 & 1 & -2 \\ 3 & 0 & 1 \end{array}\right] \longrightarrow \left[\begin{array}{cc|c} 1 & 1 & 1 \\ 0 & -1 & -4 \\ 0 & -3 & -2 \end{array}\right] \longrightarrow \left[\begin{array}{cc|c} 1 & 1 & 1 \\ 0 & -1 & -4 \\ 0 & 0 & 10 \end{array}\right].$$

Since $0 \neq 10$, $\mathbf{b}$ cannot be expressed as a linear combination of the given vectors.

■

We define the **span** of a set of vectors $\{\mathbf{u}^{(j)}\}_{j=1}^{n}$ to be the set of all possible linear combinations of the given vectors. The span is denoted by $S[\mathbf{u}^{(1)}, \ldots, \mathbf{u}^{(n)}]$ and a vector $\mathbf{b}$ is in this span if

$$\mathbf{b} = x_1\mathbf{u}^{(1)} + x_2\mathbf{u}^{(2)} + \cdots + x_n\mathbf{u}^{(n)}$$

for some scalars $x_1, \ldots, x_n$. If the matrix $\mathbf{A}$ has columns $\mathbf{u}^{(1)}, \ldots, \mathbf{u}^{(n)}$ then $\mathbf{b} \in S[\mathbf{u}^{(1)}, \ldots, \mathbf{u}^{(n)}]$ means that $\mathbf{Ax} = \mathbf{b}$ for some $\mathbf{x}$. Therefore, we have $R(\mathbf{A}) = S[\mathbf{u}^{(1)}, \ldots, \mathbf{u}^{(n)}]$. That is, the range of $L_{\mathbf{A}}$ is the span of the columns of the matrix which defines $L_{\mathbf{A}}$. If $\mathbf{u}^{(j)} \in \mathbf{R}^m$ for $1 \leq j \leq n$, it follows (since the range is a subspace) that $S[\mathbf{u}^{(1)}, \ldots, \mathbf{u}^{(n)}]$ is a subspace of $\mathbf{R}^m$.

Theorem 4.3 Let $\mathbf{A}$ be an $m \times n$ matrix which defines the linear transformation $L_{\mathbf{A}}$. Then

$$R(\mathbf{A}) = S[\mathbf{A}_1, \ldots, \mathbf{A}_n].$$

■

Next we consider the possibility that one or more of the columns of $\mathbf{A}$ are redundant in the sense that possibly one column of $\mathbf{A}$ contributes nothing new to the span. To be precise we need to introduce the idea of linear dependence. We say that a given set of m-vectors $\{\mathbf{u}^{(j)}\}_{j=1}^{n}$ is **linearly dependent** if there exist scalars $x_1, \ldots, x_n$, *not all zero,* such that

$$x_1\mathbf{u}^{(1)} + x_2\mathbf{u}^{(2)} + \cdots + x_n\mathbf{u}^{(n)} = \boldsymbol{\theta}.$$

We emphasize that at least one of the scalars is nonzero, say $x_k \neq 0$. Then we can solve for $\mathbf{u}^{(k)}$:

$$x_k\mathbf{u}^{(k)} = -\sum_{\substack{j=1 \\ j\neq k}}^{n} x_j u^{(j)}$$

and since $x_k \neq 0$,

$$\mathbf{u}^{(k)} = -\sum_{\substack{j=1 \\ j\neq k}}^{n} \frac{x_j}{x_k}\mathbf{u}^{(j)}.$$

Thus $\mathbf{u}^{(k)}$ is a linear combination of (i.e., *depends linearly upon*) the vectors $\{\mathbf{u}^{(j)}\}_{j=1, j\neq k}^{n}$. Conversely, if one of the vectors, say $\mathbf{u}^{(k)}$, is a linear combination of the others,

$$\mathbf{u}^{(k)} = \sum_{\substack{j=1 \\ j\neq k}}^{n} c_j\mathbf{u}^{(j)},$$

then $\sum_{j=1}^{n} x_j\mathbf{u}^{(j)} = \boldsymbol{\theta}$, where $x_j = -c_j$ for $j \neq k$ and $x_k = 1$. That is, $\{\mathbf{u}^{(j)}\}_{j=1}^{n}$ is linearly dependent. Therefore, $\{\mathbf{u}^{(j)}\}_{j=1}^{n}$ is linearly dependent if and only if at least one of the vectors can be expressed as a linear combination of the remaining vectors.

Example 4.8

Consider the set of vectors

$$\left\{ \begin{bmatrix} 1 \\ -1 \end{bmatrix}, \begin{bmatrix} 2 \\ 1 \end{bmatrix}, \begin{bmatrix} 1 \\ 3 \end{bmatrix} \right\}.$$

To determine if this set is linearly dependent we seek x_1, x_2, x_3 (not all zero) such that

$$x_1 \begin{bmatrix} 1 \\ -1 \end{bmatrix} + x_2 \begin{bmatrix} 2 \\ 1 \end{bmatrix} + x_3 \begin{bmatrix} 1 \\ 3 \end{bmatrix} = \begin{bmatrix} 0 \\ 0 \end{bmatrix}.$$

That is, we seek a nontrivial solution of the homogeneous matrix problem

$$\begin{bmatrix} 1 & 2 & 1 \\ -1 & 1 & 3 \end{bmatrix} \begin{bmatrix} x_1 \\ x_2 \\ x_3 \end{bmatrix} = \begin{bmatrix} 0 \\ 0 \end{bmatrix}.$$

Gaussian elimination gives the solution: $x_2 = -\frac{4}{3}x_3$ and $x_1 = \frac{5}{3}x_3$ with x_3 arbitrary. Choosing $x_3 = 3$ gives $x_2 = -4$, $x_1 = 5$ and with this choice

$$5 \begin{bmatrix} 1 \\ -1 \end{bmatrix} + (-4) \begin{bmatrix} 2 \\ 1 \end{bmatrix} + 3 \begin{bmatrix} 1 \\ 3 \end{bmatrix} = \begin{bmatrix} 0 \\ 0 \end{bmatrix}.$$

Therefore, the set is linearly dependent, and since each scalar is nonzero, any one of the vectors can be written as a linear combination of the other two. For example, the second vector can be expressed as

$$\begin{bmatrix} 2 \\ 1 \end{bmatrix} = \frac{5}{4} \begin{bmatrix} 1 \\ -1 \end{bmatrix} + \frac{3}{4} \begin{bmatrix} 1 \\ 3 \end{bmatrix}.$$

■

Before continuing, we make two important observations about Example 4.8. First we note that a given set of vectors is linearly dependent if and only if the corresponding matrix of these column vectors has a nontrivial nullspace. More precisely, let $\mathbf{u}^{(j)}$ be the jth column of $\mathbf{A}$: then $\{\mathbf{u}^{(j)}\}_{j=1}^{n}$ is linearly dependent if and only if $\mathbf{Ax} = \boldsymbol{\theta}$ for some $\mathbf{x} \neq \boldsymbol{\theta}$, that is, if and only if $N(\mathbf{A}) \neq \{\boldsymbol{\theta}\}$.

Theorem 4.4 Let $L_{\mathbf{A}}$ be the linear transformation defined by $m \times n$ matrix $\mathbf{A}$. Then $\{\mathbf{A}_j\}_{j=1}^{n}$, the set of columns of $\mathbf{A}$, is linearly dependent if and only if $N(\mathbf{A}) \neq \{\boldsymbol{\theta}\}$, that is, if and only if the nullspace contains a nonzero vector.

■

The second observation is not as straightforward. In Example 4.8 we found that the second vector could be expressed as a linear combination of the other two. Actually, we found that $\mathbf{u}^{(2)} = \frac{5}{4}\mathbf{u}^{(1)} + \frac{3}{4}\mathbf{u}^{(3)}$. This means that $\mathbf{u}^{(2)}$ is redundant and can be dropped without affecting the span.

Example 4.9

We show that on the basis of Example 4.8, $S[\mathbf{u}^{(1)}, \mathbf{u}^{(2)}, \mathbf{u}^{(3)}] = S[\mathbf{u}^{(1)}, \mathbf{u}^{(3)}]$. Clearly, the latter is a subset of the former, so it suffices to show that if $\mathbf{b} \in S[\mathbf{u}^{(1)}, \mathbf{u}^{(2)}, \mathbf{u}^{(3)}]$, then $\mathbf{b} \in S[\mathbf{u}^{(1)}, \mathbf{u}^{(3)}]$. Now suppose that

$$\mathbf{b} = x_1\mathbf{u}^{(1)} + x_2\mathbf{u}^{(2)} + x_3\mathbf{u}^{(3)}$$

for some scalars x_1, x_2, x_3. Since $\mathbf{u}^{(2)}$ is a linear combination of $\mathbf{u}^{(1)}$ and $\mathbf{u}^{(3)}$, we obtain

$$\begin{aligned}\mathbf{b} &= x_1\mathbf{u}^{(1)} + x_2(\tfrac{5}{4}\mathbf{u}^{(1)} + \tfrac{3}{4}\mathbf{u}^{(3)}) + x_3\mathbf{u}^{(3)} \\ &= (x_1 + \tfrac{5}{4}x_2)\mathbf{u}^{(1)} + (x_3 + \tfrac{3}{4}x_2)\mathbf{u}^{(3)}\end{aligned}$$

and hence $\mathbf{b} \in S[\mathbf{u}^{(1)}, \mathbf{u}^{(3)}]$.

■

In general, a vector that depends linearly on other vectors in a set can be dropped without diminishing the span of that set.

Theorem 4.5 Suppose that $\{\mathbf{u}^{(j)}\}_{j=1}^{n}$ is linearly dependent. Then for some k, $1 \le k \le n$, $\mathbf{u}^{(k)}$ is a linear combination of $\{\mathbf{u}^{(j)}\}_{j=1, j\neq k}^{n}$ and

$$S[\mathbf{u}^{(1)}, \ldots, \mathbf{u}^{(k-1)}, \mathbf{u}^{(k+1)}, \ldots, \mathbf{u}^{(n)}] = S[\mathbf{u}^{(1)}, \ldots, \mathbf{u}^{(k-1)}, \mathbf{u}^{(k)}, \mathbf{u}^{(k+1)}, \ldots, \mathbf{u}^{(n)}].$$

■

Next we give a procedure for determining if the columns of an $m \times n$ matrix are linearly dependent. The idea is to use Gaussian elimination to reduce $\mathbf{A}$ to a particular form called the echelon form. By inspection of the echelon form we can determine which columns of $\mathbf{A}$, if any, are redundant.

A matrix $\mathbf{B}$ is said to be in **echelon form** if it has the following properties:

1. For some r the first r rows of $\mathbf{B}$ are nonzero. That is, each of the first r rows contains at least one nonzero entry.
2. The first nonzero entry in a given row is called its *leading entry*. Below each leading entry is a column of zeros.
3. Each leading entry lies to the right of the leading entry of the previous row.
4. After the first r rows each row contains only zeros.

A matrix in echelon form takes the form depicted in Figure 4.3, where $*$ indicates a leading entry.

A given matrix can always be reduced to a matrix in echelon form by using a sequence of elementary row operations. The procedure is the same as we previously used in Gaussian elimination for square matrices. It is time for an example.

$$\begin{bmatrix} * & \cdots & \cdots & \cdots & \cdots & \cdots & \cdots & \cdots \\ \cdots & \cdots & \cdots & \cdots & \cdots & \cdots & \cdots & \cdots \\ 0 & 0 & 0 & * & \cdots & \cdots & \cdots & \cdots \\ 0 & 0 & 0 & 0 & 0 & * & \cdots & \cdots \\ 0 & 0 & 0 & 0 & 0 & 0 & * & \cdots \\ 0 & 0 & 0 & 0 & 0 & 0 & 0 & 0 \\ 0 & 0 & 0 & 0 & 0 & 0 & 0 & 0 \end{bmatrix}$$

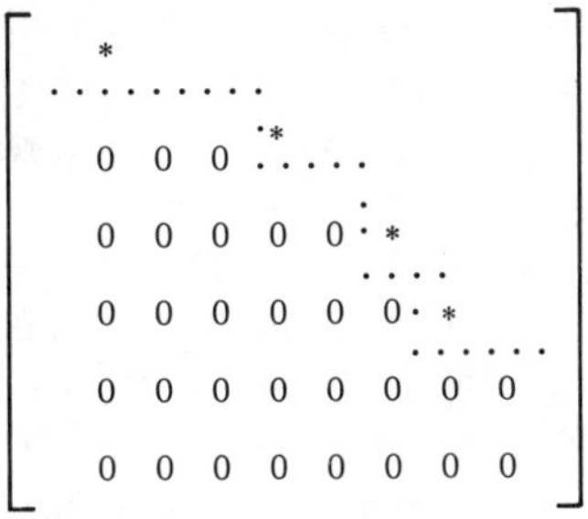

Figure 4.3

Example 4.10

We shall reduce **A** to an echelon form **B**, where

$$\mathbf{A} = \begin{bmatrix} 2 & -1 & 0 & 3 & -1 \\ 4 & -2 & 3 & -3 & -1 \\ -2 & 1 & -3 & 6 & 0 \\ 2 & 0 & 1 & -1 & 1 \end{bmatrix}.$$

We have by elementary row operations:

$$\begin{bmatrix} 2 & -1 & 0 & 3 & -1 \\ 4 & -2 & 3 & -3 & -1 \\ -2 & 1 & -3 & 6 & 0 \\ 2 & 0 & 1 & -1 & 1 \end{bmatrix} \longrightarrow \begin{bmatrix} 2 & -1 & 0 & 3 & -1 \\ 0 & 0 & 3 & -9 & 1 \\ 0 & 0 & -3 & 9 & -1 \\ 0 & 1 & 1 & -4 & 2 \end{bmatrix} \longrightarrow \begin{bmatrix} 2 & -1 & 0 & 3 & -1 \\ 0 & 1 & 1 & -4 & 2 \\ 0 & 0 & -3 & 9 & -1 \\ 0 & 0 & 3 & -9 & 1 \end{bmatrix}$$

$$\longrightarrow \begin{bmatrix} 2 & -1 & 0 & 3 & -1 \\ 0 & 1 & 1 & -4 & 2 \\ 0 & 0 & -3 & 9 & -1 \\ 0 & 0 & 0 & 0 & 0 \end{bmatrix} = \mathbf{B}.$$

Note that the leading entries (i.e., 2, 1, and -3) occupy the positions of the nonzero pivots in the elimination process.

Example 4.11

Let **A** and **C** be given by

$$\mathbf{A} = \begin{bmatrix} 3 & 7 & 1 & -2 \\ 0 & 0 & 0 & 0 \\ 0 & 0 & 4 & 6 \\ 0 & 0 & 0 & -3 \end{bmatrix} \quad \text{and} \quad \mathbf{C} = \begin{bmatrix} 3 & 7 & 1 & -2 \\ 0 & 0 & 4 & 6 \\ 0 & 0 & 0 & -3 \\ 0 & 0 & 0 & 0 \end{bmatrix}.$$

Then **A** is not in echelon form (why?) but **C** is in echelon form.

Let us return to the question of linear dependence and show how the echelon form enters the picture. When the given matrix **A** is reduced to echelon form **B**, the homogeneous problems $\mathbf{Ax} = \boldsymbol{\theta}$ and $\mathbf{Bx} = \boldsymbol{\theta}$ have exactly the same solutions (recall Theorem 2.1). This implies, by Theorem 4.4, that the columns of the $m \times n$ matrix **A** are linearly dependent if and only if the columns of **B** are linearly dependent. Due to the special structure of the echelon form we shall find that our task has been greatly simplified. The key observation is to note that each unknown in $\mathbf{Bx} = \boldsymbol{\theta}$ corresponding to a column containing a leading entry can be expressed in terms of the remaining $n - r$ unknowns (the first r rows each contain a leading entry) which may be given arbitrary values. The leading entries are the nonzero pivots used in the process of reducing **A** to **B**. Hence we call the variables corresponding to columns with leading entries **pivot variables**. Those that correspond to the remaining $n - r$ variables are called **arbitrary variables** (the terms basic variables and nonbasic variables are used in some texts).

Example 4.12

We determine if the columns of **A** are linearly dependent. First we reduce **A** to echelon form:

$$\mathbf{A} = \begin{bmatrix} 1 & 0 & -1 & 2 \\ 2 & 1 & -2 & 5 \\ -1 & -2 & 1 & -4 \end{bmatrix} \longrightarrow \begin{bmatrix} 1 & 0 & -1 & 2 \\ 0 & 1 & 0 & 1 \\ 0 & -2 & 0 & -2 \end{bmatrix}$$

$$\longrightarrow \begin{bmatrix} \textcircled{1} & 0 & -1 & 2 \\ 0 & \textcircled{1} & 0 & 1 \\ 0 & 0 & 0 & 0 \end{bmatrix} = \mathbf{B}.$$

The leading entries in **B** are circled and the corresponding pivot variables are x_1 and x_2. Solving $\mathbf{Bx} = \boldsymbol{\theta}$ gives $x_2 = -x_4$ and $x_1 = x_3 - 2x_4$. Thus the pivot variables are expressed in terms of the arbitrary variables. More important, the columns of **A** satisfy

$$(x_3 - 2x_4)\begin{bmatrix} 1 \\ 2 \\ -1 \end{bmatrix} + (-x_4)\begin{bmatrix} 0 \\ 1 \\ -2 \end{bmatrix} + x_3\begin{bmatrix} -1 \\ -2 \\ 1 \end{bmatrix} + x_4\begin{bmatrix} 2 \\ 5 \\ -4 \end{bmatrix} = \boldsymbol{\theta}$$

with x_3 and x_4 arbitrary. We are free to choose $x_3 \neq 0$ or $x_4 \neq 0$, and hence the columns of **A** are linearly dependent.

■

In this 3×4 example there can be at most three leading entries (one for each row) and hence (because $4 > 3$) there must be at least one arbitrary variable. Actually, we found two variables of each type, but this simple dimension argument shows that the columns of **A** must be dependent. In fact, we have the following general result.

Theorem 4.6. If **A** is an $m \times n$ matrix with $n > m$, then **A** has linearly dependent columns, that is, $\mathbf{Ax} = \boldsymbol{\theta}$ has a nontrivial solution.

■

If every column in an echelon form **B** for **A** has a leading entry, then there are no arbitrary variables. In this case $\mathbf{Bx} = \boldsymbol{\theta}$ and $\mathbf{Ax} = \boldsymbol{\theta}$ have only the trivial solution $\mathbf{x} = \boldsymbol{\theta}$, and therefore the columns of **A** are not linearly dependent. A set of vectors $\{\mathbf{u}^{(j)}\}_{j=1}^{n}$ that is not linearly dependent is called **linearly independent**. In other words, the set is linearly independent if the only scalars $x_1, \ldots, x_n$ satisfying

$$x_1\mathbf{u}^{(1)} + x_2\mathbf{u}^{(2)} + \cdots + x_n\mathbf{u}^{(n)} = \boldsymbol{\theta}$$

are $x_1 = x_2 = \cdots = x_n = 0$. Theorem 4.4 can then be rephrased as stating that $N(\mathbf{A}) = \{\boldsymbol{\theta}\}$ if and only if the columns of **A** are linearly independent.

We summarize our procedure for determining if a given set of m-vectors $\{\mathbf{u}^{(j)}\}_{j=1}^{n}$ is linearly independent:

(1) Form the $m \times n$ matrix **A** whose jth column is $\mathbf{u}^{(j)}$.
(2) Apply elimination to **A**, forming an echelon form **B** for **A**.
(3) If the number of pivot variables, r, is equal to n, the set is linearly independent. The set is linearly dependent if $r < n$.

We conclude this section by establishing some general results on linear independence. First we show that any set of m-vectors which contains $\boldsymbol{\theta}$ must be linearly dependent. Indeed, if the set is $\{\boldsymbol{\theta}, \mathbf{u}^{(2)}, \ldots, \mathbf{u}^{(n)}\}$, then choose $x_1 \neq 0$, $x_2 = \cdots x_n = 0$ and hence

$$\sum_{j=2}^{n} x_j\mathbf{u}^{(j)} + x_1\boldsymbol{\theta} = \boldsymbol{\theta}.$$

The set is linearly dependent since $x_1 \neq 0$. We note that this could also be established by observing that the first column of the appropriate echelon form has no leading entry and hence x_1 is an arbitrary variable.

Next, suppose that $S = \{\mathbf{u}^{(j)}\}_{j=1}^{n}$ is given and some subset of S, say $T = \{\mathbf{u}^{(j)}\}_{j=1}^{k}$, is linearly dependent. Then S must be linearly dependent. To see this, we have for some $x_1, \ldots, x_k$ (not all zero), $\sum_{j=1}^{k} x_j\mathbf{u}^{(j)} = \boldsymbol{\theta}$ with $x_i \neq 0$ for some i, $1 \leq i \leq k$. Now choose $x_{k+1} = \cdots = x_n = 0$ and we have

$$\sum_{j=1}^{k} x_j\mathbf{u}^{(j)} + \sum_{j=k+1}^{n} 0\mathbf{u}^{(j)} = \sum_{j=1}^{n} x_j\mathbf{u}^{(j)} = \boldsymbol{\theta}.$$

Therefore, we have established the following theorem.

Theorem 4.7. Let S be a given set of m-vectors.
(a) If $\boldsymbol{\theta} \in S$, then S is linearly dependent.
(b) If $T \subseteq S$ and T is linearly dependent, then S is also linearly dependent.
(c) If S is linearly independent, then every subset T of S is linearly independent. ■

Part (c) of this theorem is actually a logically equivalent rewording (called the contrapositive) of part (b).

EXERCISES 4.2

1. In each case determine if b is a linear combination of the given set of vectors.

(a) $\left\{\begin{bmatrix}1\\0\\-1\end{bmatrix}, \begin{bmatrix}2\\1\\1\end{bmatrix}, \begin{bmatrix}4\\1\\-1\end{bmatrix}\right\}$; $\mathbf{b} = \begin{bmatrix}4\\-2\\4\end{bmatrix}$ **(b)** $\left\{\begin{bmatrix}1\\2\end{bmatrix}, \begin{bmatrix}-1\\3\end{bmatrix}\right\}$; $\mathbf{b} = \begin{bmatrix}0\\1\end{bmatrix}$

(c) $\left\{\begin{bmatrix}1\\-1\\2\end{bmatrix}, \begin{bmatrix}4\\6\\7\end{bmatrix}\right\}$; $\mathbf{b} = \boldsymbol{\theta}$ **(d)** $\left\{\begin{bmatrix}1\\2\end{bmatrix}, \begin{bmatrix}4\\-2\end{bmatrix}, \begin{bmatrix}1\\1\end{bmatrix}\right\}$; $\mathbf{b} = \begin{bmatrix}0\\0\\1\end{bmatrix}$

2. Let

$$\mathbf{e}^{(1)} = \begin{bmatrix}1\\0\\0\end{bmatrix}, \quad \mathbf{e}^{(2)} = \begin{bmatrix}0\\1\\0\end{bmatrix}, \quad \text{and} \quad \mathbf{e}^{(3)} = \begin{bmatrix}0\\0\\1\end{bmatrix}.$$

Which vectors $\mathbf{b} \in \mathbf{R}^3$ are in $S[\mathbf{e}^{(1)}, \mathbf{e}^{(2)}, \mathbf{e}^{(3)}]$?

3. Suppose that $\mathbf{A}$ is the 3×3 upper triangular matrix

$$\begin{bmatrix}2 & 1 & -2\\0 & -1 & 0\\0 & 0 & 4\end{bmatrix}.$$

Which vectors $\mathbf{b} \in \mathbf{R}^3$ are in the span of the columns of $\mathbf{A}$?

4. Suppose that $\mathbf{A}$ is an $n \times n$ invertible matrix. Show that $R(\mathbf{A}) = \mathbf{R}^n$.

5. For each of the $m \times n$ matrices below, determine if the span $S[\mathbf{A}_1, \ldots, \mathbf{A}_n]$ is equal to $\mathbf{R}^m$.

(a) $\mathbf{A} = \begin{bmatrix}1 & 0\\-2 & 1\\4 & 5\end{bmatrix}$ **(b)** $\mathbf{A} = \begin{bmatrix}4 & -1 & 0\\2 & 3 & 6\end{bmatrix}$ **(c)** $\mathbf{A} = \begin{bmatrix}1 & -2\\-2 & 4\end{bmatrix}$

6. In each case determine if the set is linearly dependent.

(a) $\left\{\begin{bmatrix}1\\-6\\3\end{bmatrix}, \begin{bmatrix}2\\4\\0\end{bmatrix}, \begin{bmatrix}0\\1\\0\end{bmatrix}, \begin{bmatrix}1\\1\\1\end{bmatrix}\right\}$ **(b)** $\left\{\begin{bmatrix}0\\0\end{bmatrix}, \begin{bmatrix}1\\-7\end{bmatrix}\right\}$

(c) $\left\{\begin{bmatrix}1\\5\\-2\\0\end{bmatrix}, \begin{bmatrix}2\\1\\1\\-2\end{bmatrix}, \begin{bmatrix}1\\0\\0\\-1\end{bmatrix}\right\}$

7. Determine if one of the vectors can be expressed as a linear combination of the others.

$$\mathbf{u}^{(1)} = \begin{bmatrix}1\\1\\1\end{bmatrix}, \quad \mathbf{u}^{(2)} = \begin{bmatrix}2\\0\\-4\end{bmatrix}, \quad \mathbf{u}^{(3)} = \begin{bmatrix}0\\2\\-6\end{bmatrix}$$

8. For the vectors of Exercise 7, is it true that $S[\mathbf{u}^{(1)}, \mathbf{u}^{(2)}, \mathbf{u}^{(3)}] = S[\mathbf{u}^{(1)}, \mathbf{u}^{(2)}]$?

9. In each case reduce $\mathbf{A}$ to echelon form.

(a) $\mathbf{A} = \begin{bmatrix}6 & -1 & 2 & 1\\2 & 1 & 4 & 1\\-3 & 0 & 1 & -1\end{bmatrix}$ **(b)** $\mathbf{A} = \begin{bmatrix}2 & 3 & 1\\-4 & -6 & 2\\1 & 1 & 1\\-2 & -3 & 4\end{bmatrix}$

(c) $\mathbf{A} = \begin{bmatrix} 0 & 0 & 1 \\ 0 & 1 & 0 \\ 1 & 0 & 0 \end{bmatrix}$ (d) $\mathbf{A} = \begin{bmatrix} 1 & -3 & 4 \\ 2 & -6 & 1 \end{bmatrix}$

Identify the pivot variables in each part.

10. For each matrix of Exercise 9, determine if the columns are linearly dependent.

11. Without actually calculating a solution, show that

$$\mathbf{Ax} = \begin{bmatrix} 2 & 3 & 4 & -1 \\ 1 & 0 & 1 & 2 \\ -6 & 2 & 1 & 3 \end{bmatrix} \mathbf{x} = \boldsymbol{\theta}$$

has a nontrivial solution.

12. List all possible echelon forms for a 2×3 matrix. Indicate a leading entry by $*$.

***13.** Suppose that $\{\mathbf{u}^{(1)}, \mathbf{u}^{(2)}, \mathbf{u}^{(3)}\}$ is linearly independent. Show that $\{\mathbf{u}^{(1)} + \mathbf{u}^{(2)}, \mathbf{u}^{(1)} + \mathbf{u}^{(3)}, \mathbf{u}^{(2)} + \mathbf{u}^{(3)}\}$ is also linearly independent.

4.3 The Rank-Nullity Theorem

In Section 4.2 we found, for a given $m \times n$ matrix $\mathbf{A}$, that $R(\mathbf{A}) = S[\mathbf{A}_1, \ldots, \mathbf{A}_n]$. That is, any vector $\mathbf{b}$ in the range can be written as a linear combination of the columns of $\mathbf{A}$. Thus $\mathbf{Ax} = \mathbf{b}$ is equivalent to

$$\sum_{j=1}^{n} x_j \mathbf{A}_j = \mathbf{b}.$$

If the columns of $\mathbf{A}$ are linearly independent, the coefficients $x_1, \ldots, x_n$ are unique. Indeed, if $\mathbf{b}$ can also be written as

$$\sum_{j=1}^{n} u_j \mathbf{A}_j = \mathbf{b},$$

then, upon subtraction, we obtain

$$\sum_{j=1}^{n} (u_j - x_j)\mathbf{A}_j = \mathbf{b} - \mathbf{b} = \boldsymbol{\theta}.$$

Hence, since the columns of $\mathbf{A}$ are linearly independent, we must have $x_1 = u_1, \ldots, x_n = u_n$. Therefore, if the columns of $\mathbf{A}$ are linearly independent, then $\mathbf{x} = [x_1, \ldots, x_n]^T$ is uniquely determined. To put this another way, we recall from Theorem 4.1 that every solution of $\mathbf{Ax} = \mathbf{b}$ takes the form $\mathbf{x} = \mathbf{u} + \mathbf{z}$ with $\mathbf{Au} = \mathbf{b}$ and $\mathbf{Az} = \boldsymbol{\theta}$. Linear independence of the columns of $\mathbf{A}$ implies that $N(\mathbf{A}) = \{\boldsymbol{\theta}\}$ and hence $\mathbf{z} = \boldsymbol{\theta}$. Thus $\mathbf{x} = \mathbf{u}$ is unique.

The situation changes if $\{\mathbf{A}_j\}_{j=1}^{n}$ is linearly dependent. Then there is some redundancy in the columns. That is, at least one of the columns is a linear combination of the remaining columns. In this case uniqueness of solutions of $\mathbf{Ax} = \mathbf{b}$ is lost. We shall determine a set of columns of $\mathbf{A}$ that is both linearly independent and spans the range.

A **basis** for a subspace V of $\mathbf{R}^m$ is a set of vectors that spans V and is linearly independent.

Example 4.13

In $\mathbf{R}^3$ take $V = \mathbf{R}^3$, that is, take the subspace to be the entire space. Then a basis for V is

$$\left\{ \mathbf{e}^{(1)} = \begin{bmatrix} 1 \\ 0 \\ 0 \end{bmatrix}, \quad \mathbf{e}^{(2)} = \begin{bmatrix} 0 \\ 1 \\ 0 \end{bmatrix}, \quad \mathbf{e}^{(3)} = \begin{bmatrix} 0 \\ 0 \\ 1 \end{bmatrix} \right\}.$$

Clearly, $\{\mathbf{e}^{(j)}\}_{j=1}^{3}$ is linearly independent and any $\mathbf{b} \in \mathbf{R}^3$ can be uniquely expressed as

$$\mathbf{b} = \begin{bmatrix} b_1 \\ b_2 \\ b_3 \end{bmatrix} = b_1\mathbf{e}^{(1)} + b_2\mathbf{e}^{(2)} + b_3\mathbf{e}^{(3)}.$$

The set $\{\mathbf{e}^{(j)}\}_{j=1}^{3}$ is called the **natural basis** for $\mathbf{R}^3$.

■

Example 4.14

Consider the upper triangular matrix

$$\mathbf{A} = \begin{bmatrix} 1 & -1 & 3 \\ 0 & 2 & 2 \\ 0 & 0 & 4 \end{bmatrix}.$$

By back substitution it is easy to see that $\mathbf{Ax} = \mathbf{b}$ has a unique solution for each $\mathbf{b} \in \mathbf{R}^3$. In particular, the only solution of $\mathbf{Ax} = \boldsymbol{\theta}$ is $\mathbf{x} = \boldsymbol{\theta}$. Therefore, the columns of $\mathbf{A}$ are linearly independent and each vector $\mathbf{b} \in \mathbf{R}^3$ can be written as a linear combination of the columns of $\mathbf{A}$. That is, the columns of $\mathbf{A}$ form a basis for $\mathbf{R}^3$.

■

Example 4.15

The columns of any $n \times n$ invertible matrix $\mathbf{A}$ constitute a basis for $\mathbf{R}^n$. The unique solution of $\mathbf{Ax} = \mathbf{b}$ is $\mathbf{x} = \mathbf{A}^{-1}\mathbf{b}$. Thus every $\mathbf{b} \in \mathbf{R}^n$ can be written as $\mathbf{b} = \sum_{j=1}^{n} x_j\mathbf{A}_j$, where $\mathbf{x} = [x_1, \ldots, x_n]^T$ is given by $\mathbf{x} = \mathbf{A}^{-1}\mathbf{b}$. This shows that $\mathbf{R}^n = S[\mathbf{A}_1, \ldots, \mathbf{A}_n]$. The only solution of $\mathbf{Ax} = \boldsymbol{\theta}$ is $\mathbf{x} = \mathbf{A}^{-1}\boldsymbol{\theta} = \boldsymbol{\theta}$ and hence the nullspace consists only of the zero vector. Thus $\{\mathbf{A}_j\}_{j=1}^{n}$ is linearly independent.

■

It happens that *every basis for a subspace contains the same number of vectors*. We will not prove this, but we point it out as a fundamental result. The number of vectors in any basis for a subspace is called the **dimension** of the subspace.

We now show how to determine a basis for the range of a given $m \times n$ matrix **A**. The key is to first reduce **A** to echelon form. Before describing the procedure in general terms we consider a specific example.

Example 4.16

For

$$\mathbf{A} = \begin{bmatrix} 1 & 0 & -1 & 2 \\ 2 & 1 & -2 & 5 \\ -1 & -2 & 1 & -4 \end{bmatrix}$$

we found, in Example 4.12, the echelon form

$$\mathbf{B} = \begin{bmatrix} 1 & 0 & -1 & 2 \\ 0 & 1 & 0 & 1 \\ 0 & 0 & 0 & 0 \end{bmatrix}.$$

The columns of **A** are linearly dependent and we found that for arbitrary variables x_3 and x_4,

$$(x_3 - 2x_4)\mathbf{A}_1 + (-x_4)\mathbf{A}_2 + x_3\mathbf{A}_3 + x_4\mathbf{A}_4 = \boldsymbol{\theta}.$$

By choosing $x_4 = 1$ we can solve for $\mathbf{A}_4$ in terms of the remaining columns:

$$\mathbf{A}_4 = \mathbf{A}_2 - x_3\mathbf{A}_3 + (2 - x_3)\mathbf{A}_1.$$

Thus $R(\mathbf{A}) = S[\mathbf{A}_1, \mathbf{A}_2, \mathbf{A}_3, \mathbf{A}_4] = S[\mathbf{A}_1, \mathbf{A}_2, \mathbf{A}_3]$. The fourth column of **A** can be deleted without diminishing the range. The remaining columns are still linearly dependent since, for $x_4 = 0$,

$$x_3\mathbf{A}_1 + 0\mathbf{A}_2 + x_3\mathbf{A}_3 = \boldsymbol{\theta}$$

and hence $\mathbf{A}_3 = -\mathbf{A}_1$. Thus $S[\mathbf{A}_1, \mathbf{A}_2] = S[\mathbf{A}_1, \mathbf{A}_2, \mathbf{A}_3]$ and the range is spanned by $\{\mathbf{A}_1, \mathbf{A}_2\}$. Moreover, $\{\mathbf{A}_1, \mathbf{A}_2\}$ is linearly independent since deleting $\mathbf{A}_3$ *and* $\mathbf{A}_4$ corresponds to $x_3 = x_4 = 0$, which implies that $x_1 = x_3 - 2x_4 = 0$, $x_2 = -x_4 = 0$. Therefore, $\{\mathbf{A}_1, \mathbf{A}_2\}$ is a basis for the range, and the dimension of $R(\mathbf{A})$ is 2.

■

As illustrated in Example 4.16, we determine a basis for the range by successively deleting redundant column vectors corresponding to the arbitrary variables. This does not diminish the range and the columns of **A** corresponding to the pivot variables constitute a basis for the range. Thus the number of vectors in a basis for the range, that is, the dimension of $R(\mathbf{A})$, is equal to the number of pivot variables and is called the **rank** of **A**. We denote the rank of **A** by $r(\mathbf{A})$.

Theorem 4.8. Suppose that **A** is an $m \times n$ matrix. Let **B** be an echelon form for **A** with r leading entries. Then $r(\mathbf{A}) = r$ and the r columns of **A** corresponding to the pivot variables constitute a basis for $R(\mathbf{A})$. Every $\mathbf{b} \in R(\mathbf{A})$ can be uniquely expressed as a linear combination of the basis vectors.

■

With the aid of another example, the procedure for determining the rank and a basis for the range should become clear.

Example 4.17

Consider the 5 × 6 matrix

$$\mathbf{A} = \begin{bmatrix} 2 & 4 & 4 & -2 & 4 & 4 \\ 1 & 2 & 2 & 0 & 1 & 3 \\ 0 & 0 & 1 & -1 & 1 & 0 \\ -2 & -4 & -3 & 2 & -4 & -2 \\ 1 & 2 & 1 & -2 & 3 & 2 \end{bmatrix}.$$

Since there are more columns than rows, we know, by Theorem 4.6, that the columns of $\mathbf{A}$ are linearly dependent. Thus $r(\mathbf{A}) < 6$. To determine the rank and a basis for the range we reduce $\mathbf{A}$ to echelon form. We compute

$$\begin{bmatrix} 2 & 4 & 4 & -2 & 4 & 4 \\ 1 & 2 & 2 & 0 & 1 & 3 \\ 0 & 0 & 1 & -1 & 1 & 0 \\ -2 & -4 & -3 & 2 & -4 & -2 \\ 1 & 2 & 1 & -2 & 3 & 2 \end{bmatrix} \longrightarrow \begin{bmatrix} 2 & 4 & 4 & -2 & 4 & 4 \\ 0 & 0 & 0 & 1 & -1 & 1 \\ 0 & 0 & 1 & -1 & 1 & 0 \\ 0 & 0 & 1 & 0 & 0 & 2 \\ 0 & 0 & -1 & -1 & 1 & 0 \end{bmatrix}$$

and interchange the second and fourth rows:

$$\longrightarrow \begin{bmatrix} 2 & 4 & 4 & -2 & 4 & 4 \\ 0 & 0 & 1 & 0 & 0 & 2 \\ 0 & 0 & 1 & -1 & 1 & 0 \\ 0 & 0 & 0 & 1 & -1 & 1 \\ 0 & 0 & -1 & -1 & 1 & 0 \end{bmatrix}$$

$$\longrightarrow \begin{bmatrix} 2 & 4 & 4 & -2 & 4 & 4 \\ 0 & 0 & 1 & 0 & 0 & 2 \\ 0 & 0 & 0 & -1 & 1 & -2 \\ 0 & 0 & 0 & 1 & -1 & 1 \\ 0 & 0 & 0 & -1 & 1 & 2 \end{bmatrix}$$

$$\longrightarrow \begin{bmatrix} 2 & 4 & 4 & -2 & 4 & 4 \\ 0 & 0 & 1 & 0 & 0 & 2 \\ 0 & 0 & 0 & -1 & 1 & -2 \\ 0 & 0 & 0 & 0 & 0 & -1 \\ 0 & 0 & 0 & 0 & 0 & 4 \end{bmatrix}$$

$$\longrightarrow \begin{bmatrix} 2 & 4 & 4 & -2 & 4 & 4 \\ 0 & 0 & 1 & 0 & 0 & 2 \\ 0 & 0 & 0 & -1 & 1 & -2 \\ 0 & 0 & 0 & 0 & 0 & -1 \\ 0 & 0 & 0 & 0 & 0 & 0 \end{bmatrix}$$

Thus there are four leading entries, corresponding to the pivot variables x_1, x_3, x_4, and x_6. We have $r(\mathbf{A}) = 4$ and hence $R(\mathbf{A})$ is a four-dimensional subspace of $\mathbf{R}^5$. A basis for the range of $\mathbf{A}$ is given by

$$\{\mathbf{A}_1, \mathbf{A}_3, \mathbf{A}_4, \mathbf{A}_6\}.$$

■

Next we carry out a similar procedure for the nullspace. The number of vectors in a basis for the nullspace, that is, the dimension of $N(\mathbf{A})$, is called the **nullity** of $\mathbf{A}$ and is denoted by $n(\mathbf{A})$. Again the key is to examine an echelon form for $\mathbf{A}$. Let us return to the matrix of Examples 4.12 and 4.16.

Example 4.18

The matrix $\mathbf{A}$ and an echelon form $\mathbf{B}$ are as follows;

$$\mathbf{A} = \begin{bmatrix} 1 & 0 & -1 & 2 \\ 2 & 1 & -2 & 5 \\ -1 & 2 & 1 & -4 \end{bmatrix}, \qquad \mathbf{B} = \begin{bmatrix} 1 & 0 & -1 & 2 \\ 0 & 1 & 0 & 1 \\ 0 & 0 & 0 & 0 \end{bmatrix}.$$

A vector $\mathbf{x}$ is in the nullspace if $\mathbf{Ax} = \boldsymbol{\theta}$ or equivalently, if $\mathbf{Bx} = \boldsymbol{\theta}$. We found that $x_2 = -x_4$ and $x_1 = x_3 - 2x_4$. Therefore, $\mathbf{x}$ is given by

$$\mathbf{x} = \begin{bmatrix} x_3 - 2x_4 \\ -x_4 \\ x_3 \\ x_4 \end{bmatrix} = x_3 \begin{bmatrix} 1 \\ 0 \\ 1 \\ 0 \end{bmatrix} + x_4 \begin{bmatrix} -2 \\ -1 \\ 0 \\ 1 \end{bmatrix}.$$

If we set $x_3 = 1$ and $x_4 = 0$, we find a particular vector in the nullspace, that is,

$$\mathbf{z}^{(1)} = \begin{bmatrix} 1 \\ 0 \\ 1 \\ 0 \end{bmatrix} \quad \text{satisfies} \quad \mathbf{Az}^{(1)} = \boldsymbol{\theta}.$$

Similarly, for $x_4 = 1$ and $x_3 = 0$, the vector

$$\mathbf{z}^{(2)} = \begin{bmatrix} -2 \\ -1 \\ 0 \\ 1 \end{bmatrix} \quad \text{satisfies} \quad \mathbf{Az}^{(2)} = \boldsymbol{\theta}.$$

Moreover, $\mathbf{x} = x_3\mathbf{z}^{(1)} + x_4\mathbf{z}^{(2)}$ and hence the nullspace is spanned by $\{\mathbf{z}^{(1)}, \mathbf{z}^{(2)}\}$. By considering the third and fourth components, it is easy to see that $\{\mathbf{z}^{(1)}, \mathbf{z}^{(2)}\}$ is linearly independent. Thus $n(\mathbf{A}) = 2$ and $\{\mathbf{z}^{(1)}, \mathbf{z}^{(2)}\}$ is a basis for the nullspace. Note also that the nullity is equal to the number of arbitrary variables.

■

Example 4.19

We find bases for the range and nullspace of

$$\mathbf{A} = \begin{bmatrix} 1 & 2 & 3 & 4 & 5 \\ 0 & 1 & 1 & 1 & 0 \\ -3 & -6 & -9 & -2 & 5 \\ 2 & 4 & 6 & 1 & -4 \end{bmatrix}.$$

First we reduce to echelon form

$$\mathbf{A} \longrightarrow \begin{bmatrix} 1 & 2 & 3 & 4 & 5 \\ 0 & 1 & 1 & 1 & 0 \\ 0 & 0 & 0 & 10 & 20 \\ 0 & 0 & 0 & -7 & -14 \end{bmatrix} \longrightarrow \begin{bmatrix} 1 & 2 & 3 & 4 & 5 \\ 0 & 1 & 1 & 1 & 0 \\ 0 & 0 & 0 & 10 & 20 \\ 0 & 0 & 0 & 0 & 0 \end{bmatrix} = \mathbf{B}.$$

Thus $r(\mathbf{A}) = 3$ and $n(\mathbf{A}) = 5 - 3 = 2$. A basis for the range is $\{\mathbf{A}_1, \mathbf{A}_2, \mathbf{A}_4\}$. To determine the nullspace we solve $\mathbf{Bx} = \boldsymbol{\theta}$ by back substitution and find $x_4 = -2x_5$, $x_2 = -x_3 - x_4 = 2x_5 - x_3$, and $x_1 = -(2x_2 + 3x_3 + 4x_4 + 5x_5) = -x_3 - x_5$. Therefore, any $\mathbf{x}$ in the nullspace is given by

$$\mathbf{x} = \begin{bmatrix} -x_3 - x_5 \\ x_5 - x_3 \\ x_3 \\ -x_5 \\ x_5 \end{bmatrix} = x_3 \begin{bmatrix} -1 \\ -1 \\ 1 \\ 0 \\ 0 \end{bmatrix} + x_5 \begin{bmatrix} -1 \\ 1 \\ 0 \\ -1 \\ 1 \end{bmatrix} = x_3\mathbf{z}^{(1)} + x_5\mathbf{z}^{(2)},$$

where

$$\mathbf{z}^{(1)} = \begin{bmatrix} -1 \\ -1 \\ 1 \\ 0 \\ 0 \end{bmatrix} \quad \text{and} \quad \mathbf{z}^{(2)} = \begin{bmatrix} -1 \\ 1 \\ 0 \\ -1 \\ 1 \end{bmatrix}.$$

The set $\{\mathbf{z}^{(1)}, \mathbf{z}^{(2)}\}$ spans the nullspace and is linearly independent. Thus $\{\mathbf{z}^{(1)}, \mathbf{z}^{(2)}\}$ is a basis for the nullspace.

■

The method of Examples 4.18 and 4.19 can always be used to find the nullity of $m \times n$ matrix $\mathbf{A}$. From the echelon form we find that the pivot variables can be expressed in terms of the $n - r(\mathbf{A})$ arbitrary variables. Thus each vector in the nullspace depends only on these arbitrary variables. By setting one arbitrary variable equal to 1 and the rest equal to zero we obtain a vector in a basis for the nullspace. Following this process for each arbitrary variable we obtain all the vectors in a basis for the nullspace.

It is important to remember that the number of pivot variables plus the number of arbitrary variables is equal to the number of columns of $\mathbf{A}$. Since the first is equal to $r(\mathbf{A})$ and the second is equal to $n(\mathbf{A})$, we obtain the **rank-nullity theorem**.

Theorem 4.9. If $\mathbf{A}$ is an $m \times n$ matrix, then

$$r(\mathbf{A}) + n(\mathbf{A}) = n \quad \text{and} \quad r(\mathbf{A}) \leq m.$$

■

On the basis of the rank-nullity theorem we can answer many of the questions relating to the existence and uniqueness of solutions for linear systems of equations. We consider three cases: (1) $m > n$, (2) $m < n$, and (3) $m = n$. In case (1) we find $m > n \geq r(\mathbf{A})$ and hence $R(\mathbf{A}) \neq \mathbf{R}^m$. Thus for some $\mathbf{b} \in \mathbf{R}^m$, $\mathbf{Ax} = \mathbf{b}$ has no solution and hence we cannot be assured of existence. In case (2) we have $n(\mathbf{A}) = n - r(\mathbf{A}) \geq n - m > 0$, and hence $N(\mathbf{A})$ contains at least one nonzero vector. Therefore, no solution of $\mathbf{Ax} = \mathbf{b}$ is unique. Consequently, for existence and uniqueness of solutions for all $\mathbf{b} \in \mathbf{R}^m$ it is *necessary* that $m = n$. For $m = n$ we have $r(\mathbf{A}) = m$ if and only if $n(\mathbf{A}) = 0$. Thus for square matrices uniqueness is equivalent to existence of a solution for every $\mathbf{b} \in \mathbf{R}^m$, that is, $R(\mathbf{A}) = \mathbf{R}^m$ if and only if $N(\mathbf{A}) = \{\boldsymbol{\theta}\}$.

We summarize much of our previous discussion in the next theorem, which gives several equivalent properties of square matrices.

Theorem 4.10. Suppose that A is an $n \times n$ matrix. If any one of the following properties is true, then all of them are true.

(a) $\mathbf{A}$ is invertible.
(b) $\det \mathbf{A} \neq 0$.
(c) The columns of $\mathbf{A}$ form a linearly independent set.
(d) $R(\mathbf{A}) = \mathbf{R}^n$, that is, $r(\mathbf{A}) = n$.
(e) $N(\mathbf{A}) = \{\boldsymbol{\theta}\}$, that is $n(\mathbf{A}) = 0$.
(f) For each $\mathbf{b} \in \mathbf{R}^n$, $\mathbf{Ax} = \mathbf{b}$ has at least one solution.
(g) $\mathbf{Ax} = \mathbf{b}$ has at most one solution.
(h) The columns of $\mathbf{A}$ form a basis for $\mathbf{R}^n$.

■

As a consequence of property (*h*) any set of n linearly independent n-vectors is a basis for $\mathbf{R}^n$.

EXERCISES 4.3

1. Let V be the subspace of $\mathbf{R}^3$ given by

$$V = \{\mathbf{x} \in \mathbf{R}^3 : x_1 + x_2 + x_3 = 0\}.$$

Show that a basis for V is

$$\left\{ \begin{bmatrix} 1 \\ 0 \\ -1 \end{bmatrix}, \begin{bmatrix} 0 \\ 1 \\ -1 \end{bmatrix} \right\}.$$

Describe V geometrically.

2. Do the columns of $\mathbf{A}$ form a basis for $\mathbf{R}^3$? Justify your answer.

$$\mathbf{A} = \begin{bmatrix} -1 & 0 & -3 \\ 1 & 1 & 1 \\ 2 & 3 & 0 \end{bmatrix}.$$

***3.** Suppose that $\{\mathbf{u}^{(1)}, \mathbf{u}^{(2)}, \mathbf{u}^{(3)}\}$ is a basis for $\mathbf{R}^3$. Show that another basis is $\{\mathbf{u}^{(1)} + \mathbf{u}^{(2)}, \mathbf{u}^{(1)} + \mathbf{u}^{(3)}, \mathbf{u}^{(2)} + \mathbf{u}^{(3)}\}$.

4. What is the largest possible rank of a 3×4 matrix? Of a 5×4 matrix? Of an $m \times n$ matrix?

5. Let $\mathbf{A}$ be given by

$$\mathbf{A} = \begin{bmatrix} 2 & -1 & 6 & 1 \\ 4 & 5 & 2 & 0 \\ -2 & 3 & -1 & 1 \end{bmatrix}.$$

Find $r(\mathbf{A})$ and a basis for the range.

6. What is the rank of the $n \times n$ identity matrix?

7. Suppose that $\mathbf{A}$ has the echelon form

$$\mathbf{B} = \begin{bmatrix} 0 & -3 & 2 & 6 \\ 0 & 0 & 0 & 2 \\ 0 & 0 & 0 & 0 \\ 0 & 0 & 0 & 0 \end{bmatrix}.$$

What is $r(\mathbf{A})$? What is $n(\mathbf{A})$? Find a basis for the nullspace.

8. Give an example of matrices $\mathbf{A}$ and $\mathbf{B}$ such that $r(\mathbf{A}) = r(\mathbf{B}) = 3$ and $r(\mathbf{A} + \mathbf{B}) = 1$.

9. Can you find a 3×4 matrix $\mathbf{A}$ with $r(\mathbf{A}) = 4$? Explain your answer.

10. Find the rank and a basis for the range of

$$\mathbf{A} = \begin{bmatrix} 1 & 2 & 0 & 1 & 1 \\ 1 & 1 & 2 & 5 & 4 \\ 2 & 3 & 10 & 7 & 5 \\ 1 & 2 & 4 & 1 & 1 \end{bmatrix}.$$

11. Find the nullity and a basis for the nullspace for the matrix of Example 4.17.

12. Suppose that $\mathbf{A}$ is an $m \times n$ matrix. In each of the following cases, can you determine if $m > n$, $m < n$, or $m = n$?

(a) $\mathbf{Ax} = \mathbf{b}$ has 0 or infinitely many solutions, depending on $\mathbf{b}$.
(b) $\mathbf{Ax} = \mathbf{b}$ has 0 or 1 solution, depending on $\mathbf{b}$.
(c) $\mathbf{Ax} = \mathbf{b}$ has exactly one solution for each $\mathbf{b}$.
(d) $\mathbf{Ax} = \boldsymbol{\theta}$ has a solution $\mathbf{x} \neq \boldsymbol{\theta}$.

4.4 Error-Correcting Codes

Computers receive, store, and transmit information in strings of **bits** (binary digits), that is, in some fixed number of 0's and 1's. For example, when a user depresses the key Z on a remote terminal keyboard, a message consisting of say, 111001 is then transmitted to the computer. The remote terminal and the computer each follow a dictionary which uniquely associates the binary string 111001 with the letter Z. The main problem is to transmit messages in such a way that transmission errors can be detected and corrected. In 1950, Hamming discovered a method whereby any single transmission error can be detected and corrected. The key is to code the message with some redundancy (through parity checks). We shall consider a particular Hamming code, called the (7, 4) code, and examine its relationship to some of the ideas that have been introduced in the chapter.

We denote by $\mathbf{Z}_2$ the set of binary digits, that is, $\mathbf{Z}_2 = \{0, 1\}$. Only messages consisting of four bits will be considered. Such a message can be viewed as a vector:

$$\mathbf{b} = \begin{bmatrix} b_1 \\ b_2 \\ b_3 \\ b_4 \end{bmatrix} \quad \text{with } b_i \in \mathbf{Z}_2, \quad i = 1, 2, 3, 4.$$

For example, the message 1011 is viewed as the vector

$$\begin{bmatrix} 1 \\ 0 \\ 1 \\ 1 \end{bmatrix}.$$

Let $\mathbf{Z}_2^4$ denote the **message space** consisting of all possible messages. Therefore, $\mathbf{Z}_2^4$ consists of those 4-vectors whose entries are in $\mathbf{Z}_2$. Clearly, $\mathbf{Z}_2^4$ consists of exactly $2^4 = 16$ vectors:

$$\begin{bmatrix} 0 \\ 0 \\ 0 \\ 0 \end{bmatrix}, \begin{bmatrix} 1 \\ 0 \\ 0 \\ 0 \end{bmatrix}, \begin{bmatrix} 0 \\ 1 \\ 0 \\ 0 \end{bmatrix}, \begin{bmatrix} 0 \\ 0 \\ 1 \\ 0 \end{bmatrix}, \begin{bmatrix} 0 \\ 0 \\ 0 \\ 1 \end{bmatrix}, \begin{bmatrix} 1 \\ 1 \\ 0 \\ 0 \end{bmatrix}, \begin{bmatrix} 1 \\ 0 \\ 1 \\ 0 \end{bmatrix}, \begin{bmatrix} 1 \\ 0 \\ 0 \\ 1 \end{bmatrix}$$

$$\begin{bmatrix} 0 \\ 1 \\ 1 \\ 0 \end{bmatrix}, \begin{bmatrix} 0 \\ 1 \\ 0 \\ 1 \end{bmatrix}, \begin{bmatrix} 0 \\ 0 \\ 1 \\ 1 \end{bmatrix}, \begin{bmatrix} 1 \\ 1 \\ 1 \\ 0 \end{bmatrix}, \begin{bmatrix} 1 \\ 0 \\ 1 \\ 1 \end{bmatrix}, \begin{bmatrix} 1 \\ 1 \\ 0 \\ 1 \end{bmatrix}, \begin{bmatrix} 0 \\ 1 \\ 1 \\ 1 \end{bmatrix}, \begin{bmatrix} 1 \\ 1 \\ 1 \\ 1 \end{bmatrix}.$$

Unfortunately, the usual rules of vector addition and scalar multiplication do not apply in $\mathbf{Z}_2^4$.

Example 4.20

Let

$$\mathbf{u} = \begin{bmatrix} 1 \\ 0 \\ 1 \\ 0 \end{bmatrix}, \quad \mathbf{v} = \begin{bmatrix} 1 \\ 1 \\ 1 \\ 1 \end{bmatrix}, \quad \text{and} \quad \alpha = 3.$$

Then $\mathbf{u}, \mathbf{v} \in \mathbf{Z}_2^4$ but

$$\mathbf{u} + \mathbf{v} = \begin{bmatrix} 2 \\ 1 \\ 2 \\ 1 \end{bmatrix} \notin \mathbf{Z}_2^4 \quad \text{and} \quad \alpha\mathbf{u} = \begin{bmatrix} 3 \\ 0 \\ 3 \\ 0 \end{bmatrix} \notin \mathbf{Z}_2^4,$$

since 2 and 3 (respectively) are not binary digits.

■

We introduce some new rules of arithmetic so that $\mathbf{Z}_2^4$ is closed under vector addition and scalar multiplication. Scalar multiplication is easy—we only allow scalars α in $\mathbf{Z}_2$. Then if $\mathbf{u} \in \mathbf{Z}_2^4$ and $\alpha \in \mathbf{Z}_2$, we have $\alpha\mathbf{u} = \mathbf{u}$ if $\alpha = 1$ and $\alpha\mathbf{u} = \boldsymbol{\theta}$ if $\alpha = 0$. In either case, $\alpha\mathbf{u} \in \mathbf{Z}_2^4$ for $\alpha \in \mathbf{Z}_2$ and $\mathbf{u} \in \mathbf{Z}_2^4$. For addition we use the table

+	0	1
0	0	1
1	1	0

which defines a new type of addition, called *addition modulo 2*. Therefore,

$$0 + 0 = 0$$

$$0 + 1 = 1$$

$$1 + 0 = 1$$

$$1 + 1 = 0.$$

As long as we remember that $1 + 1 = 0$ we should not have difficulty with modulo 2 addition.

Example 4.21

For the vectors $\mathbf{u}$, $\mathbf{v}$ of Example 4.20, we have

$$\mathbf{u} + \mathbf{v} = \begin{bmatrix} 1+1 \\ 0+1 \\ 1+1 \\ 0+1 \end{bmatrix} = \begin{bmatrix} 0 \\ 1 \\ 0 \\ 1 \end{bmatrix}.$$

■

In this section all calculations will be done with scalars from $\mathbf{Z}_2$ and addition modulo 2.

Suppose that we have a string of bits, say $b_1 b_2 \cdots b_n$; then the sum

$$p = \sum_{k=1}^{n} b_k$$

has the value $p = 0$ or $p = 1$. The value of $p = 1$ if the number of bits b_k that equal 1 is odd and $p = 0$ if the number of bits that equal 1 is even. We say that the **parity** of $b_1 b_2 \cdots b_n$ is even if $p = 0$ and odd if $p = 1$.

Example 4.22

The parity of 1010 is even since $1 + 0 + 1 + 0 = 0$ and the parity of 1101 is odd since $1 + 1 + 0 + 1 = 1$.

■

Hamming's (7, 4) code consists in appending to each message in $\mathbf{Z}_2^4$ three parity checks, as follows. Suppose that $\mathbf{b} \in \mathbf{Z}_2^4$ is a given message; then create a codeword in $\mathbf{Z}_2^7$ by appending the parity checks

$$b_5 = b_1 + b_2 + b_3$$
$$b_6 = b_1 + b_2 + b_4$$
$$b_7 = b_2 + b_3 + b_4.$$

Thus a **valid codeword** is a vector in $\mathbf{Z}_2^7$, say

$$\mathbf{b} = \begin{bmatrix} b_1 \\ b_2 \\ b_3 \\ b_4 \\ b_5 \\ b_6 \\ b_7 \end{bmatrix} \begin{matrix} \left.\begin{matrix} \\ \\ \\ \\ \end{matrix}\right\} \text{message} \\ \left.\begin{matrix} \\ \\ \\ \end{matrix}\right\} \text{parity checks} \end{matrix}$$

with b_5, b_6, b_7 given above.

Example 4.23

For the messages **u**, **v** from Example 4.20 the corresponding codewords are

$$\begin{bmatrix} 1 \\ 0 \\ 1 \\ 0 \\ 1+0+1 \\ 1+1+0 \\ 0+1+0 \end{bmatrix} = \begin{bmatrix} 1 \\ 0 \\ 1 \\ 0 \\ 0 \\ 0 \\ 1 \end{bmatrix} \quad \text{and} \quad \begin{bmatrix} 1 \\ 1 \\ 1 \\ 1 \\ 1+1+1 \\ 1+1+1 \\ 1+1+1 \end{bmatrix} = \begin{bmatrix} 1 \\ 1 \\ 1 \\ 1 \\ 1 \\ 1 \\ 1 \end{bmatrix}.$$

The vector

$$\mathbf{b} = \begin{bmatrix} 1 \\ 1 \\ 0 \\ 1 \\ 0 \\ 0 \\ 0 \end{bmatrix}$$

is not a valid codeword since $b_1 + b_2 + b_4 = 1 + 1 + 1 = 1$ but $b_6 = 0$. ■

A basis for $\mathbf{Z}_2^4$ is $\{\mathbf{e}^{(j)}\}_{j=1}^4$, where

$$\mathbf{e}^{(1)} = \begin{bmatrix} 1 \\ 0 \\ 0 \\ 0 \end{bmatrix}, \quad \mathbf{e}^{(2)} = \begin{bmatrix} 0 \\ 1 \\ 0 \\ 0 \end{bmatrix}, \quad \mathbf{e}^{(3)} = \begin{bmatrix} 0 \\ 0 \\ 1 \\ 0 \end{bmatrix}, \quad \mathbf{e}^{(4)} = \begin{bmatrix} 0 \\ 0 \\ 0 \\ 1 \end{bmatrix}.$$

The corresponding codewords are easy to determine and we use them as columns of the following matrix:

$$\mathbf{M} = \begin{bmatrix} 1 & 0 & 0 & 0 \\ 0 & 1 & 0 & 0 \\ 0 & 0 & 1 & 0 \\ 0 & 0 & 0 & 1 \\ 1 & 1 & 1 & 0 \\ 1 & 1 & 0 & 1 \\ 0 & 1 & 1 & 1 \end{bmatrix} \begin{matrix} \left.\vphantom{\begin{matrix}1\\1\\1\\1\end{matrix}}\right\} \text{message} \\ \left.\vphantom{\begin{matrix}1\\1\\1\end{matrix}}\right\} \text{parity checks} \end{matrix}$$

If $\mathbf{b} \in \mathbf{Z}_2^4$ is any message, then we have

$$\mathbf{Mb} = \begin{bmatrix} b_1 \\ b_2 \\ b_3 \\ b_4 \\ b_1 + b_2 + b_3 \\ b_1 + b_2 + b_4 \\ b_2 + b_3 + b_4 \end{bmatrix},$$

which is a valid codeword. $\mathbf{M}$ is called the **encoding matrix** and the codeword space is $R(\mathbf{M})$, the range of the linear transformation defined by $\mathbf{M}$. Thus the codeword space is the subspace of $\mathbf{Z}_2^7$ given by $R(\mathbf{M})$. The message space is mapped to the codeword space by $L_{\mathbf{M}}$: $\mathbf{Z}_2^4 \to \mathbf{Z}_2^7$.

We assume that during transmission, *at most one error is introduced.* This is a severe restriction, but we mention that more sophisticated codes have been devised which can detect and correct multiple errors. We show that a valid codeword in $\mathbf{Z}_2^7$ must be in the nullspace of a particular linear transformation $L_{\mathbf{H}}$, that is, $\mathbf{u} \in \mathbf{Z}_2^7$ is a valid codeword only if $\mathbf{Hu} = \boldsymbol{\theta}$. To discover $\mathbf{H}$ we consider for given $\mathbf{u} \in \mathbf{Z}_2^7$ the following: If $\mathbf{u}$ is a valid codeword, then

$$\overbrace{u_1 + u_2 + u_3}^{u_5} + u_5 = u_5 + u_5 = 0$$

$$\overbrace{u_1 + u_2 + u_4}^{u_6} + u_6 = u_6 + u_6 = 0$$

$$\overbrace{u_2 + u_3 + u_4}^{u_7} + u_7 = u_7 + u_7 = 0.$$

(Recall that for any $b \in \mathbf{Z}_2$ we have $b + b = 0$.) It follows that a valid codeword $\mathbf{u} \in \mathbf{Z}_2^7$ is in the nullspace of $L_{\mathbf{H}}$, where

$$\mathbf{H} = \begin{bmatrix} 1 & 1 & 1 & 0 & 1 & 0 & 0 \\ 1 & 1 & 0 & 1 & 0 & 1 & 0 \\ 0 & 1 & 1 & 1 & 0 & 0 & 1 \end{bmatrix}.$$

Indeed, we have for $\mathbf{w} \in \mathbf{Z}_2^7$,

$$\mathbf{Hw} = \begin{bmatrix} w_1 + w_2 + w_3 + w_5 \\ w_1 + w_2 + w_4 + w_6 \\ w_2 + w_3 + w_4 + w_7 \end{bmatrix}$$

and $\mathbf{Hw} = \boldsymbol{\theta}$ if $\mathbf{w}$ is a valid codeword, that is, if $\mathbf{w} \in R(\mathbf{M})$. It follows that $R(\mathbf{M}) \subseteq N(\mathbf{H})$. In fact, it is true that $R(\mathbf{M}) = N(\mathbf{H})$, but we postpone demonstrating this until later. $\mathbf{H}$ is called the **syndrome matrix.**

Example 4.24

Suppose that $\mathbf{w} = [1 \quad 1 \quad 0 \quad 1 \quad 1 \quad 0 \quad 0]^{\mathrm{T}}$; then

$$\mathbf{Hw} = \begin{bmatrix} 1 + 1 + 0 + 1 \\ 1 + 1 + 1 + 0 \\ 1 + 0 + 1 + 0 \end{bmatrix} = \begin{bmatrix} 1 \\ 1 \\ 0 \end{bmatrix} \neq \boldsymbol{\theta}$$

and hence $\mathbf{w}$ is not a valid codeword.

■

It is clear that *error detection* is easy—one simply checks $\mathbf{Hw} = \mathbf{v}$. If $\mathbf{v} = \boldsymbol{\theta}$, then $\mathbf{w}$ is a valid codeword, and if $\mathbf{v} \neq \boldsymbol{\theta}$, then $\mathbf{w}$ is not a valid codeword, that is, an error has occurred. Error correction is a bit more subtle but again $\mathbf{H}$ is the key. Suppose that $\mathbf{u}$ is a valid codeword but that a single error, say in the jth component, is introduced during transmission. Then $\hat{\mathbf{u}} = \mathbf{u} + \boldsymbol{\epsilon}^{(j)}$ is received, where $\boldsymbol{\epsilon}^{(j)}$ denotes one of the standard basis vectors for $\mathbf{Z}_2^7$. What we would like to do is determine the value of j so that the error can be corrected. Now

$$\mathbf{H}\hat{\mathbf{u}} = \mathbf{Hu} + \mathbf{H}\boldsymbol{\epsilon}^{(j)} = \boldsymbol{\theta} + \mathbf{H}\boldsymbol{\epsilon}^{(j)} = \mathbf{H}\boldsymbol{\epsilon}^{(j)}$$

and since $\mathbf{H}\boldsymbol{\epsilon}^{(j)} = j$th column of $\mathbf{H}$, we know which component of $\hat{\mathbf{u}}$ is in error.

Example 4.25

Suppose that

$$\hat{\mathbf{u}} = \begin{bmatrix} 1 \\ 1 \\ 0 \\ 1 \\ 1 \\ 0 \\ 0 \end{bmatrix};$$

then

$$\mathbf{H}\hat{\mathbf{u}} = \begin{bmatrix} 1 \\ 1 \\ 0 \end{bmatrix} = \text{first column of } \mathbf{H}.$$

Therefore, the first entry of $\hat{\mathbf{u}}$ is in error. The correct codeword is $\mathbf{u} = [0\ \ 1\ \ 0\ \ 1\ \ 1\ \ 0\ \ 0]^T$ and $\hat{\mathbf{u}} = \mathbf{u} + \boldsymbol{\epsilon}^{(1)}$.

■

Thus if only one error is present in some $\mathbf{u} \in \mathbf{Z}_2^7$, we have $\mathbf{Hu} \neq \boldsymbol{\theta}$ and $\mathbf{Hu}$ is some column of $\mathbf{H}$, that is, $\mathbf{Hu} = \mathbf{H}_j$ for some j. Moreover, u_j (the jth entry of $\mathbf{u}$) is in error.

Let us summarize the (7, 4) Hamming code procedure. Suppose that $\mathbf{b} \in \mathbf{Z}_2^4$ is the message to be transmitted and at most one error is made in transmission. Then $L_{\mathbf{M}}(\mathbf{b}) = \mathbf{Mb} = \mathbf{u} \in \mathbf{Z}_2^7$ is the corresponding codeword to be transmitted. If $\hat{\mathbf{u}} \in \mathbf{Z}_2^7$ is the binary string received, we check $L_{\mathbf{H}}(\hat{\mathbf{u}}) = \mathbf{H}\hat{\mathbf{u}} = \mathbf{v} \in \mathbf{Z}_2^3$. If $\mathbf{v} = \boldsymbol{\theta}$, then $\mathbf{u} = \hat{\mathbf{u}}$ and the correct message is received. Otherwise, $\mathbf{v} = \mathbf{H}_j$ for some j and the jth entry of $\hat{\mathbf{u}}$ is in error. After error correction we take the correct message to consist of the first four entries of the resultant vector. We emphasize that only one transmission error is presumed. The following is an example of the entire process.

Example 4.26

Suppose that we want to transmit the message

$$\mathbf{b} = \begin{bmatrix} 0 \\ 1 \\ 1 \\ 0 \end{bmatrix}.$$

Then $\mathbf{Mb} = \mathbf{u} = [0\ \ 1\ \ 1\ \ 0\ \ 0\ \ 1\ \ 0]^T$. If $\hat{\mathbf{u}} = [0\ \ 1\ \ 1\ \ 1\ \ 0\ \ 1\ \ 0]^T$ is the binary string received, we check:

$$\mathbf{H}\hat{\mathbf{u}} = \begin{bmatrix} 0 \\ 1 \\ 1 \end{bmatrix} = \mathbf{v} \neq \boldsymbol{\theta}.$$

Since $\mathbf{v}$ = 4th column of $\mathbf{H} = \mathbf{H}_4$, we correct the fourth entry of $\hat{\mathbf{u}}$ (from 1 to 0):

$$\hat{\mathbf{u}} \longrightarrow \begin{bmatrix} 0 \\ 1 \\ 1 \\ 0 \\ 0 \\ 1 \\ 0 \end{bmatrix} \quad \left.\begin{matrix} \\ \\ \\ \\ \end{matrix}\right\} \text{correct message (first four entries)}$$

■

Finally, we demonstrate our earlier contention that the range of $L_{\mathbf{M}}$ is equal to the nullspace of $L_{\mathbf{H}}$. Recall that $R(\mathbf{M}) \subseteq N(\mathbf{H})$ and hence we must show that $N(\mathbf{H}) \subseteq R(\mathbf{M})$. First note that $L_{\mathbf{M}}$: $\mathbf{Z}_2^4 \rightarrow \mathbf{Z}_2^7$ and $L_{\mathbf{H}}$: $\mathbf{Z}_2^7 \rightarrow \mathbf{Z}_2^3$; hence $R(\mathbf{M})$ and $N(\mathbf{H})$ are both subspaces of $\mathbf{Z}_2^7$. In fact, we will show that the columns of $\mathbf{M}$, $\{\mathbf{M}_j\}_{j=1}^4$, constitute a basis for $N(\mathbf{H})$, thereby showing that $N(\mathbf{H}) \subseteq R(\mathbf{M})$.

To see that $\{\mathbf{M}_j\}_{j=1}^4$ is a basis for $N(\mathbf{H})$, we first note that if $\alpha, \beta \in \mathbf{Z}_2$ satisfy $\alpha + \beta = 0$, then $\alpha = \beta$. Now if $\mathbf{w} \in N(\mathbf{H})$, then

$$\mathbf{Hw} = \begin{bmatrix} w_1 + w_2 + w_3 + w_5 \\ w_1 + w_2 + w_4 + w_5 \\ w_2 + w_3 + w_4 + w_7 \end{bmatrix} = \boldsymbol{\theta}$$

and hence by our previous observation

$$w_5 = w_1 + w_2 + w_3$$
$$w_6 = w_1 + w_2 + w_4$$
$$w_7 = w_2 + w_3 + w_4.$$

But this implies that

$$\mathbf{w} = \begin{bmatrix} w_1 \\ w_2 \\ w_3 \\ w_4 \\ w_1 + w_2 + w_3 \\ w_1 + w_2 + w_4 \\ w_2 + w_3 + w_4 \end{bmatrix} = \mathbf{M}\begin{bmatrix} w_1 \\ w_2 \\ w_3 \\ w_4 \end{bmatrix} = w_1\mathbf{M}_1 + w_2\mathbf{M}_2 + w_3\mathbf{M}_3 + w_4\mathbf{M}_4$$

and hence $\mathbf{w} \in S[\mathbf{M}_1, \mathbf{M}_2, \mathbf{M}_3, \mathbf{M}_4]$. Since $\{\mathbf{M}_j\}_{j=1}^4$ is linearly independent (see Exercise 10) and spans the nullspace of $L_\mathbf{H}$, it is a basis for $N(\mathbf{H})$.

EXERCISES 4.4

1. Using modulo 2 addition, calculate each of the following.

(a) $\begin{bmatrix} 1 \\ 1 \\ 1 \\ 1 \end{bmatrix} + \begin{bmatrix} 0 \\ 1 \\ 1 \\ 0 \end{bmatrix}$ **(b)** $\begin{bmatrix} 1 & 0 & 1 & 1 \\ 0 & 1 & 0 & 1 \\ 1 & 0 & 0 & 1 \end{bmatrix}\begin{bmatrix} 1 \\ 0 \\ 1 \\ 1 \end{bmatrix}$

(c) $\begin{bmatrix} b & 1 \\ c & 0 \end{bmatrix}\begin{bmatrix} 0 \\ 1 \end{bmatrix}$ where $b, c \in \mathbf{Z}_2$

2. Fill in the remaining entries of the following addition table for vectors in $\mathbf{Z}_2^2$.

$+$	$[0, 0]^T$	$[0, 1]^T$	$[1, 0]^T$	$[1, 1]^T$
$[0, 0]^T$				$[1, 1]^T$
$[0, 1]^T$		$[0, 0]^T$		
$[1, 0]^T$				
$[1, 1]^T$			$[0, 1]^T$	

3. What is the parity of each binary string?
   (a) 1001 **(b)** 00110 **(c)** 1101101
4. List all the valid codewords for the (7, 4) Hamming code.
5. If $\mathbf{u}, \mathbf{v} \in \mathbf{Z}_2^7$, let $H(\mathbf{u}, \mathbf{v})$ = number of bits in which $\mathbf{u}$ and $\mathbf{v}$ differ. From the list in Exercise 4, can you find two codewords, $\mathbf{u}$ and $\mathbf{v}$, with $H(\mathbf{u}, \mathbf{v}) \le 2$?
6. Suppose that a valid codeword $\mathbf{u} \in \mathbf{Z}_2^7$ is transmitted. A transmission error $\boldsymbol{\epsilon}^{(j)}$ gives $\hat{\mathbf{u}} = \mathbf{u} + \boldsymbol{\epsilon}^{(j)}$ as the received binary string. Show that the error is corrected by adding $\boldsymbol{\epsilon}^{(j)}$ to $\hat{\mathbf{u}}$, that is, $\mathbf{u} = \hat{\mathbf{u}} + \boldsymbol{\epsilon}^{(j)}$.
7. Suppose that $\mathbf{u}, \mathbf{v} \in \mathbf{Z}_2^7$ are valid codewords in the Hamming (7, 4) code. Show that $\mathbf{u} + \mathbf{v}$ is also a valid codeword. Give an example to verify this.
8. Show that the syndrome matrix $\mathbf{H}$ of the text cannot detect multiple transmision errors. [*Hint:* Consider $\hat{\mathbf{u}} = \mathbf{u} + \boldsymbol{\epsilon}^{(i)} + \boldsymbol{\epsilon}^{(j)}$, where $\mathbf{u}$ is a valid codeword.]
9. Suppose that we use the parity checks

$$b_5 = b_1 + b_2 + b_3$$

$$b_6 = b_1 + b_3 + b_4$$

$$b_7 = b_1 + b_2 + b_4.$$

   Find the corresponding encoding and syndrome matrices for the resultant (7, 4) Hamming code.
10. Find echelon forms for the encoding and syndrome matrices of the text. Show that $r(\mathbf{M}) = n(\mathbf{H}) = 4$.
11. Suppose that a (7, 4) Hamming code is used and the bit string $[1, 0, 1, 0, 0, 0, 1]^{\mathrm{T}}$ is received. Which bit is in error?

GLOSSARY

arbitrary variable A variable corresponding to a column of an echelon form which contains no leading entry.

basis A basis for a subspace V is a linearly independent set of vectors that span V.

dimension The number of vectors in any basis for a subspace.

echelon form A form, analogous to upper triangular, to which each rectangular matrix can be reduced using elementary row operations.

linear combination A linear combination of vectors $\{\mathbf{u}^{(1)}, \cdots, \mathbf{u}^{(m)}\}$ is a sum of the form

$$\alpha_1 \mathbf{u}^{(1)} + \cdots + \alpha_m \mathbf{u}^{(m)},$$

where $\alpha_1, \ldots, \alpha_m$ are scalars.

linear dependence A set of vectors is linearly dependent if one of the vectors can be expressed as a linear combination of the other vectors in the set.

linear independence A set of vectors $\{\mathbf{u}^{(1)}, \ldots, \mathbf{u}^{(k)}\}$ is linearly independent if the only scalars satisfying

$$\alpha_1 \mathbf{u}^{(1)} + \cdots + \alpha_k \mathbf{u}^{(k)} = \boldsymbol{\theta}$$

are $\alpha_1 = \cdots = \alpha_k = 0$.

linear transformation (defined by $m \times n$ matrix $\mathbf{A}$) Associated with $m \times n$ matrix $\mathbf{A}$ is the function $L_{\mathbf{A}}$ defined by $L_{\mathbf{A}}(\mathbf{x}) = \mathbf{A}\mathbf{x}$. The transformation $L_{\mathbf{A}}$ is linear, that is, $L_{\mathbf{A}}(\alpha\mathbf{x} + \beta\mathbf{v}) = \alpha L_{\mathbf{A}}(\mathbf{x}) + \beta L_{\mathbf{A}}(\mathbf{v})$ for all n-vectors $\mathbf{x}$ and $\mathbf{v}$ and scalars α and β.

nullity The nullity of a matrix $\mathbf{A}$ is the number of vectors in a basis for $N(\mathbf{A})$.

nullspace The set of all vectors that are mapped to the zero vector by a linear transformation.

pivot variable A variable corresponding to a column of an echelon form that contains a leading entry.

$\mathbf{R}^n$ The set of all n-vectors.

range The set of all values of a linear transformation.

rank The rank of a matrix $\mathbf{A}$ is the number of vectors in a basis for $R(\mathbf{A})$.

span The span of a set of vectors S is the subspace of all vectors that are linear combinations of vectors of S.

subspace A subset of $\mathbf{R}^n$ that is closed under vector addition and scalar multiplication.

NOTES AND COMMENTS

In more theoretical books on linear algebra the notions of basis, linear dependence, linear transformation, and subspace are presented in a more abstract framework—that of a vector space. See, for example, the book *Linear Algebra* (New York: Harcourt Brace Jovanovich, 1977) by Michael O'Nan.

An introduction to error-correcting codes can be found in the text *Discrete Mathematics and Applied Modern Algebra* by Henry B. Laufer (Boston: Prindle, Weber & Schmidt, 1984). A more advanced treatment is given in *Introduction to the Theory of Error-Correcting Codes* by Vera Ples (New York: Wiley-Interscience, 1982).

5

Geometrical Notions

We now introduce some geometrical aspects of vectors and matrices. In particular we meet for the first time the important notion of inner product and learn of the crucial relationship between the inner product and the transpose matrix. A method of producing perpendicular vectors, the Gram–Schmidt process, is studied and interpreted as a matrix factorization method. Finally, several important new classes of matrices are introduced and complex numbers make their first appearance.

5.1 Vectors in the Plane

The reader is probably familiar with the representation of directed segments in the plane as ordered pairs of numbers. Such directed segments are called **vectors** and we represent them as 2-vectors. Two directed segments are considered equivalent if they are parallel, have the same length, and point in the same direction. For example, the column vector

$$\mathbf{x} = \begin{bmatrix} 3 \\ 2 \end{bmatrix}$$

is used to represent each of the vectors in Figure 5.1.

The directed segment whose "tail" is at the origin is called the standard representation of the vector.

The length (or **norm**) of a vector $\mathbf{x}$ is denoted $\|\mathbf{x}\|$ and is given by

$$\|\mathbf{x}\| = \sqrt{x_1^2 + x_2^2}. \tag{1}$$

Of course, this is just a restatement of the *Pythagorean theorem* (see Figure 5.2).

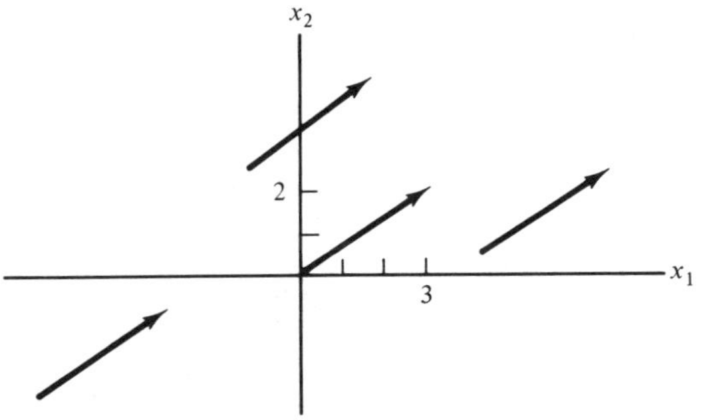

Figure 5.1

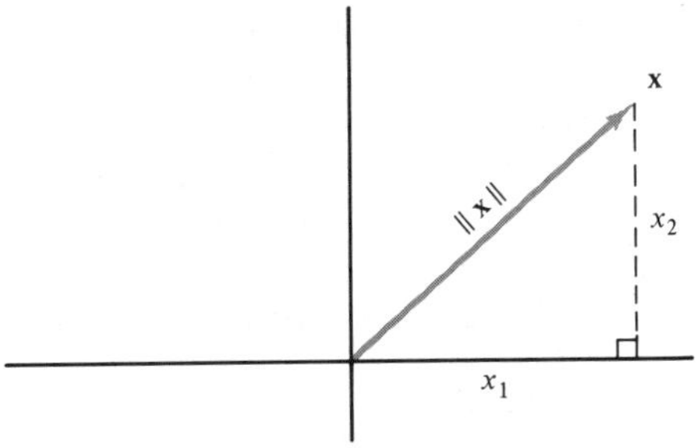

Figure 5.2

It is easy to give geometrical interpretations to the operations of multiplication of a vector by a scalar and vector addition. For example,

$$\|t\mathbf{x}\| = \sqrt{(tx_1)^2 + (tx_2)^2} = |t|\sqrt{x_1^2 + x_2^2} = |t|\,\|\mathbf{x}\|;$$

therefore, multiplication by the scalar t has the effect of changing the length of a vector by the factor $|t|$ and, depending on the sign of t, perhaps reversing its direction (see Figure 5.3). Vector addition and subtraction are accomplished geometrically by the "tail-to-head" method (see Figure 5.4).

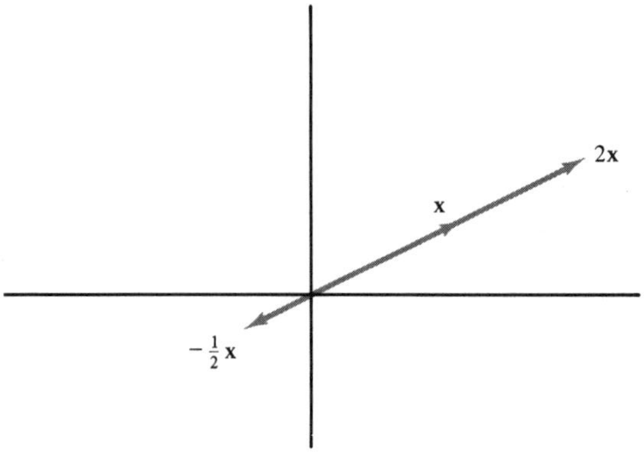

Figure 5.3

The **inner product** of two vectors $\mathbf{x}$ and $\mathbf{y}$ is the number $\langle \mathbf{x}, \mathbf{y} \rangle$ given by

$$\langle \mathbf{x}, \mathbf{y} \rangle = x_1 y_1 + x_2 y_2, \tag{2}$$

which the reader may recognize as the "dot product" of elementary vector analysis. There is an important connection between the norm and inner product which follows immediately from (1) and (2), namely

$$\|\mathbf{x}\|^2 = x_1^2 + x_2^2 = \langle \mathbf{x}, \mathbf{x} \rangle. \tag{3}$$

We also note two further properties of the inner product which follow directly from (2), specifically

$$\langle \mathbf{x}, \mathbf{y} \rangle = \langle \mathbf{y}, \mathbf{x} \rangle \tag{4}$$

and

$$\langle \alpha\mathbf{x} + \beta\mathbf{y}, \mathbf{z} \rangle = \alpha\langle \mathbf{x}, \mathbf{z} \rangle + \beta\langle \mathbf{y}, \mathbf{z} \rangle \tag{5}$$

for any vectors $\mathbf{x}$, $\mathbf{y}$, $\mathbf{z}$ and scalars α, β.

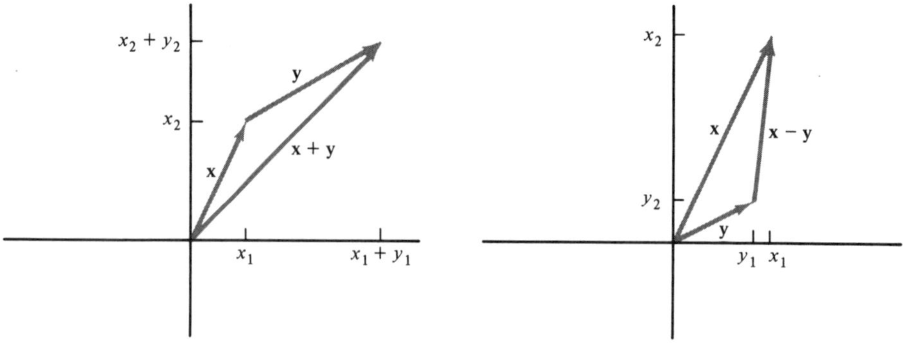

Figure 5.4

Example 5.1

Suppose that

$$\mathbf{x} = \begin{bmatrix} 3 \\ -1 \end{bmatrix}, \quad \mathbf{y} = \begin{bmatrix} 2 \\ 1 \end{bmatrix}, \quad \mathbf{z} = \begin{bmatrix} -1 \\ 1 \end{bmatrix}.$$

Then

$$\begin{aligned}
\langle \mathbf{x}, \mathbf{y} \rangle &= 3 \cdot 2 + (-1) \cdot 1 = 5, \\
\langle \mathbf{y}, \mathbf{x} \rangle &= 2 \cdot 3 + 1 \cdot (-1) = 5, \\
\|\mathbf{x}\| &= \sqrt{9 + 1} = \sqrt{10}, \\
\|\mathbf{x}\|^2 &= 10 = 3 \cdot 3 + (-1)(-1) = \langle \mathbf{x}, \mathbf{x} \rangle, \\
2\mathbf{x} + 3\mathbf{y} &= \begin{bmatrix} 12 \\ 1 \end{bmatrix}, \\
\langle 2\mathbf{x} + 3\mathbf{y}, \mathbf{z} \rangle &= 12 \cdot (-1) + 1 \cdot 1 = -11, \\
2\langle \mathbf{x}, \mathbf{z} \rangle &= 2\{3 \cdot (-1) + (-1) \cdot 1\} = -8, \\
3\langle \mathbf{y}, \mathbf{z} \rangle &= 3\{2 \cdot (-1) + 1 \cdot 1\} = -3.
\end{aligned}$$

■

The *law of cosines* (see Exercise 4) gives additional insight into the geometrical nature of the inner product. In applying the law of cosines to the triangle in Figure 5.5, we find that

$$\|\mathbf{x} - \mathbf{y}\|^2 = \|\mathbf{x}\|^2 + \|\mathbf{y}\|^2 - 2\|\mathbf{x}\|\|\mathbf{y}\| \cos \theta.$$

However, by (3)–(5), we have

$$\begin{aligned}
\|\mathbf{x} - \mathbf{y}\|^2 &= \langle \mathbf{x} - \mathbf{y}, \mathbf{x} - \mathbf{y} \rangle \\
&= \langle \mathbf{x}, \mathbf{x} - \mathbf{y} \rangle - \langle \mathbf{y}, \mathbf{x} - \mathbf{y} \rangle \\
&= \langle \mathbf{x}, \mathbf{x} \rangle - \langle \mathbf{x}, \mathbf{y} \rangle - \langle \mathbf{y}, \mathbf{x} \rangle + \langle \mathbf{y}, \mathbf{y} \rangle \\
&= \|\mathbf{x}\|^2 - 2\langle \mathbf{x}, \mathbf{y} \rangle + \|\mathbf{y}\|^2,
\end{aligned}$$

and hence

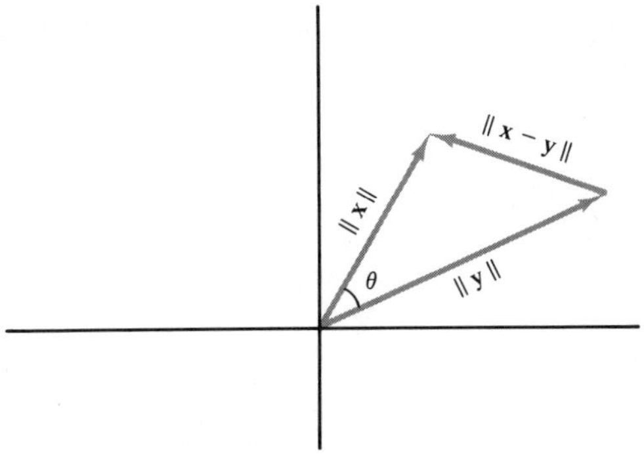

Figure 5.5

$$\langle \mathbf{x}, \mathbf{y} \rangle = \|\mathbf{x}\|\|\mathbf{y}\| \cos \theta.$$

From this we see (since $|\cos \theta| \leq 1$) that

$$|\langle \mathbf{x}, \mathbf{y} \rangle| \leq \|\mathbf{x}\|\|\mathbf{y}\|, \tag{6}$$

a fact known as the **Cauchy–Schwarz inequality.** We also see that the vectors $\mathbf{x}$ and $\mathbf{y}$ are perpendicular, denoted $\mathbf{x} \perp \mathbf{y}$, if and only if $\langle \mathbf{x}, \mathbf{y} \rangle = 0$.

Example 5.2

If

$$\mathbf{x} = \begin{bmatrix} -1 \\ 3 \end{bmatrix}, \qquad \mathbf{y} = \begin{bmatrix} 1 \\ 1 \end{bmatrix},$$

then

$$|\langle \mathbf{x}, \mathbf{y} \rangle| = |(-1) \cdot 1 + 3 \cdot 0| = 1 \leq \sqrt{10} \cdot 1 = \|\mathbf{x}\|\|\mathbf{y}\|.$$

■

The inner product is very useful in computing the projection $\mathbf{P}_{\mathbf{y}}\mathbf{x}$ of a vector $\mathbf{x}$ onto the line containing the vector $\mathbf{y}$, as illustrated in Figure 5.6. The vector $\mathbf{P}_{\mathbf{y}}\mathbf{x}$ is defined to be that scalar multiple of $\mathbf{y}$ such that $(\mathbf{x} - \mathbf{P}_{\mathbf{y}}\mathbf{x}) \perp \mathbf{y}$ (see Figure 5.6). If we set $\mathbf{P}_{\mathbf{y}}\mathbf{x} = \lambda \mathbf{y}$, we see that the scalar λ must be determined such that

$$(\mathbf{x} - \lambda \mathbf{y}) \perp \mathbf{y}, \qquad \text{that is,} \quad \langle \mathbf{x} - \lambda \mathbf{y}, \mathbf{y} \rangle = 0.$$

But this gives

$$\langle \mathbf{x}, \mathbf{y} \rangle - \lambda \langle \mathbf{y}, \mathbf{y} \rangle = 0$$

or

$$\langle \mathbf{x}, \mathbf{y} \rangle = \lambda \|\mathbf{y}\|^2,$$

and hence $\lambda = \langle \mathbf{x}, \mathbf{y} \rangle / \|\mathbf{y}\|^2$. That is,

$$\mathbf{P}_{\mathbf{y}}\mathbf{x} = \frac{\langle \mathbf{x}, \mathbf{y} \rangle}{\|\mathbf{y}\|^2} \mathbf{y}. \tag{7}$$

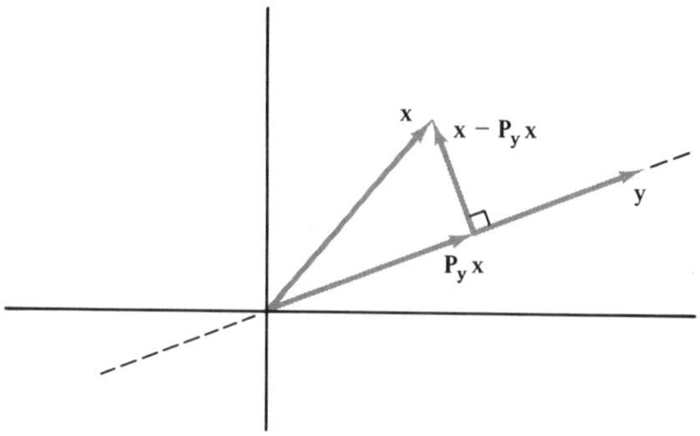

Figure 5.6

Example 5.3

If

$$\mathbf{x} = \begin{bmatrix} -1 \\ 2 \end{bmatrix} \quad \text{and} \quad \mathbf{y} = \begin{bmatrix} 3 \\ 4 \end{bmatrix},$$

then

$$\langle \mathbf{x}, \mathbf{y} \rangle = (-1) \cdot 3 + 2 \cdot 4 = 5$$

and

$$\|\mathbf{y}\|^2 = 3 \cdot 3 + 4 \cdot 4 = 25.$$

Therefore,

$$P_{\mathbf{y}}\mathbf{x} = \frac{5}{25}\mathbf{y} = \frac{1}{5}\begin{bmatrix} 3 \\ 4 \end{bmatrix}.$$

Note that $\mathbf{x} - P_{\mathbf{y}}\mathbf{x} = [-\frac{8}{5} \ \ \frac{6}{5}]^T$ and $\langle \mathbf{x} - P_{\mathbf{y}}\mathbf{x}, \mathbf{y} \rangle = 0$. ■

EXERCISES 5.1

1. Suppose that

$$\mathbf{x} = \begin{bmatrix} 2 \\ 1 \end{bmatrix}, \quad \mathbf{y} = \begin{bmatrix} -1 \\ 3 \end{bmatrix}.$$

Find $\|\mathbf{x}\|$, $\|-2\mathbf{x}\|$, $\|\mathbf{y}\|$, $\|\mathbf{x} + \mathbf{y}\|$, $\|\mathbf{x} - \mathbf{y}\|$, and $\langle \mathbf{x}, \mathbf{y} \rangle$.

2. Let

$$\mathbf{x} = \begin{bmatrix} -1 \\ 2 \end{bmatrix}, \quad \mathbf{y} = \begin{bmatrix} 6 \\ 3 \end{bmatrix}, \quad \text{and} \quad \mathbf{z} = \begin{bmatrix} \frac{1}{2} \\ -1 \end{bmatrix}.$$

Show that $\mathbf{x} \perp \mathbf{y}$, and $\mathbf{y} \perp \mathbf{z}$. Is $\mathbf{x} \perp \mathbf{z}$?

3. Let $\mathbf{x}$, $\mathbf{y}$, and $\mathbf{z}$ be the vectors in Exercise 2. Compute $\langle 2\mathbf{x} + 7\mathbf{y}, \mathbf{z} \rangle$. Show that $\langle s\mathbf{x} + t\mathbf{y}, \mathbf{z} \rangle = s\langle \mathbf{x}, \mathbf{z} \rangle$ for any scalars s and t.

4. Prove the law of cosines: $C^2 = A^2 + B^2 - 2AB\cos\theta$.

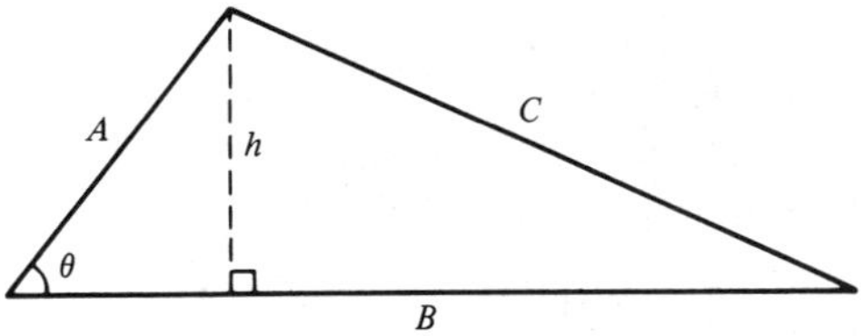

(*Hint:* Note that $h = A\sin\theta$. Apply the Pythagorean theorem as indicated.)

5. Show that the Pythagorean theorem is a special case of the law of cosines.

6. Let $\mathbf{x}$ and $\mathbf{y}$ be the vectors in Exercise 1. Draw a picture to represent the vectors $\mathbf{x}$, $\mathbf{y}$, $\mathbf{x} + \mathbf{y}$, and $\mathbf{x} - \mathbf{y}$ geometrically. Compute $\|\mathbf{x} + \mathbf{y}\|^2 + \|\mathbf{x} - \mathbf{y}\|^2$ and $2(\|\mathbf{x}\|^2 + \|\mathbf{y}\|^2)$. Can you interpret the result geometrically?

7. Suppose that

$$\mathbf{x} = \begin{bmatrix} -1 \\ 2 \end{bmatrix} \quad \text{and} \quad \mathbf{y} = \begin{bmatrix} 3 \\ 4 \end{bmatrix}.$$

Show that there is no nonzero 2-vector $\mathbf{z}$ such that $\mathbf{z} \perp \mathbf{x}$ and $\mathbf{z} \perp \mathbf{y}$.

8. For

$$\mathbf{x} = \begin{bmatrix} -1 \\ 1 \end{bmatrix} \quad \text{and} \quad \mathbf{y} = \begin{bmatrix} 2 \\ 3 \end{bmatrix},$$

find $\mathbf{P}_{\mathbf{y}}\mathbf{x}$ and $\mathbf{P}_{\mathbf{x}}\mathbf{y}$. Give a geometrical representation of these vectors.

9. Let $\mathbf{x}$ and $\mathbf{y}$ be the vectors in Exercise 8 and let

$$\mathbf{z} = \begin{bmatrix} 3 \\ 7 \end{bmatrix}.$$

Verify that $\mathbf{P}_{\mathbf{z}}(\mathbf{x} + \mathbf{y}) = \mathbf{P}_{\mathbf{z}}\mathbf{x} + \mathbf{P}_{\mathbf{z}}\mathbf{y}$. Illustrate this geometrically.

10. For $\mathbf{x}$ and $\mathbf{z}$ in Exercise 9, verify that $\mathbf{P}_{\mathbf{z}}(3\mathbf{x}) = 3\mathbf{P}_{\mathbf{z}}\mathbf{x}$ and $\mathbf{P}_{\mathbf{z}}(-2\mathbf{x}) = -2\mathbf{P}_{\mathbf{z}}\mathbf{x}$. What can you say about $\mathbf{P}_{\mathbf{z}}(s\mathbf{x})$, where s is any scalar?

11. Is it true that $\mathbf{P}_{t\mathbf{y}}\mathbf{x} = \mathbf{P}_{\mathbf{y}}\mathbf{x}$ for each nonzero scalar t?

12. Suppose that $\mathbf{y}$ is a nonzero 2-vector and let

$$\mathbf{A} = \frac{1}{\|\mathbf{y}\|^2} \begin{bmatrix} y_1^2 & y_1 y_2 \\ y_1 y_2 & y_2^2 \end{bmatrix}.$$

Verify that $\mathbf{A}\mathbf{x} = \mathbf{P}_{\mathbf{y}}\mathbf{x}$.

13. Suppose that

$$\mathbf{y} = \begin{bmatrix} 2 \\ -2 \end{bmatrix} \quad \text{and} \quad \mathbf{x} = \begin{bmatrix} 1 \\ t \end{bmatrix}.$$

Find t if

$$\mathbf{P}_{\mathbf{y}}\mathbf{x} = \begin{bmatrix} -1 \\ 1 \end{bmatrix}.$$

14. Verify that $\|\mathbf{x}\|^2 = \|\mathbf{P}_{\mathbf{y}}\mathbf{x}\|^2 + \|\mathbf{x} - \mathbf{P}_{\mathbf{y}}\mathbf{x}\|^2$ for any nonzero vector $\mathbf{y}$. (*Hint:* $\|\mathbf{x} - \mathbf{P}_{\mathbf{y}}\mathbf{x}\|^2 = \langle \mathbf{x} - \mathbf{P}_{\mathbf{y}}\mathbf{x}, \mathbf{x} - \mathbf{P}_{\mathbf{y}}\mathbf{x} \rangle$.) Interpret this geometrically.

5.2 Vectors in Higher Dimensions

The notions discussed in Section 5.1 may be naturally extended to n-vectors. Without further ado we make the following definitions for the inner product and corresponding norm of n-vectors $\mathbf{x}$, $\mathbf{y}$:

$$\langle \mathbf{x}, \mathbf{y} \rangle = x_1 y_1 + x_2 y_2 + \cdots + x_n y_n \tag{8}$$

$$\|\mathbf{x}\| = \sqrt{x_1^2 + x_2^2 + \cdots + x_n^2}. \tag{9}$$

Example 5.4

Suppose that $\mathbf{x} = [0 \quad -1 \quad 2 \quad 4 \quad 3]^{\mathrm{T}}$ and $\mathbf{y} = [7 \quad 1 \quad -1 \quad 3 \quad 0]^{\mathrm{T}}$; then

$$\langle \mathbf{x}, \mathbf{y} \rangle = (0)(7) + (-1)(1) + 2(-1) + 4(3) + 3(0) = 9$$

and

$$\|\mathbf{x}\| = \{0^2 + (-1)^2 + 2^2 + 4^2 + 3^2\}^{1/2} = \sqrt{30}\,.$$

■

Some basic properties of the inner product are given in the following theorem.

Theorem 5.1. If $\mathbf{x}$, $\mathbf{y}$, $\mathbf{z}$ are n-vectors and α, β are scalars, then
(a) $\langle \mathbf{x}, \mathbf{y}\rangle = \langle \mathbf{y}, \mathbf{x}\rangle$.
(b) $\langle \alpha\mathbf{x} + \beta\mathbf{y}, \mathbf{z}\rangle = \alpha\langle \mathbf{x}, \mathbf{z}\rangle + \beta\langle \mathbf{y}, \mathbf{z}\rangle$.
(c) $\langle \mathbf{x}, \mathbf{x}\rangle = \|\mathbf{x}\|^2$.
(d) $|\langle \mathbf{x}, \mathbf{y}\rangle| \le \|\mathbf{x}\|\|\mathbf{y}\|$.

■

Properties (a)–(c) are immediate and obvious consequences of the definitions. The inequality (d) is again called the Cauchy–Schwarz inequality and is not as obvious as in the plane; a proof is outlined in the exercises. In the following theorem we state the essential properties of the vector norm.

Theorem 5.2. If $\mathbf{x}$ and $\mathbf{y}$ are n-vectors and α is a scalar, then
(a) $\|\mathbf{x}\| \ge 0$ for all $\mathbf{x}$ and $\|\mathbf{x}\| = 0$ if and only if $\mathbf{x} = \boldsymbol{\theta}$.
(b) $\|\alpha\mathbf{x}\| = |\alpha|\|\mathbf{x}\|$.
(c) $\|\mathbf{x} + \mathbf{y}\| \le \|\mathbf{x}\| + \|\mathbf{y}\|$.
(d) $\|\mathbf{x} + \mathbf{y}\|^2 = \|\mathbf{x}\|^2 + \|\mathbf{y}\|^2$ if and only if $\langle \mathbf{x}, \mathbf{y}\rangle = 0$.

■

Because of their geometrical interpretations (see Figure 5.4), (c) and (d) are called the **triangle inequality** and **Pythagorean theorem,** respectively. Part (d) follows from the identity

$$\begin{aligned}\|\mathbf{x} + \mathbf{y}\|^2 &= \langle \mathbf{x} + \mathbf{y}, \mathbf{x} + \mathbf{y}\rangle = \langle \mathbf{x}, \mathbf{x}\rangle + 2\langle \mathbf{x}, \mathbf{y}\rangle + \langle \mathbf{y}, \mathbf{y}\rangle \\ &= \|\mathbf{x}\|^2 + 2\langle \mathbf{x}, \mathbf{y}\rangle + \|\mathbf{y}\|^2.\end{aligned}$$

The same identity, together with the Cauchy–Schwarz inequality, gives the triangle inequality:

$$\begin{aligned}\|\mathbf{x} + \mathbf{y}\|^2 &\le \|\mathbf{x}\|^2 + 2\|\mathbf{x}\|\|\mathbf{y}\| + \|\mathbf{y}\|^2 \\ &= (\|\mathbf{x}\| + \|\mathbf{y}\|)^2.\end{aligned}$$

Our next theorem points out a very important connection between the inner product and the transpose of a matrix. This is not surprising since the inner product can be defined in terms of transposes by $\langle \mathbf{x}, \mathbf{y}\rangle = \mathbf{x}^{\mathrm{T}}\mathbf{y}$. Indeed, if $\mathbf{x}$ and $\mathbf{y}$ are n-vectors, that is, $n \times 1$ matrices, then by the definition of matrix multiplication $\mathbf{x}^{\mathrm{T}}\mathbf{y}$ is the 1×1 matrix (i.e., scalar) given by

$$\mathbf{x}^{\mathrm{T}}\mathbf{y} = x_1y_1 + x_2y_2 + \cdots + x_ny_n = \langle \mathbf{x}, \mathbf{y}\rangle.$$

Theorem 5.3. Suppose that $\mathbf{A}$ is an $m \times n$ matrix, $\mathbf{x}$ is an n-vector, and $\mathbf{y}$ is an m-vector. Then $\langle \mathbf{A}\mathbf{x}, \mathbf{y}\rangle = \langle \mathbf{x}, \mathbf{A}^{\mathrm{T}}\mathbf{y}\rangle$.

■

Note that the inner product on the left is an inner product of m-vectors, while that on the right is an inner product of n-vectors. The proof of Theorem 5.3 is a simple computation using properties of the transpose:

$$\langle \mathbf{Ax}, \mathbf{y} \rangle = (\mathbf{Ax})^T\mathbf{y} = (\mathbf{x}^T\mathbf{A}^T)\mathbf{y} = \mathbf{x}^T(\mathbf{A}^T\mathbf{y}) = \langle \mathbf{x}, \mathbf{A}^T\mathbf{y} \rangle.$$

Example 5.5

Let

$$\mathbf{A} = \begin{bmatrix} 3 & 2 & 1 \\ -1 & 0 & 1 \end{bmatrix}, \qquad \mathbf{x} = \begin{bmatrix} 1 \\ -1 \\ 2 \end{bmatrix}, \qquad \mathbf{y} = \begin{bmatrix} -2 \\ 1 \end{bmatrix}.$$

Then

$$\mathbf{Ax} = \begin{bmatrix} 3 \\ 1 \end{bmatrix}, \qquad \mathbf{A}^T = \begin{bmatrix} 3 & -1 \\ 2 & 0 \\ 1 & 1 \end{bmatrix}, \quad \text{and} \quad \mathbf{A}^T\mathbf{y} = \begin{bmatrix} -7 \\ -4 \\ -1 \end{bmatrix}.$$

Therefore,

$$\langle \mathbf{Ax}, \mathbf{y} \rangle = 3 \cdot (-2) + 1 \cdot 1 = -5$$

and

$$\langle \mathbf{x}, \mathbf{A}^T\mathbf{y} \rangle = 1 \cdot (-7) + (-1) \cdot (-4) + 2(-1) = -5.$$

■

It is often necessary to be able to express a given vector in $\mathbf{R}^n$ as a linear combination of the vectors in a given basis for $\mathbf{R}^n$. If $\{\mathbf{u}^{(j)}\}_{j=1}^n$ is a basis for $\mathbf{R}^n$ and $\mathbf{y}$ is a given n-vector, then to express $\mathbf{y}$ as a linear combination of the basis vectors requires the solution of

$$\mathbf{y} = \sum_{j=1}^{n} c_j \mathbf{u}^{(j)}$$

or the equivalent matrix problem

$$\mathbf{Ac} = \mathbf{y},$$

where $\mathbf{A}_j = \mathbf{u}^{(j)}$ and $\mathbf{c} = [c_1, \ldots, c_n]^T$.

The system above can be solved by Gaussian elimination to obtain the coefficients c_j of the "expansion" of $\mathbf{y}$ in terms of the basis $\{\mathbf{u}^{(j)}\}_{j=1}^n$. However, for a certain special type of basis, called an orthonormal basis, the calculation of the coefficients c_j is a relatively simple matter.

Again we write $\mathbf{x} \perp \mathbf{y}$ if $\langle \mathbf{x}, \mathbf{y} \rangle = 0$. In two dimensions we have termed such vectors perpendicular. However, for higher dimensions the term **orthogonal** is generally used to used to describe vectors $\mathbf{x}$ and $\mathbf{y}$ with $\mathbf{x} \perp \mathbf{y}$.

We say that a set $\{\mathbf{x}^{(1)}, \ldots, \mathbf{x}^{(k)}\}$ of nonzero n-vectors is orthogonal if $\mathbf{x}^{(i)} \perp \mathbf{x}^{(j)}$ for $i \neq j$.

Example 5.6

The set

$$\left\{ \mathbf{x}^{(1)} = \begin{bmatrix} 0 \\ 1 \\ 0 \end{bmatrix}, \quad \mathbf{x}^{(2)} = \begin{bmatrix} 1 \\ 0 \\ -1 \end{bmatrix}, \quad \mathbf{x}^{(3)} = \begin{bmatrix} 1 \\ 0 \\ 1 \end{bmatrix} \right\}$$

is orthogonal since

$$\langle \mathbf{x}^{(1)}, \mathbf{x}^{(2)} \rangle = 0 \cdot 1 + 1 \cdot 0 + 0 \cdot (-1) = 0,$$

$$\langle \mathbf{x}^{(1)}, \mathbf{x}^{(3)} \rangle = 0 \cdot 1 + 1 \cdot 0 + 0 \cdot 1 \quad = 0,$$

and

$$\langle \mathbf{x}^{(2)}, \mathbf{x}^{(3)} \rangle = 1 \cdot 1 + 0 \cdot 0 + (-1) \cdot 1 = 0.$$

■

An orthogonal set is automatically linearly independent. To see this, suppose that

$$c_1 \mathbf{x}^{(1)} + c_2 \mathbf{x}^{(2)} + \cdots + c_k \mathbf{x}^{(k)} = \boldsymbol{\theta}.$$

Then

$$\begin{aligned} 0 &= \langle c_1 \mathbf{x}^{(1)} + \cdots + c_k \mathbf{x}^{(k)}, \mathbf{x}^{(j)} \rangle \\ &= c_1 \langle \mathbf{x}^{(1)}, \mathbf{x}^{(j)} \rangle + \cdots + c_j \langle \mathbf{x}^{(j)}, \mathbf{x}^{(j)} \rangle + \cdots + c_k \langle \mathbf{x}^{(k)}, \mathbf{x}^{(j)} \rangle \\ &= c_j \, \| \mathbf{x}^{(j)} \|^2. \end{aligned}$$

Since $\mathbf{x}^{(j)} \neq \boldsymbol{\theta}$, we find that $c_j = 0$ for each j. That is, the set $\{\mathbf{x}^{(1)}, \ldots, \mathbf{x}^{(k)}\}$ is linearly independent.

If, in addition to being orthogonal, each vector in the set has norm 1, we say that the set is **orthonormal**. An orthonormal set has a very convenient property with respect to expansions. Indeed, if $\{\mathbf{x}^{(1)}, \ldots, \mathbf{x}^{(k)}\}$ is orthonormal and $\mathbf{y} \in S[\mathbf{x}^{(1)}, \ldots, \mathbf{x}^{(k)}]$, then for some coefficients $c_1, c_2, \ldots, c_k$,

$$\mathbf{y} = c_1 \mathbf{x}^{(1)} + c_2 \mathbf{x}^{(2)} + \cdots + c_k \mathbf{x}^{(k)}.$$

The coefficients c_j are easy to compute since

$$\begin{aligned} \langle \mathbf{y}, \mathbf{x}^{(j)} \rangle &= \langle c_1 \mathbf{x}^{(1)} + c_2 \mathbf{x}^{(2)} + \cdots + c_k \mathbf{x}^{(k)}, \mathbf{x}^{(j)} \rangle \\ &= \sum_{i=1}^{k} c_i \langle \mathbf{x}^{(i)}, \mathbf{x}^{(j)} \rangle \\ &= c_j \langle \mathbf{x}^{(j)}, \mathbf{x}^{(j)} \rangle = c_j 1 = c_j. \end{aligned}$$

To put it another way, if $\mathbf{y}$ is a linear combination of orthonormal vectors $\{\mathbf{x}^{(1)}, \ldots, \mathbf{x}^{(k)}\}$, then the coefficient of $\mathbf{x}^{(j)}$ in this combination is $\langle \mathbf{y}, \mathbf{x}^{(j)} \rangle$.

Example 5.7

If

$$\mathbf{x}^{(1)} = \begin{bmatrix} 0 \\ 1 \\ 0 \end{bmatrix}, \qquad \mathbf{x}^{(2)} = \frac{1}{\sqrt{2}}\begin{bmatrix} 1 \\ 0 \\ -1 \end{bmatrix}, \qquad \mathbf{x}^{(3)} = \frac{1}{\sqrt{2}}\begin{bmatrix} 1 \\ 0 \\ 1 \end{bmatrix},$$

then $\{\mathbf{x}^{(1)}, \mathbf{x}^{(2)}, \mathbf{x}^{(3)}\}$ is an orthonormal basis for $\boldsymbol{R}^3$. Given a 3-vector $\mathbf{y}$, say

$$\mathbf{y} = \begin{bmatrix} 1 \\ -1 \\ 2 \end{bmatrix},$$

then we may write $\mathbf{y} = c_1\mathbf{x}^{(1)} + c_2\mathbf{x}^{(2)} + c_3\mathbf{x}^{(3)}$, where $c_1 = \langle \mathbf{y}, \mathbf{x}^{(1)} \rangle = -1$, $c_2 = \langle \mathbf{y}, \mathbf{x}^{(2)} \rangle = -1/\sqrt{2}$, and $c_3 = \langle \mathbf{y}, \mathbf{x}^{(3)} \rangle = 3/\sqrt{2}$. Indeed

$$c_1\mathbf{x}^{(1)} + c_2\mathbf{x}^{(2)} + c_3\mathbf{x}^{(3)}$$

$$= -1\begin{bmatrix} 0 \\ 1 \\ 0 \end{bmatrix} - \frac{1}{\sqrt{2}} \cdot \frac{1}{\sqrt{2}}\begin{bmatrix} 1 \\ 0 \\ -1 \end{bmatrix} + \frac{3}{\sqrt{2}}\frac{1}{\sqrt{2}}\begin{bmatrix} 1 \\ 0 \\ 1 \end{bmatrix} = \begin{bmatrix} -\frac{1}{2} + \frac{3}{2} \\ -1 \\ \frac{1}{2} + \frac{3}{2} \end{bmatrix}$$

$$= \begin{bmatrix} 1 \\ -1 \\ 2 \end{bmatrix} = \mathbf{y}.$$

■

There is a constructive method, called the **Gram–Schmidt** process, of transforming a given linearly independent set of vectors into an orthonormal set which has the same span as does the original set of vectors. Suppose that $\{\mathbf{x}^{(1)}, \mathbf{x}^{(2)}, \ldots, \mathbf{x}^{(k)}\}$ is a linearly independent set of vectors and we wish to produce from them an orthonormal set $\{\mathbf{q}^{(1)}, \mathbf{q}^{(2)}, \ldots, \mathbf{q}^{(k)}\}$. To begin, we set $\mathbf{y}^{(1)} = \mathbf{x}^{(1)}$ and then normalize:

$$\mathbf{y}^{(1)} = \mathbf{x}^{(1)}, \qquad \mathbf{q}^{(1)} = \frac{\mathbf{y}^{(1)}}{\|\mathbf{y}^{(1)}\|}.$$

Next we seek a vector $\mathbf{y}^{(2)}$ which is orthogonal to $\mathbf{q}^{(1)}$. The key to this is Figure 5.6. From it we see that

$$\mathbf{x}^{(2)} - \mathbf{P}_{\mathbf{q}^{(1)}}\mathbf{x}^{(2)} = \mathbf{x}^{(2)} - \langle \mathbf{x}^{(2)}, \mathbf{q}^{(1)} \rangle \mathbf{q}^{(1)}$$

is orthogonal to $\mathbf{q}^{(1)}$. We therefore set

$$\mathbf{y}^{(2)} = \mathbf{x}^{(2)} - \langle \mathbf{x}^{(2)}, \mathbf{q}^{(1)} \rangle \mathbf{q}^{(1)}, \qquad \mathbf{q}^{(2)} = \frac{\mathbf{y}^{(2)}}{\|\mathbf{y}^{(2)}\|}.$$

Continuing in this way, we next set

$$\mathbf{y}^{(3)} = \mathbf{x}^{(3)} - \langle \mathbf{x}^{(3)}, \mathbf{q}^{(1)} \rangle \mathbf{q}^{(1)} - \langle \mathbf{x}^{(3)}, \mathbf{q}^{(2)} \rangle \mathbf{q}^{(2)}, \qquad \mathbf{q}^{(3)} = \frac{\mathbf{y}^{(3)}}{\|\mathbf{y}^{(3)}\|}.$$

Then note that

$$\langle \mathbf{y}^{(3)}, \mathbf{q}^{(1)} \rangle = \langle \mathbf{x}^{(3)}, \mathbf{q}^{(1)} \rangle - \langle \mathbf{x}^{(3)}, \mathbf{q}^{(1)} \rangle - 0 = 0$$

and

$$\langle \mathbf{y}^{(3)}, \mathbf{q}^{(2)} \rangle = \langle \mathbf{x}^{(3)}, \mathbf{q}^{(2)} \rangle - 0 - \langle \mathbf{x}^{(3)}, \mathbf{q}^{(2)} \rangle = 0$$

and hence $\{\mathbf{q}^{(1)}, \mathbf{q}^{(2)}, \mathbf{q}^{(3)}\}$ is orthonormal. In general, we set

$$\mathbf{y}^{(j)} = \mathbf{x}^{(j)} - \langle \mathbf{x}^{(j)}, \mathbf{q}^{(1)} \rangle \mathbf{q}^{(1)} - \cdots - \langle \mathbf{x}^{(j)}, \mathbf{q}^{(j-1)} \rangle \mathbf{q}^{(j-1)}$$

and $\mathbf{q}^{(j)} = \mathbf{y}^{(j)}/\|\mathbf{y}^{(j)}\|$.

Example 5.8

We apply the Gram–Schmidt process to the vectors

$$\mathbf{x}^{(1)} = \begin{bmatrix} 2 \\ 1 \\ 0 \end{bmatrix}, \qquad \mathbf{x}^{(2)} = \begin{bmatrix} 0 \\ 1 \\ 1 \end{bmatrix}, \qquad \mathbf{x}^{(3)} = \begin{bmatrix} 2 \\ 0 \\ 2 \end{bmatrix}.$$

Then

$$\mathbf{q}^{(1)} = \frac{1}{\sqrt{5}} \begin{bmatrix} 2 \\ 1 \\ 0 \end{bmatrix}$$

$$\mathbf{y}^{(2)} = \mathbf{x}^{(2)} - \langle \mathbf{x}^{(2)}, \mathbf{q}^{(1)} \rangle \mathbf{q}^{(1)} = \mathbf{x}^{(2)} - \frac{1}{\sqrt{5}} \mathbf{q}^{(1)} = \frac{1}{5} \begin{bmatrix} -2 \\ 4 \\ 5 \end{bmatrix}$$

$$\mathbf{q}^{(2)} = \frac{\mathbf{y}^{(2)}}{\|\mathbf{y}^{(2)}\|} = \frac{1}{3\sqrt{5}} \begin{bmatrix} -2 \\ 4 \\ 5 \end{bmatrix}$$

$$\begin{aligned} \mathbf{y}^{(3)} &= \mathbf{x}^{(3)} - \langle \mathbf{x}^{(3)}, \mathbf{q}^{(1)} \rangle \mathbf{q}^{(1)} - \langle \mathbf{x}^{(3)}, \mathbf{q}^{(2)} \rangle \mathbf{q}^{(2)} \\ &= \mathbf{x}^{(3)} - \frac{4}{\sqrt{5}} \mathbf{q}^{(1)} - \frac{2}{\sqrt{5}} \mathbf{q}^{(2)} \\ &= \frac{1}{3} \begin{bmatrix} 2 \\ -4 \\ 4 \end{bmatrix}, \end{aligned}$$

$$\mathbf{q}^{(3)} = \frac{\mathbf{y}^{(3)}}{\|\mathbf{y}^{(3)}\|} = \frac{1}{6} \begin{bmatrix} 2 \\ -4 \\ 4 \end{bmatrix}.$$

■

It should be noted that each $\mathbf{q}^{(j)}$ is a linear combination of the $\mathbf{x}^{(i)}$ with $i \le j$ and also that each $\mathbf{x}^{(j)}$ is a linear combination of the $\mathbf{q}^{(i)}$ with $i \le j$. We summarize this process in the following theorem.

Theorem 5.4. Suppose that $\{\mathbf{x}^{(j)}\}_{j=1}^{k}$ is a given set of linearly independent vectors. Define the vectors $\{\mathbf{q}^{(j)}\}_{j=1}^{k}$ by

$$\mathbf{q}^{(1)} = \frac{\mathbf{x}^{(1)}}{\|\mathbf{x}^{(1)}\|}$$

and for $j = 2, \ldots, k$,

$$\mathbf{q}^{(j)} = \frac{\mathbf{x}^{(j)} - \sum_{i=1}^{j-1} \langle \mathbf{x}^{(j)}, \mathbf{q}^{(i)} \rangle \mathbf{q}^{(i)}}{\left\| \mathbf{x}^{(j)} - \sum_{i=1}^{j-1} \langle \mathbf{x}^{(j)}, \mathbf{q}^{(i)} \rangle \mathbf{q}^{(i)} \right\|}.$$

Then $\{\mathbf{q}^{(j)}\}_{j=1}^{k}$ is an orthonormal set and for each j with $1 \le j \le k$,

$$S[\mathbf{x}^{(1)}, \ldots, \mathbf{x}^{(j)}] = S[\mathbf{q}^{(1)}, \ldots, \mathbf{q}^{(j)}].$$

■

EXERCISES 5.2

1. Let $\mathbf{x} = [1 \quad -2 \quad 4 \quad 1 \quad 5]^{\mathrm{T}}$ and $\mathbf{y} = [-2 \quad 2 \quad 3 \quad 6 \quad 0]^{\mathrm{T}}$. Compute $\langle \mathbf{x}, \mathbf{y} \rangle$, $\|\mathbf{x}\|^2$, $\|\mathbf{y}\|^2$, and $\|\mathbf{x} + \mathbf{y}\|^2$.
2. Suppose that $\mathbf{x}$ and $\mathbf{y}$ are orthogonal n-vectors. Verify $\|\mathbf{x} - \mathbf{y}\|^2 = \|\mathbf{x}\|^2 + \|\mathbf{y}\|^2$. (*Hint:* $\|\mathbf{x} - \mathbf{y}\|^2 = \langle \mathbf{x} - \mathbf{y}, \mathbf{x} - \mathbf{y} \rangle$.)
3. Verify the *parallelogram law* for n-vectors:

$$\|\mathbf{x} + \mathbf{y}\|^2 + \|\mathbf{x} - \mathbf{y}\|^2 = 2\|\mathbf{x}\|^2 + 2\|\mathbf{y}\|^2.$$

   What is the geometric significance of this law? [*Hint:* Use part (c) Theorem 5.1.]
4. Verify that $4\langle \mathbf{x}, \mathbf{y} \rangle = \|\mathbf{x} + \mathbf{y}\|^2 - \|\mathbf{x} - \mathbf{y}\|^2$.
5. Show that if $\mathbf{x}$ and $\mathbf{y}$ are linearly dependent, then $|\langle \mathbf{x}, \mathbf{y} \rangle| = \|\mathbf{x}\|\|\mathbf{y}\|$.
6. **(a)** Suppose that $\mathbf{x}$ and $\mathbf{y}$ are n-vectors. Verify that for any number t

$$0 \le \|t\mathbf{x} + \mathbf{y}\|^2 = \|\mathbf{x}\|^2 t^2 + 2\langle \mathbf{x}, \mathbf{y} \rangle t + \|\mathbf{y}\|^2.$$

   (b) Suppose that for given real numbers A, B, C and all real numbers t,

$$0 \le At^2 + 2Bt + C.$$

   Show that $B^2 - AC \le 0$. (*Hint:* Use the quadratic formula.)

   ***(c)** Use parts (a) and (b) to establish the Cauchy–Schwarz inequality.
7. Let

$$\mathbf{A} = \begin{bmatrix} 1 & 2 & -1 \\ 4 & 6 & 1 \end{bmatrix}, \qquad \mathbf{x} = [5, \quad -2, \quad 1]^{\mathrm{T}}, \quad \text{and} \quad \mathbf{y} = [-3, \quad 1]^{\mathrm{T}}.$$

   Compute $\langle \mathbf{A}\mathbf{x}, \mathbf{y} \rangle$ and $\langle \mathbf{x}, \mathbf{A}^{\mathrm{T}}\mathbf{y} \rangle$.
8. Let

$$\mathbf{A} = \begin{bmatrix} 1 & -1 \\ 2 & 0 \\ 1 & 4 \end{bmatrix}, \qquad \mathbf{x} = \begin{bmatrix} 2 \\ -1 \end{bmatrix}, \qquad \mathbf{y} = \begin{bmatrix} -1 \\ 3 \\ 1 \end{bmatrix}.$$

   Verify that $\langle \mathbf{A}\mathbf{x}, \mathbf{y} \rangle = \langle \mathbf{x}, \mathbf{A}^{\mathrm{T}}\mathbf{y} \rangle$.

9. Let

$$\mathbf{A} = \begin{bmatrix} \frac{1}{\sqrt{2}} & 0 \\ 0 & 1 \\ \frac{1}{\sqrt{2}} & 0 \end{bmatrix}.$$

Show that $\mathbf{A}^T\mathbf{A} = \mathbf{I}_2$.

10. Suppose that $\mathbf{A}$ is an $n \times n$ matrix whose column vectors $\{\mathbf{A}_1, \mathbf{A}_2, \ldots, \mathbf{A}_n\}$ form an orthonormal set. Compute $\mathbf{A}^T\mathbf{A}$.

11. Suppose that $\mathbf{x} = [t \quad 1 \quad 6t \quad 3]^T$ and $\mathbf{y} = [t \quad -t \quad 1 \quad 2]^T$. Find all values of t for which $\mathbf{x} \perp \mathbf{y}$.

12. Let

$$\mathbf{A} = \begin{bmatrix} 0 & 1 \\ -1 & 0 \end{bmatrix}.$$

Verify that $\mathbf{x} \perp \mathbf{Ax}$ for every 2-vector $\mathbf{x}$.

13. Show that if $\mathbf{y}$ is an n-vector and $\langle \mathbf{x}, \mathbf{y} \rangle = 0$ for all n-vectors $\mathbf{x}$, then $\mathbf{y} = \boldsymbol{\theta}$. [*Hint:* Set $\mathbf{x} = \mathbf{y}$ and use part (a) of Theorem 5.2.]

***14.** Suppose that $\mathbf{A}$, $\mathbf{B}$ are matrices and $\mathbf{x}$, $\mathbf{y}$ are vectors of the appropriate sizes, then $\langle \mathbf{ABx}, \mathbf{y} \rangle = \langle \mathbf{Bx}, \mathbf{A}^T\mathbf{y} \rangle = \langle \mathbf{x}, \mathbf{B}^T\mathbf{A}^T\mathbf{y} \rangle$ and $\langle \mathbf{ABx}, \mathbf{y} \rangle = \langle \mathbf{x}, (\mathbf{AB})^T\mathbf{y} \rangle$. Use this to give an alternative proof of Theorem 1.3.

15. Suppose that $\mathbf{A}^T\mathbf{y} = \boldsymbol{\theta}$. Show that $\mathbf{y} \perp \mathbf{Ax}$ for any vector $\mathbf{x}$ of the appropriate size.

16. Orthogonalize the vectors

$$\begin{bmatrix} 3 \\ 4 \end{bmatrix}, \quad \begin{bmatrix} 4 \\ 3 \end{bmatrix}$$

by using the Gram–Schmidt process.

17. Use the Gram–Schmidt process to orthogonalize the vectors

$$\begin{bmatrix} 1 \\ 0 \\ 1 \\ 0 \end{bmatrix}, \quad \begin{bmatrix} 2 \\ 1 \\ 0 \\ 0 \end{bmatrix}, \quad \begin{bmatrix} 1 \\ 0 \\ 1 \\ 1 \end{bmatrix}.$$

18. Try to orthogonalize the vectors

$$\begin{bmatrix} 2 \\ 1 \\ 1 \end{bmatrix}, \quad \begin{bmatrix} 1 \\ 0 \\ 1 \end{bmatrix}, \quad \begin{bmatrix} -2 \\ 1 \\ 2 \end{bmatrix}, \quad \begin{bmatrix} 1 \\ 1 \\ 1 \end{bmatrix}.$$

What happens? Why?

19. Express $\mathbf{y}$ as a linear combination of the vectors in the orthonormal set S.

(a) $\mathbf{y} = \begin{bmatrix} 3 \\ 4 \end{bmatrix}, \quad S = \left\{ \frac{1}{\sqrt{2}} \begin{bmatrix} 1 \\ 1 \end{bmatrix}, \frac{1}{\sqrt{2}} \begin{bmatrix} -1 \\ 1 \end{bmatrix} \right\}$

(b) $\mathbf{y} = \begin{bmatrix} 1 \\ -1 \\ 1 \end{bmatrix}, \quad S = \left\{ \frac{1}{\sqrt{5}} \begin{bmatrix} 2 \\ 1 \\ 0 \end{bmatrix}, \frac{1}{3\sqrt{5}} \begin{bmatrix} -2 \\ 4 \\ 5 \end{bmatrix}, \frac{1}{6} \begin{bmatrix} 2 \\ -4 \\ 4 \end{bmatrix} \right\}$

(c) $\mathbf{y} = \begin{bmatrix} 5 \\ 3 \\ -1 \\ 3 \end{bmatrix}, \quad S = \left\{ \frac{1}{5} \begin{bmatrix} 1 \\ 2 \\ -2 \\ 4 \end{bmatrix}, \frac{1}{\sqrt{14}} \begin{bmatrix} 0 \\ -1 \\ 3 \\ 2 \end{bmatrix}, \frac{1}{\sqrt{19}} \begin{bmatrix} 4 \\ 1 \\ 1 \\ -1 \end{bmatrix} \right\}$

5.3 Orthogonal Matrices and QR Factorization

In an elementary calculus course the reader may have already encountered the notion of a rotation of the coordinate plane about the origin. The situation is illustrated for a counterclockwise rotation of θ radians in Figure 5.7. Under this rotation a vector $\mathbf{x} = [x_1, x_2]^T$ is transformed into a vector $\mathbf{x}' = [x_1', x_2']^T$. To see the relationship between these vectors it is perhaps easiest to notice that the standard unit vectors $[1, 0]^T$ and $[0, 1]^T$ are transformed as follows:

$$\begin{bmatrix} 1 \\ 0 \end{bmatrix} \longrightarrow \begin{bmatrix} \cos\theta \\ \sin\theta \end{bmatrix}, \qquad \begin{bmatrix} 0 \\ 1 \end{bmatrix} \longrightarrow \begin{bmatrix} -\sin\theta \\ \cos\theta \end{bmatrix}.$$

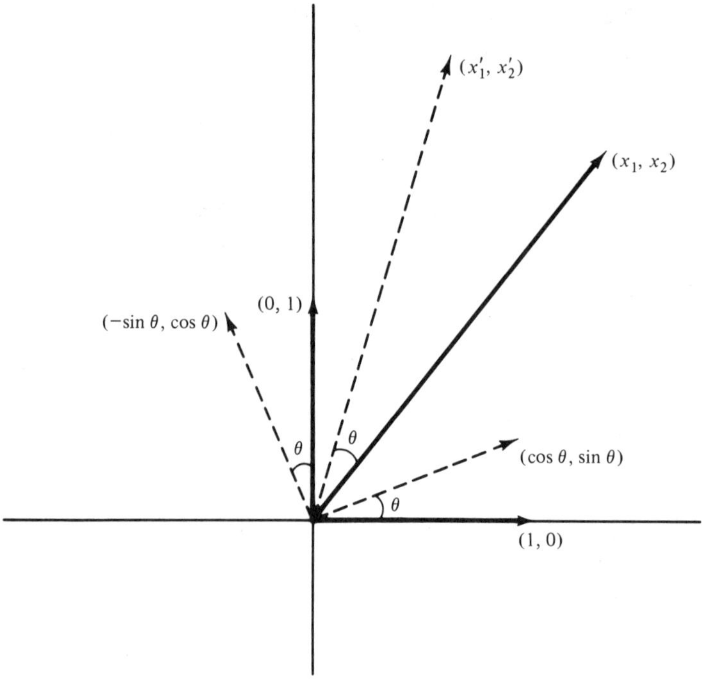

Figure 5.7

A glance at Figure 5.7 should then convince you that

$$\begin{bmatrix} x_1 \\ x_2 \end{bmatrix} = x_1 \begin{bmatrix} 1 \\ 0 \end{bmatrix} + x_2 \begin{bmatrix} 0 \\ 1 \end{bmatrix} \longrightarrow x_1 \begin{bmatrix} \cos\theta \\ \sin\theta \end{bmatrix} + x_2 \begin{bmatrix} -\sin\theta \\ \cos\theta \end{bmatrix} = \begin{bmatrix} x_1' \\ x_2' \end{bmatrix}$$

$$= \begin{bmatrix} \cos\theta & -\sin\theta \\ \sin\theta & \cos\theta \end{bmatrix} \begin{bmatrix} x_1 \\ x_2 \end{bmatrix}$$

or equivalently that $\mathbf{x}' = \mathbf{R}_\theta \mathbf{x}$, where $\mathbf{R}_\theta$ is the "rotation matrix" given by

$$\mathbf{R}_\theta = \begin{bmatrix} \cos\theta & -\sin\theta \\ \sin\theta & \cos\theta \end{bmatrix}.$$

Example 5.9

To rotate the vector

$$\mathbf{x} = \begin{bmatrix} 1 \\ -1 \end{bmatrix}$$

through $\pi/4$ radians, we compute $\mathbf{x}' = \mathbf{R}_{\pi/4}\mathbf{x}$. We find

$$\mathbf{x}' = \begin{bmatrix} \frac{\sqrt{2}}{2} & \frac{-\sqrt{2}}{2} \\ \frac{\sqrt{2}}{2} & \frac{\sqrt{2}}{2} \end{bmatrix} \begin{bmatrix} 1 \\ -1 \end{bmatrix} = \begin{bmatrix} \sqrt{2} \\ 0 \end{bmatrix}.$$

■

It should be clear that this transformation can be "undone" simply by rotating through an angle of $-\theta$, that is

$$\mathbf{R}_\theta^{-1} = \mathbf{R}_{-\theta}.$$

Indeed,

$$\mathbf{R}_{-\theta} = \begin{bmatrix} \cos\theta & \sin\theta \\ -\sin\theta & \cos\theta \end{bmatrix}$$

and it is easy to verify (using $\cos^2\theta + \sin^2\theta = 1$) that

$$\mathbf{R}_\theta \mathbf{R}_{-\theta} = \mathbf{R}_{-\theta}\mathbf{R}_\theta = \mathbf{I},$$

and hence $\mathbf{R}_{-\theta} = \mathbf{R}_\theta^{-1}$. The interesting thing is that $\mathbf{R}_\theta^{-1}$ is the transpose of $\mathbf{R}_\theta$: $\mathbf{R}_\theta^{-1} = \mathbf{R}_\theta^{\mathrm{T}}$.

In general we will call a square matrix $\mathbf{Q}$ **orthogonal** if

$$\mathbf{Q}^{\mathrm{T}}\mathbf{Q} = \mathbf{Q}\mathbf{Q}^{\mathrm{T}} = \mathbf{I},$$

That is, $\mathbf{Q}^{-1} = \mathbf{Q}^{\mathrm{T}}$. Denote by $\mathbf{Q}_i$ the ith column vector of an orthogonal matrix $\mathbf{Q}$, then note that

$$\langle \mathbf{Q}_i, \mathbf{Q}_j \rangle = \mathbf{Q}_i^{\mathrm{T}}\mathbf{Q}_j = \delta_{ij}.$$

That is, *orthogonal matrices are those square matrices whose column vectors are orthonormal.*

Example 5.10

The matrix

$$\mathbf{Q} = \begin{bmatrix} \frac{1}{\sqrt{2}} & 0 & \frac{1}{-\sqrt{2}} \\ 0 & 1 & 0 \\ \frac{1}{\sqrt{2}} & 0 & \frac{1}{\sqrt{2}} \end{bmatrix}$$

is orthogonal since its column vectors are orthonormal.
Indeed,

$$\mathbf{Q}^{\mathrm{T}}\mathbf{Q} = \begin{bmatrix} \frac{1}{\sqrt{2}} & 0 & \frac{1}{\sqrt{2}} \\ 0 & 1 & 0 \\ \frac{1}{-\sqrt{2}} & 0 & \frac{1}{\sqrt{2}} \end{bmatrix} \begin{bmatrix} \frac{1}{\sqrt{2}} & 0 & \frac{1}{-\sqrt{2}} \\ 0 & 1 & 0 \\ \frac{1}{\sqrt{2}} & 0 & \frac{1}{\sqrt{2}} \end{bmatrix} = \begin{bmatrix} 1 & 0 & 0 \\ 0 & 1 & 0 \\ 0 & 0 & 1 \end{bmatrix}$$

and similarly, $\mathbf{Q}\mathbf{Q}^{\mathrm{T}} = \mathbf{I}$.

■

Orthogonal matrices share many of the properties of the rotation matrices discussed above. For example, rotations *preserve the length of vectors and angles between vectors* (and hence inner products); the same is true of orthogonal matrices.

Theorem 5.5. If $\mathbf{Q}$ is orthogonal, then $\langle \mathbf{Qx}, \mathbf{Qy} \rangle = \langle \mathbf{x}, \mathbf{y} \rangle$ and $\|\mathbf{Qx}\| = \|\mathbf{x}\|$.

Proof. Simply note that, by Theorem 5.3,

$$\langle \mathbf{Qx}, \mathbf{Qy} \rangle = \langle \mathbf{x}, \mathbf{Q}^{\mathrm{T}}\mathbf{Qy} \rangle = \langle \mathbf{x}, \mathbf{y} \rangle.$$

Replacing $\mathbf{y}$ by $\mathbf{x}$ then gives

$$\|\mathbf{Qx}\|^2 = \langle \mathbf{x}, \mathbf{x} \rangle = \|\mathbf{x}\|^2.$$

■

It turns out that the Gram–Schmidt process has an interesting and useful interpretation as a matrix factorization method. Let us consider the Gram–Schmidt process as applied to the linearly independent column vectors of the matrix

$$\mathbf{A} = \begin{bmatrix} 1 & 2 \\ 1 & 1 \end{bmatrix}.$$

To begin with, we find

$$\mathbf{q}^{(1)} = \frac{1}{\sqrt{2}}\mathbf{A}_1 \quad \text{or, equivalently,} \quad \mathbf{A}_1 = \sqrt{2}\,\mathbf{q}^{(1)}.$$

Also,

$$\mathbf{y}^{(2)} = \mathbf{A}_2 - \langle \mathbf{A}_2, \mathbf{q}^{(1)} \rangle \mathbf{q}^{(1)}$$
$$= \mathbf{A}_2 - \frac{3}{\sqrt{2}}\mathbf{q}^{(1)} = \frac{1}{2}\begin{bmatrix} 1 \\ -1 \end{bmatrix}, \qquad \|\mathbf{y}^{(2)}\| = \frac{1}{\sqrt{2}}.$$

Therefore,

$$\mathbf{q}^{(2)} = \sqrt{2}\,\mathbf{A}_2 - 3\mathbf{q}^{(1)} \quad \text{or, equivalently, } \mathbf{A}_2 = \frac{3}{\sqrt{2}}\mathbf{q}^{(1)} + \frac{1}{\sqrt{2}}\mathbf{q}^{(2)}.$$

Combining this with the equation for $\mathbf{A}_1$, we see that

$$[\mathbf{A}_1 \quad \mathbf{A}_2] = [\mathbf{q}^{(1)} \quad \mathbf{q}^{(2)}]\begin{bmatrix} \sqrt{2} & \frac{3}{\sqrt{2}} \\ 0 & \frac{1}{\sqrt{2}} \end{bmatrix},$$

that is,

$$\mathbf{A} = \mathbf{QR},$$

where

$$\mathbf{Q} = [\mathbf{q}^{(1)} \quad \mathbf{q}^{(2)}] = \begin{bmatrix} \frac{1}{\sqrt{2}} & \frac{1}{\sqrt{2}} \\ \frac{1}{\sqrt{2}} & -\frac{1}{\sqrt{2}} \end{bmatrix}$$

and

$$\mathbf{R} = \begin{bmatrix} \sqrt{2} & \frac{3}{\sqrt{2}} \\ 0 & \frac{1}{\sqrt{2}} \end{bmatrix}.$$

The important thing to notice is that the matrix $\mathbf{Q}$ formed in this way always has orthonormal columns and the matrix $\mathbf{R}$ is invertible and upper triangular since $\mathbf{A}_j$ depends only on $\mathbf{q}^{(i)}$ for $i \leq j$. It should be clear that this interpretation of the Gram–Schmidt process is valid for any matrix with linearly independent columns.

Theorem 5.6. If $\mathbf{A}$ is a matrix with linearly independent columns, then $\mathbf{A}$ can be factored as $\mathbf{A} = \mathbf{QR}$, where $\mathbf{Q}$ has orthonormal columns and $\mathbf{R}$ is an invertible upper triangular matrix.

■

It is important to notice that the matrix $\mathbf{A}$ in this theorem need not be square. We illustrate in the following examples.

Example 5.11

We will find the QR factorization of

$$\mathbf{A} = \begin{bmatrix} 1 & 0 \\ 1 & 1 \\ 0 & 1 \end{bmatrix}.$$

Applying the Gram–Schmidt procedure, we find that

$$\mathbf{q}^{(1)} = \frac{1}{\sqrt{2}}\mathbf{A}_1 = \frac{1}{\sqrt{2}}\begin{bmatrix} 1 \\ 1 \\ 0 \end{bmatrix}.$$

$$\mathbf{y}^{(2)} = \mathbf{A}_2 - \langle \mathbf{A}_2, \mathbf{q}^{(1)} \rangle \mathbf{q}^{(1)} = \begin{bmatrix} 0 \\ 1 \\ 1 \end{bmatrix} - \frac{1}{2}\begin{bmatrix} 1 \\ 1 \\ 0 \end{bmatrix} = \begin{bmatrix} -\frac{1}{2} \\ \frac{1}{2} \\ 1 \end{bmatrix}$$

$$\mathbf{q}^{(2)} = \frac{\mathbf{y}^{(2)}}{\|\mathbf{y}^{(2)}\|} = \sqrt{\frac{2}{3}}\mathbf{A}_2 - \frac{1}{\sqrt{3}}\mathbf{q}^{(1)} = \begin{bmatrix} -\dfrac{1}{\sqrt{6}} \\ \dfrac{1}{\sqrt{6}} \\ \dfrac{\sqrt{2}}{\sqrt{3}} \end{bmatrix}.$$

Therefore,

$$\mathbf{A}_1 = \sqrt{2}\,\mathbf{q}^{(1)} \qquad \mathbf{A}_2 = \frac{1}{\sqrt{2}}\mathbf{q}^{(1)} + \sqrt{\frac{3}{2}}\,\mathbf{q}^{(2)},$$

and hence

$$\mathbf{A} = \begin{bmatrix} \dfrac{1}{\sqrt{2}} & \dfrac{-1}{\sqrt{6}} \\ \dfrac{1}{\sqrt{2}} & \dfrac{1}{\sqrt{6}} \\ 0 & \sqrt{\dfrac{2}{3}} \end{bmatrix} \begin{bmatrix} \sqrt{2} & \dfrac{1}{\sqrt{2}} \\ 0 & \sqrt{\dfrac{3}{2}} \end{bmatrix} = \mathbf{QR}.$$

■

Example 5.12

We will find the QR factorization of

$$\mathbf{A} = \begin{bmatrix} 1 & 1 & 5 \\ 2 & 1 & 2 \\ -2 & 1 & 2 \\ 4 & 6 & 5 \end{bmatrix}.$$

By the Gram–Schmidt process we find

$$\mathbf{q}^{(1)} = \frac{\mathbf{A}_1}{\|\mathbf{A}_1\|} = \frac{1}{5}\mathbf{A}_1 = \frac{1}{5}\begin{bmatrix} 1 \\ 2 \\ -2 \\ 4 \end{bmatrix}$$

$$\mathbf{y}^{(2)} = \mathbf{A}_2 - \langle \mathbf{A}_2, \mathbf{q}^{(1)} \rangle \mathbf{q}^{(1)} = \mathbf{A}_2 - \langle \mathbf{A}_2, \mathbf{A}_1 \rangle \frac{\mathbf{A}_1}{25} = \mathbf{A}_2 - \mathbf{A}_1$$

$$\mathbf{q}^{(2)} = \frac{\mathbf{y}^{(2)}}{\|\mathbf{y}^{(2)}\|} = \frac{\mathbf{A}_2 - \mathbf{A}_1}{\sqrt{14}} = \frac{1}{\sqrt{14}}\begin{bmatrix} 0 \\ -1 \\ 3 \\ 2 \end{bmatrix}$$

$$\begin{aligned}\mathbf{y}^{(3)} &= \mathbf{A}_3 - \langle \mathbf{A}_3, \mathbf{q}^{(1)} \rangle \mathbf{q}^{(1)} - \langle \mathbf{A}_3, \mathbf{q}^{(2)} \rangle \mathbf{q}^{(2)} = \mathbf{A}_3 - 5\mathbf{q}^{(1)} - \sqrt{14}\,\mathbf{q}^{(2)} \\ &= \mathbf{A}_3 - \mathbf{A}_2\end{aligned}$$

and

$$\mathbf{q}^{(3)} = \frac{\mathbf{A}_3 - \mathbf{A}_2}{\|\mathbf{A}_3 - \mathbf{A}_2\|} = \frac{1}{\sqrt{19}}\begin{bmatrix} 4 \\ 1 \\ 1 \\ -1 \end{bmatrix}.$$

Solving for the $\mathbf{A}_i$'s in terms of the $\mathbf{q}^{(i)}$'s gives

$$\begin{aligned}\mathbf{A}_1 &= 5\mathbf{q}^{(1)} \\ \mathbf{A}_2 &= \sqrt{14}\,\mathbf{q}^{(2)} + \mathbf{A}_1 = 5\mathbf{q}^{(1)} + \sqrt{14}\,\mathbf{q}^{(2)} \\ \mathbf{A}_3 &= \sqrt{19}\,\mathbf{q}^{(3)} + \mathbf{A}_2 = 5\mathbf{q}^{(1)} + \sqrt{14}\,\mathbf{q}^{(2)} + \sqrt{19}\,\mathbf{q}^{(3)},\end{aligned}$$

and hence

$$\mathbf{A} = \begin{bmatrix} 1 & 1 & 5 \\ 2 & 1 & 2 \\ -2 & 1 & 2 \\ 4 & 6 & 5 \end{bmatrix} = \begin{bmatrix} \frac{1}{5} & 0 & \frac{4}{\sqrt{19}} \\ \frac{2}{5} & \frac{-1}{\sqrt{14}} & \frac{1}{\sqrt{19}} \\ -\frac{2}{5} & \frac{3}{\sqrt{14}} & \frac{1}{\sqrt{19}} \\ \frac{4}{5} & \frac{2}{\sqrt{14}} & -\frac{1}{\sqrt{19}} \end{bmatrix} \begin{bmatrix} 5 & 5 & 5 \\ 0 & \sqrt{14} & \sqrt{14} \\ 0 & 0 & \sqrt{19} \end{bmatrix} = \mathbf{QR}.$$

■

EXERCISES 5.3

1. Compute $\mathbf{R}_{\pi/4}\mathbf{y}$ and $\mathbf{R}_{-\pi/4}\mathbf{y}$, where $\mathbf{y} = [1 \quad 2]^T$. Draw a geometrical illustration of these vectors.
2. Show that the rotation matrix satisfies $\mathbf{R}_\theta \mathbf{R}_\phi = \mathbf{R}_{\theta+\phi}$.
3. Verify directly that $\langle \mathbf{R}_\theta \mathbf{x}, \mathbf{R}_\theta \mathbf{y} \rangle = \langle \mathbf{x}, \mathbf{y} \rangle$.
4. Compute $\det \mathbf{R}_\theta$.
5. Show that if $\mathbf{Q}$ is orthogonal, then $\det \mathbf{Q} = \pm 1$. (*Hint:* $1 = \det \mathbf{Q}^T\mathbf{Q} = \det \mathbf{Q}^T \det \mathbf{Q}$.)
6. Let

$$\mathbf{Q} = \frac{1}{2}\begin{bmatrix} 1 & -1 & -1 & -1 \\ -1 & 1 & -1 & -1 \\ -1 & -1 & 1 & -1 \\ -1 & -1 & -1 & 1 \end{bmatrix}.$$

   (a) Verify that $\mathbf{Q}$ is orthogonal.
   (b) Compute $\|\mathbf{x}\|$, $\|\mathbf{Qx}\|$, $\|\mathbf{y}\|$, $\|\mathbf{Qy}\|$, $\langle \mathbf{x}, \mathbf{y} \rangle$, and $\langle \mathbf{Qx}, \mathbf{Qy} \rangle$, where $\mathbf{x} = [2 \quad 1 \quad -1 \quad 0]^T$ and $\mathbf{y} = [1 \quad -3 \quad 1 \quad -1]^T$.
7. Suppose that $\mathbf{Q}$ is an orthogonal matrix and λ is a number satisfying $\mathbf{Qx} = \lambda \mathbf{x}$ for some $\mathbf{x} \neq \boldsymbol{\theta}$. Show that $|\lambda| = 1$.
8. (a)* Show that if $\mathbf{u}$ is a unit vector (i.e., $\|\mathbf{u}\| = 1$), then $\mathbf{Q} = \mathbf{I} - 2\mathbf{u}\mathbf{u}^T$ is an orthogonal matrix. (This matrix is called a *Householder transformation*.)
   (b) Compute $\mathbf{Q}$ if $\mathbf{u} = (1/\sqrt{2})[1 \quad 1]^T$.
9. Suppose that $\mathbf{Q}$ and $\mathbf{S}$ are $n \times n$ orthogonal matrices. Is the product matrix $\mathbf{QS}$ orthogonal? [Remember that $(\mathbf{QS})^T = \mathbf{S}^T\mathbf{Q}^T$.]
10. Find a *QR* factorization of the matrix

$$\mathbf{A} = \begin{bmatrix} 1 & 1 & \dfrac{2}{\sqrt{2}} \\ -1 & 0 & 0 \\ 0 & 0 & 1 \end{bmatrix}.$$

11. Find a *QR* factorization of the matrix.

$$\mathbf{A} = \begin{bmatrix} 1 & 0 \\ 1 & 1 \\ 1 & 2 \end{bmatrix}.$$

12. Find a *QR* factorization of the matrix

$$\mathbf{A} = \begin{bmatrix} 1 & 0 & 0 \\ 0 & 1 & 1 \\ 0 & 0 & 1 \\ 1 & 1 & 0 \end{bmatrix}.$$

13. Suppose that the matrix $\mathbf{A}$ has a *QR* factorization $\mathbf{A} = \mathbf{QR}$ and the system $\mathbf{Ax} = \mathbf{b}$ is consistent. Show that the solution of this system is given by $\mathbf{x} = \mathbf{R}^{-1}\mathbf{Q}^T\mathbf{b}$. In particular, note that the solution is unique.

14. Use the technique indicated in Exercise 13 to solve $\mathbf{Ax} = \mathbf{b}$, where $\mathbf{b} = [1 \;\; 5 \;\; 3 \;\; 3]^{\mathrm{T}}$ and $\mathbf{A}$ is given in Exercise 12. Note that $\mathbf{R}^{-1}$ need not be computed, as $\mathbf{Rx} = \mathbf{Q}^{\mathrm{T}}\mathbf{b}$ can be solved by back substitution.

5.4 Adjoint and Unitary Matrices

Until now we have dealt exclusively with real scalars. However, almost all of what we have developed remains true if we allow complex scalars, that is, numbers of the form $a + bi$, where a and b are real numbers and i is an "imaginary" number satisfying $i^2 = -1$. We assume that the reader has encountered complex numbers in the past and is aware that the usual arithmetic laws (commutativity, associativity, distributivity) hold for complex numbers.

If $z = a + bi$ is a complex number, its **conjugate** is the number $\overline{z} = a - bi$. Notice that $\overline{\overline{z}} = z$ and a complex number z is real (i.e., $b = 0$) if and only if $z = \overline{z}$. The **modulus** of a complex number $z = a + bi$ is the real number $|z|$ defined by

$$|z| = \sqrt{z\overline{z}} = \sqrt{a^2 + b^2}.$$

The modulus plays a role for complex numbers which is exactly analogous to that played by the absolute value for real numbers.

Example 5.13

We prove two simple properties for conjugation, namely $\overline{z + w} = \overline{z} + \overline{w}$ and $\overline{zw} = \overline{z}\,\overline{w}$.

If $z = a + bi$ and $w = c + di$, then $\overline{z} = a - bi$, $\overline{w} = c - di$. Therefore,

$$\begin{aligned}\overline{z + w} &= \overline{(a + c) + (b + d)i} = (a + c) - (b + d)i \\ &= (a - bi) + (c - di) = \overline{z} + \overline{w}.\end{aligned}$$

Also,

$$\begin{aligned}\overline{zw} &= \overline{(a + bi)(c + di)} = \overline{(ac - bd) + (ad + bc)i} \\ &= (ac - bd) - (ad + bc)i.\end{aligned}$$

while

$$\begin{aligned}\overline{z}\,\overline{w} &= (a - bi)(c - di) = ac + bdi^2 - adi - bci \\ &= (ac - bd) - (ad + bc)i = \overline{zw}.\end{aligned}$$

■

For n-vectors with complex entries, we define the **complex inner product** by

$$\langle \mathbf{x}, \mathbf{y} \rangle = \sum_{j=1}^{n} x_j \overline{y}_j. \tag{10}$$

Note that since real numbers are equal to their conjugates, this agrees with our previous definition of inner product for vectors which have real components. This inner product also has a norm associated with it given by

$$\|\mathbf{x}\|^2 = \sum_{j=1}^{n} x_j\bar{x}_j = \sum_{j=1}^{n} |x_j|^2.$$

The complex inner product has all of the properties of the real inner product with one important exception:

$$\langle \mathbf{x}, \mathbf{y} \rangle = \overline{\langle \mathbf{y}, \mathbf{x} \rangle}. \tag{11}$$

To see this, note that

$$\begin{aligned}\overline{\langle \mathbf{y}, \mathbf{x} \rangle} &= \overline{\sum_{j=1}^{n} y_j\bar{x}_j} = \sum_{i=1}^{n} \bar{y}_j\bar{\bar{x}}_j \\ &= \sum_{j=1}^{n} x_j\bar{y}_j = \langle \mathbf{x}, \mathbf{y} \rangle.\end{aligned}$$

An important consequence of this fact is

$$\langle \mathbf{x}, \lambda\mathbf{y} \rangle = \bar{\lambda}\langle \mathbf{x}, \mathbf{y} \rangle.$$

Indeed, by (11),

$$\langle \mathbf{x}, \lambda\mathbf{y} \rangle = \overline{\langle \lambda\mathbf{y}, \mathbf{x} \rangle} = \overline{\lambda\langle \mathbf{y}, \mathbf{x} \rangle} = \bar{\lambda}\overline{\langle \mathbf{y}, \mathbf{x} \rangle} = \bar{\lambda}\langle \mathbf{x}, \mathbf{y} \rangle. \tag{12}$$

Example 5.14

Suppose that

$$\mathbf{x} = \begin{bmatrix} 1+i \\ i \end{bmatrix}, \qquad \mathbf{y} = \begin{bmatrix} 1 \\ -i \end{bmatrix}.$$

Then

$$\begin{aligned}\|\mathbf{x}\|^2 &= |1 + i|^2 + |i|^2 = 1 + 1 + 1 = 3 \\ \langle \mathbf{x}, \mathbf{y} \rangle &= (1 + i)\cdot 1 + i(i) = i \\ \langle \mathbf{y}, \mathbf{x} \rangle &= 1(1 - i) + (-i)(-i) = -i = \overline{\langle \mathbf{x}, \mathbf{y} \rangle}.\end{aligned}$$

Also, if $\lambda = 2 - i$, then

$$\lambda\mathbf{y} = \begin{bmatrix} 2-i \\ -1-2i \end{bmatrix}, \qquad \langle \mathbf{x}, \lambda\mathbf{y} \rangle = (1 + i)(2 + i) + i(-1 + 2i) = -1 + 2i$$

$$\bar{\lambda}\langle \mathbf{x}, \mathbf{y} \rangle = (2 + i)(i) = -1 + 2i.$$

■

If **A** is an $m \times n$ matrix with complex entries, then we define its **adjoint** to be the $n \times m$ matrix $\mathbf{A}^*$ whose entries are given by

$$a_{ij}^* = \overline{a_{ji}}.$$

To put it another way, the entries of $\mathbf{A}^*$ are the conjugates of the entries of $\mathbf{A}^T$. In particular, if $\mathbf{A}$ has only real entries, then $\mathbf{A}^* = \mathbf{A}^T$.

Example 5.15

If

$$\mathbf{A} = \begin{bmatrix} 1-i & i & 2 \\ 3i & 4 & 1-2i \end{bmatrix}, \quad \text{then} \quad \mathbf{A}^* = \begin{bmatrix} 1+i & -3i \\ -i & 4 \\ 2 & 1+2i \end{bmatrix}.$$

■

The adjoint matrix has properties analogous to that of the transpose. For example,

$$\mathbf{A}^{**} = \mathbf{A}, \qquad (\mathbf{A} + \mathbf{B})^* = \mathbf{A}^* + \mathbf{B}^*, \qquad (\mathbf{AB})^* = \mathbf{B}^*\mathbf{A}^*.$$

But note that

$$(\lambda \mathbf{A})^* = \bar{\lambda}\mathbf{A}^*.$$

Example 5.16

Let

$$\mathbf{A} = \begin{bmatrix} 1-i & i & 2 \\ 3i & 4 & 1-2i \end{bmatrix} \quad \text{and} \quad \lambda = 1 + i.$$

Then

$$\lambda\mathbf{A} = \begin{bmatrix} 2 & -1+i & 2+2i \\ -3+3i & 4+4i & 3-i \end{bmatrix}$$

and

$$(\lambda\mathbf{A})^* = \begin{bmatrix} 2 & -3-3i \\ -1-i & 4-4i \\ 2-2i & 3+i \end{bmatrix}.$$

Also,

$$\bar{\lambda}\mathbf{A}^* = (1-i)\begin{bmatrix} 1+i & -3i \\ -i & 4 \\ 2 & 1+2i \end{bmatrix} = \begin{bmatrix} 2 & -3-3i \\ -1-i & 4-4i \\ 2-2i & 3+i \end{bmatrix} = (\lambda\mathbf{A})^*.$$

■

Most important, the adjoint behaves with respect to the complex inner product in exactly the same way that the transpose behaves with respect to the real inner product (see Theorem 5.3).

Theorem 5.7. $\langle \mathbf{Ax}, \mathbf{y} \rangle = \langle \mathbf{x}, \mathbf{A}^*\mathbf{y} \rangle$.

■

Example 5.17

Suppose that $\mathbf{A}$ is the matrix of Example 5.16 with

$$\mathbf{x} = \begin{bmatrix} i \\ 1+2i \\ 1 \end{bmatrix}, \quad \text{and} \quad \mathbf{y} = \begin{bmatrix} -i \\ 2-i \end{bmatrix}.$$

Then

$$\mathbf{Ax} = \begin{bmatrix} 1+2i \\ 2+6i \end{bmatrix}, \qquad \mathbf{A^*y} = \begin{bmatrix} -2-7i \\ 7-4i \\ 4+i \end{bmatrix},$$

and hence

$$\langle \mathbf{Ax}, \mathbf{y} \rangle = (1+2i)(+i) + (2+6i)(2+i) = -4+15i$$
$$\langle \mathbf{x}, \mathbf{A^*y} \rangle = i(-2+7i) + (1+2i)(7+4i) + 1 \cdot (4-i) = -4+15i.$$

■

A matrix $\mathbf{A}$ is said to be **self-adjoint** if $\mathbf{A} = \mathbf{A}^*$. It follows that self-adjoint matrices are square and real matrices are self-adjoint if and only if they are symmetric.

We say that a square matrix is **unitary** if its columns are orthonormal with respect to the complex inner product. If $\mathbf{Q}$ is an $n \times n$ unitary matrix, then

$$\delta_{ij} = \langle \mathbf{Q}_i, \mathbf{Q}_j \rangle = \sum_{k=1}^{n} q_{ki}\overline{q}_{kj} = \sum_{k=1}^{n} \overline{q}^{\mathrm{T}}_{jk} q_{ki} = (\mathbf{Q^*Q})_{ji};$$

therefore, $\mathbf{Q^*Q} = \mathbf{I}$ and similarly, $\mathbf{QQ^*} = \mathbf{I}$. To put it another way, $\mathbf{Q}$ is unitary if and only if $\mathbf{Q}^{-1} = \mathbf{Q}^*$. Unitary matrices are therefore just the complex analogues of orthogonal matrices.

Example 5.18

The matrix

$$\mathbf{A} = \begin{bmatrix} -1 & 1+i \\ 1-i & 2 \end{bmatrix}$$

is self-adjoint since $\mathbf{A} = \mathbf{A}^*$.

$$\mathbf{Q} = \frac{1}{\sqrt{3}} \begin{bmatrix} 1-i & \frac{1}{\sqrt{2}}(1-i) \\ -i & \sqrt{2}i \end{bmatrix}$$

is unitary because

$$\mathbf{Q}^*\mathbf{Q} = \frac{1}{3}\begin{bmatrix} 1+i & i \\ \frac{1}{\sqrt{2}}(1+i) & -\sqrt{2}i \end{bmatrix}\begin{bmatrix} 1-i & \frac{1}{\sqrt{2}}(1-i) \\ -1 & \sqrt{2}i \end{bmatrix} = \frac{1}{3}\begin{bmatrix} 3 & 0 \\ 0 & 3 \end{bmatrix} = \mathbf{I},$$

and similarly, $\mathbf{QQ}^* = \mathbf{I}$.

■

Finally, we note that a unitary matrix has the same properties with respect to the complex inner product and associated norm that an orthogonal matrix has with respect to the real inner product (see Theorem 5.5).

Theorem 5.8. If $\mathbf{Q}$ is a unitary matrix, then $\langle \mathbf{Qx}, \mathbf{Qy} \rangle = \langle \mathbf{x}, \mathbf{y} \rangle$ and $\|\mathbf{Qx}\| = \|\mathbf{x}\|$.

Proof. This follows from Theorem 5.7 since $\langle \mathbf{Qx}, \mathbf{Qy} \rangle = \langle \mathbf{x}, \mathbf{Q}^*\mathbf{Qy} \rangle = \langle \mathbf{x}, \mathbf{y} \rangle$. Setting $\mathbf{y} = \mathbf{x}$ then gives $\|\mathbf{Qx}\|^2 = \langle \mathbf{x}, \mathbf{x} \rangle = \|\mathbf{x}\|^2$.

■

EXERCISES 5.4

1. Show that $|\bar{z}| = |z|$ for any complex number z.

2. Let $\mathbf{x} = [2-i \quad 1+i \quad -4-2i \quad -i]^{\mathrm{T}}$ and $\mathbf{y} = [i \quad 3-2i \quad 1+3i \quad 1-i]^{\mathrm{T}}$. Compute $\langle \mathbf{x}, \mathbf{y} \rangle$ and $\langle \mathbf{y}, \mathbf{x} \rangle$.

3. Let

$$\mathbf{A} = \begin{bmatrix} 1-2i & i \\ 7+2i & 2-3i \end{bmatrix} \quad \text{and} \quad \mathbf{B} = \begin{bmatrix} i & 4 \\ 2-i & 1+4i \end{bmatrix}.$$

Verify that $(\mathbf{A} + \mathbf{B})^* = \mathbf{A}^* + \mathbf{B}^*$.

4. **(a)** Let $\mathbf{A}$ be the matrix in Exercise 3. Compute $(\mathbf{A}^*)^*$.
(b) Suppose that $\mathbf{A}$ is any complex matrix. How is $(\mathbf{A}^*)^*$ related to $\mathbf{A}$?

5. Suppose that

$$\mathbf{A} = \begin{bmatrix} i & 2 \\ -i & 1 \\ 1 & 1+i \end{bmatrix}, \quad \mathbf{x} = \begin{bmatrix} -i \\ 1 \end{bmatrix}, \quad \mathbf{y} = \begin{bmatrix} 1-i \\ i \\ 0 \end{bmatrix}.$$

Verify that $\langle \mathbf{Ax}, \mathbf{y} \rangle = \langle \mathbf{x}, \mathbf{A}^*\mathbf{y} \rangle$.

6. Show that the diagonal entries of a self-adjoint matrix are real.

7. **(a)** Let

$$\mathbf{A} = \begin{bmatrix} 2 & 1-i \\ 1+i & 3 \end{bmatrix} \quad \text{and} \quad \mathbf{B} = \begin{bmatrix} -1 & 2+i \\ 2-i & 1 \end{bmatrix}.$$

Show that $\mathbf{A}$ and $\mathbf{B}$ are self-adjoint.
(b) Compute $\mathbf{AB}$ and $(\mathbf{AB})^*$.
(c) Is the product of self-adjoint matrices necessarily self-adjoint?

8. Suppose that

$$\mathbf{A} = \begin{bmatrix} i \\ 2+2i \end{bmatrix} \quad \text{and} \quad \mathbf{B} = [1-i \quad 2i].$$

Verify that $(\mathbf{AB})^* = \mathbf{B}^*\mathbf{A}^*$.

9. Is the product of two $n \times n$ unitary matrices also unitary? (See Exercise 9 of Section 5.3.)

10. Verify that the complex inner product satisfies

$$4\langle \mathbf{x}, \mathbf{y}\rangle = \|\mathbf{x}+\mathbf{y}\|^2 - \|\mathbf{x}-\mathbf{y}\|^2 + i\,\|\mathbf{x}+i\mathbf{y}\|^2 - i\,\|\mathbf{x}-i\mathbf{y}\|^2.$$

11. Verify that for any square matrix $\mathbf{B}$,

$$\mathbf{B} = \tfrac{1}{2}(\mathbf{B}+\mathbf{B}^*) + \tfrac{1}{2}(\mathbf{B}-\mathbf{B}^*).$$

12. Show that any square matrix $\mathbf{B}$ can be written $\mathbf{B} = \mathbf{C} + \mathbf{D}$, where $\mathbf{C} = \mathbf{C}^*$ and $\mathbf{D} = -\mathbf{D}^*$. (*Hint*: See Exercise 11.)

13. Show directly [i.e., without recourse to (11)] that $\langle \mathbf{x}, \lambda\mathbf{y}\rangle = \bar{\lambda}\langle \mathbf{x}, \mathbf{y}\rangle$.

14. The matrices

$$\mathbf{D}(1) = \begin{bmatrix} 0 & 0 & 0 & 1 \\ 0 & 0 & 1 & 0 \\ 0 & 1 & 0 & 0 \\ 1 & 0 & 0 & 0 \end{bmatrix}, \qquad \mathbf{D}(2) = \begin{bmatrix} 0 & 0 & 0 & i \\ 0 & 0 & -i & 0 \\ 0 & i & 0 & 0 \\ -i & 0 & 0 & 0 \end{bmatrix}$$

$$\mathbf{D}(3) = \begin{bmatrix} 0 & 0 & 1 & 0 \\ 0 & 0 & 0 & -1 \\ 1 & 0 & 0 & 0 \\ 0 & -1 & 0 & 0 \end{bmatrix}, \qquad \mathbf{D}(4) = \begin{bmatrix} 1 & 0 & 0 & 0 \\ 0 & 1 & 0 & 0 \\ 0 & 0 & -1 & 0 \\ 0 & 0 & 0 & -1 \end{bmatrix},$$

occur in quantum mechanics and are called the Dirac spin matrices.

(a) Verify that $\mathbf{D}(k)$ is self-adjoint, $k = 1, 2, 3, 4$.

(b) Show that

$$\mathbf{D}(m)\mathbf{D}(n) + \mathbf{D}(n)\mathbf{D}(m) = 2\delta_{mn}\mathbf{I}, \qquad m, n = 1, 2, 3, 4.$$

15. Verify that the matrix

$$\mathbf{Q} = \begin{bmatrix} \dfrac{-i}{\sqrt{2}} & 0 & \dfrac{i}{\sqrt{2}} \\ 0 & i & 0 \\ \dfrac{i}{\sqrt{2}} & 0 & \dfrac{i}{\sqrt{2}} \end{bmatrix}$$

is a unitary matrix.

16. Suppose that $\mathbf{A}$ is a real orthogonal matrix. Verify that

$$\mathbf{Q} = \frac{1+i}{\sqrt{2}}\mathbf{A}$$

is a unitary matrix.

17. Suppose that

$$\mathbf{A} = \begin{bmatrix} 0 & 0 & a & ia \\ 0 & 0 & ia & a \\ a & ia & 0 & 0 \\ ia & a & 0 & 0 \end{bmatrix}$$

and **A** is unitary. What are the possible values of a?

18. Prove Theorem 5.7. (*Hint:* Follow the proof of Theorem 5.3.)

5.5. Alternative Norms and Digital Codes

One could take parts (a–c) of Theorem 5.2 as characteristic of "length" and define a *norm* or "generalized length" to be a function $\|\cdot\|$ that satisfies these properties. The norm which we defined previously is then called the *Euclidean norm* because of its natural association with Euclidean geometry. There are, however, many other possibilities. For example, the function $\|\cdot\|_1$ defined on n-vectors $\mathbf{x}$ by the rule

$$\|\mathbf{x}\|_1 = \sum_{i=1}^{n} |x_i|$$

satisfies parts (a)–(c) of Theorem 5.2 and hence may be considered as an alternative notion of length for n-vectors. This norm, called the l_1 *norm*, has all the usual properties of length, except for the important fact that it has no inner product associated with it.

Example 5.19.

Suppose that

$$\mathbf{x} = \begin{bmatrix} 1 \\ -1 \end{bmatrix}, \qquad \mathbf{y} = \begin{bmatrix} 2 \\ 1 \end{bmatrix}.$$

Then $\|\mathbf{x} + \mathbf{y}\|_1^2 = (3 + 0)^2 = 9$, $\|\mathbf{x} - \mathbf{y}\|_1^2 = (1 + 2)^2 = 9$, $\|\mathbf{x}\|_1^2 = (1 + 1)^2 = 4$, $\|\mathbf{y}\|^2 = (2 + 1)^2 = 9$. Therefore, $\|\mathbf{x} + \mathbf{y}\|_1^2 + \|\mathbf{x} - \mathbf{y}\|_1^2 \neq 2\|\mathbf{x}\|_1^2 + 2\|\mathbf{y}\|_1^2$. That is, the l_1 norm does not satisfy the parallelogram law and hence is not associated with an inner product, as is the Euclidean norm (see Exercise 3 of Section 5.2).

■

The l_1 norm does have a certain advantage over the Euclidean norm in that it is easier to compute. An even easier norm to compute is the l_∞ *norm* or *max norm*, which is defined on n-vectors by

$$\|\mathbf{x}\|_\infty = \max_{1 \le i \le n} |x_i|.$$

Again it is a very easy matter to verify that parts (a)–(c) of Theorem 5.2 are satisfied for this norm.

The l_1 norm plays a useful role in the theory of digital codes. In the final section of Chapter 4 we saw that errors in digital messages could be detected and corrected by appending to the message certain parity check bits. We now show that an alternative way of detecting and correcting errors in a digital message is to spread the allowable messages far enough apart in the l_1 norm. By a digital message of length n, we mean an n-vector (which we will write as a row vector) of 0's and 1's. For example,

$$[0 \quad 1 \quad 0 \quad 1 \quad 1 \quad 0]$$

is a digital message of length 6. By interpreting a "dot" as 0 and a "dash" as 1, a Morse code message could, for example, be viewed as a (long) digital message. We will consider a situation in which digital messages of some fixed length are transmitted. The *Hamming distance* between two such messages is defined to be the l_1 norm of their difference. For example, if

$$\mathbf{x} = [0 \quad 1 \quad 0 \quad 1 \quad 1 \quad 0]$$

and

$$\mathbf{y} = [1 \quad 0 \quad 0 \quad 1 \quad 0 \quad 1],$$

then their Hamming distance is

$$\|\mathbf{x} - \mathbf{y}\|_1 = 1 + 1 + 1 + 1 = 4.$$

To put it another way, the Hamming distance between two messages is the total number of entries in which the messages disagree [this is the function $H(\mathbf{u},\mathbf{v})$ defined in Exercise 5 of Section 4.4].

We will suppose that not all messages have specific meanings. Only special messages called *codewords* will have meanings. For example, there are $2^6 = 64$ possible messages of length 6, but only a relatively small number of these messages may be chosen as codewords to convey specific meanings. The reason for building such redundancy into the system is that no transmission channel is error-free. A codeword that is transmitted may be garbled in the process and received as a message that is not a codeword. If this occurs, the receiver would like to do two things: *detect* that a transmission error has occurred and *correct* the message to the codeword that was sent. For this to be possible, the codewords must not be too close together (in the sense of Hamming distance).

Example 5.20

Consider messages of length 3 and suppose the codewords are $[1 \quad 0 \quad 0]$, $[0 \quad 0 \quad 0]$, and $[1 \quad 0 \quad 1]$. If $[1 \quad 0 \quad 0]$ is transmitted and a single error is made causing $[0 \quad 0 \quad 0]$ to be received, this error is not detected and the codeword $[0 \quad 0 \quad 0]$ is assumed to have been sent. If $[0 \quad 0 \quad 0]$ is sent and a single error in transmission occurs so that $[0 \quad 0 \quad 1]$ is received, the message $[0 \quad 0 \quad 1]$ is interpreted as an error since it is not a codeword. However, assuming a single error, it is not possible to correct this message using the Hamming distance since each of the codewords $[0 \quad 0 \quad 0]$ and $[1 \quad 0 \quad 1]$ are of Hamming distance 1 from the received message. Therefore, it is impossible for the receiver to decide which of these codewords was sent.

■

Example 5.20 shows that in order to detect and correct errors in digital codes by use of the Hamming distance, the codewords must be spread sufficiently far apart (with respect to the Hamming distance). In fact, if it is to be possible to detect all patterns of k or fewer errors in a message of length n ($k < n$), the Hamming distance between any two codewords must clearly be at least $k + 1$. Moreover, to detect *and* correct any pattern of k or fewer errors, the distance between any two codewords must be at least $2k + 1$ (and hence the number of errors k must satisfy $2k < n$).

Example 5.21

Consider digital messages of length 6 and suppose that there are three codewords,

$$\mathbf{c}^{(1)} = [1 \quad 0 \quad 1 \quad 0 \quad 1 \quad 0]$$
$$\mathbf{c}^{(2)} = [0 \quad 1 \quad 0 \quad 0 \quad 1 \quad 0]$$
$$\mathbf{c}^{(3)} = [0 \quad 1 \quad 0 \quad 1 \quad 0 \quad 1].$$

Each of these codewords is 3 units distant from every other codeword; therefore, it is possible to detect 2 or fewer errors. Also it is possible to detect and correct any single error.

For example, if the codeword $\mathbf{c}^{(1)}$ is sent and errors are made in the first and third entries, then the message

$$[0 \quad 0 \quad 0 \quad 0 \quad 1 \quad 0]$$

is received. This is immediately recognized by the receiver as an error since it is not a codeword. However, if a third error were made in entry 2, the codeword $\mathbf{c}^{(2)}$ would be received and no error would be detected.

If it is assumed that no more than one error is committed in transmission, then errors may be both detected and corrected. For example, if the codeword $\mathbf{c}^{(1)}$ is transmitted and $[0 \quad 0 \quad 1 \quad 0 \quad 1 \quad 0]$ is received, the received message has Hamming distance 1, 2, 5 from the codewords $\mathbf{c}^{(1)}$, $\mathbf{c}^{(2)}$, $\mathbf{c}^{(3)}$, respectively. The message is therefore corrected to the codeword $\mathbf{c}^{(1)}$.

■

EXERCISES 5.5

1. Show that the norm $\|\cdot\|_\infty$ does not satisfy the parallelogram law. (*Hint:* See Example 5.13.)
2. Verify that for n-vectors $\mathbf{x}$, $\|\mathbf{x}\|_\infty \le \|\mathbf{x}\|_1 \le n\|\mathbf{x}\|_\infty$.
3. **(a)** Suppose that

$$\begin{bmatrix} 2 & -1 \\ 1 & 4 \end{bmatrix}.$$

Verify that $\|\mathbf{Ax}\|_\infty \le 5\,\|\mathbf{x}\|_\infty$ for each 2-vector $\mathbf{x}$.

(b) Suppose that $\mathbf{A}$ is a 2×2 matrix. Verify that $\|\mathbf{Ax}\|_\infty \le M\|\mathbf{x}\|_\infty$ for each 2-vector $\mathbf{x}$, where

$$M = \max_{1\le i\le 2} \sum_{j=1}^{2} |a_{ij}|.$$

4. (a) Verify that for each 2-vector $\mathbf{x}$, $\|\mathbf{Ax}\|_1 \le 5\,\|\mathbf{x}\|_1$, where $\mathbf{A}$ is the matrix in part (a) of Exercise 3.

(b) Suppose that $\mathbf{A}$ is a 2×2 matrix. Verify that $\|\mathbf{Ax}\|_1 \le M\,\|\mathbf{x}\|_1$ for each 2-vector $\mathbf{x}$, where

$$M = \max_{1\le j\le 2} \sum_{i=1}^{2} |a_{ij}|.$$

5. Sketch the picture of the unit sphere in the l_2 norm, that is, the set

$$\left\{\begin{bmatrix} x_1 \\ x_2 \end{bmatrix} : x_1^2 + x_2^2 = 1\right\}.$$

6. Sketch the picture of the "unit sphere" in the l_1 norm, that is, the set

$$\left\{\begin{bmatrix} x_1 \\ x_2 \end{bmatrix} : |x_1| + |x_2| = 1\right\}.$$

7. Sketch the picture of the "unit sphere" in the l_∞ norm, that is, the set

$$\left\{\begin{bmatrix} x_1 \\ x_2 \end{bmatrix} : \max\{|x_1|, |x_2|\} = 1\right\}.$$

8. Construct eight codewords for digital messages of length 4 in such a way that any single error in transmission can be detected.

9. Construct two codewords for digital messages of length 4 in such a way that any single error in transmission can be detected and corrected.

5.6 Computer Graphics

We now give an application of matrices to computer graphics. Our aim is to view a simple figure consisting of a finite number of points (called vertices) connected by straight-line segments, such as the wedge-shaped object in Figure 5.8, from various aspects as projected onto a video display monitor. We shall represent points in an *xyz* coordinate system and view only their projections onto the *xy* plane. This corresponds to taking the surface of the video screen as the *xy* plane and viewing it from a vantage point on the *z* axis (see Figure 5.9). Therefore, the *z* coordinate of a point will be invisible to the viewer.

An object will be represented by embedding it in the *xyz* coordinate system so that the origin is in its interior and specifying the coordinates of its vertices. [Of course, the connections between the vertices must also be specified, for example, by using a communications matrix (see Section 1.4), but we will not dwell on this aspect of the problem.] For example, the object in Figure 5.8 could be represented by the *coordinate matrix*

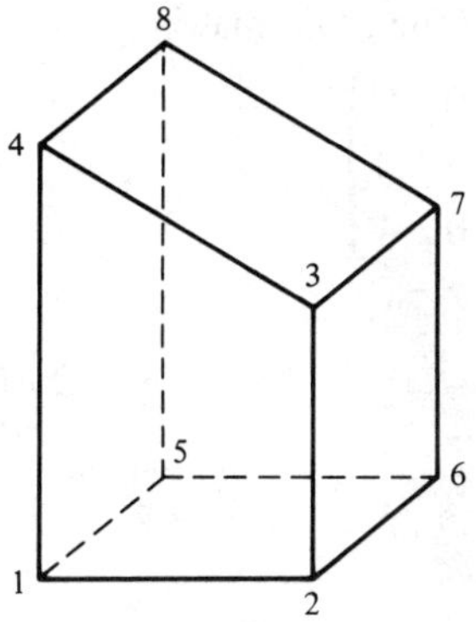

Figure 5.8

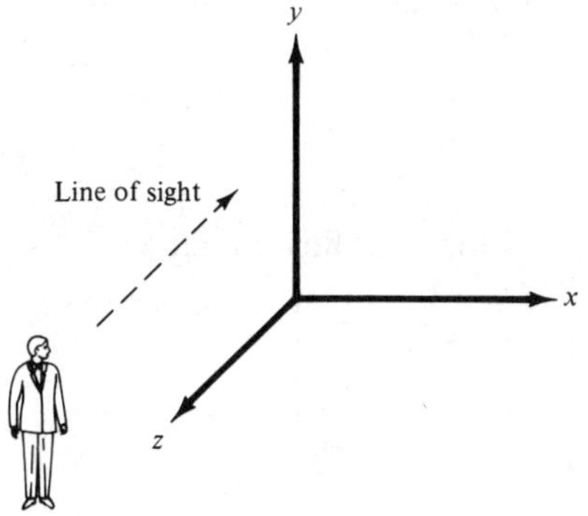

Figure 5.9

$$\begin{bmatrix} -1 & 1 & 1 & -1 & -1 & 1 & 1 & -1 \\ -1 & -1 & 1 & 2 & -1 & -1 & 1 & 2 \\ .5 & .5 & .5 & .5 & -.5 & -.5 & -.5 & -.5 \end{bmatrix}. \tag{13}$$

The ith column vector of this matrix corresponds to the ith vertex of the object. This coordinate matrix gives the view of the object in Figure 5.10 (remember that our line of sight is down the z axis and hence points with the same xy coordinates but different z coordinates are indistinguishable).

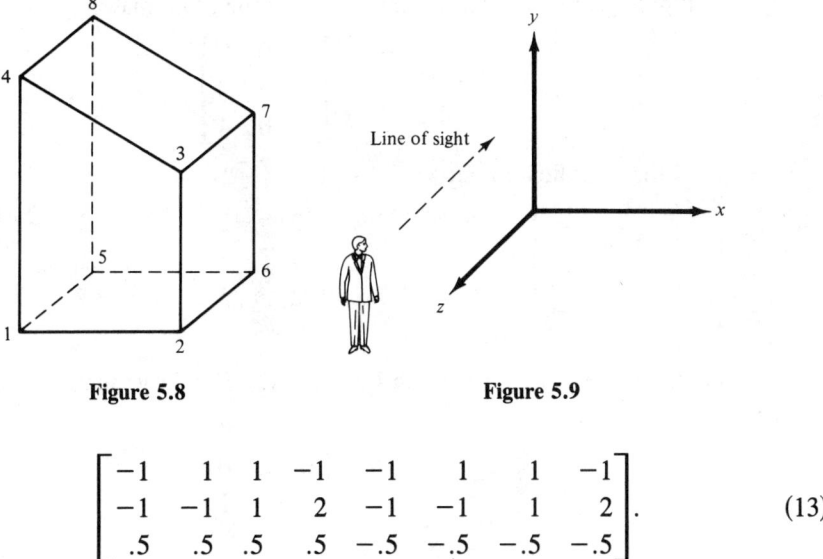

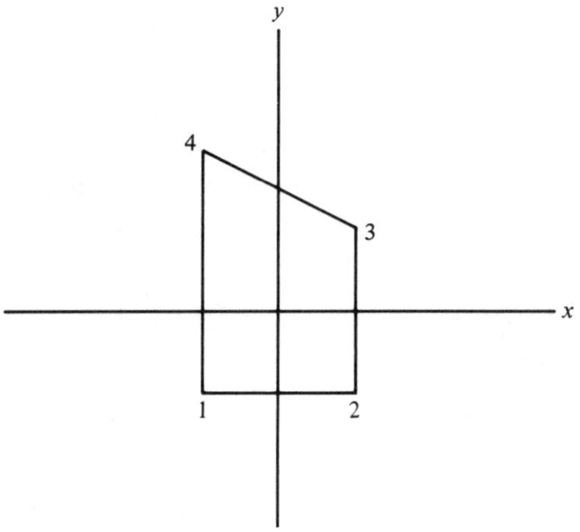

Figure 5.10

One of the simplest ways to modify a view is to make a change of scale. For example, we could increase the extent for the object in the x direction simply by multiplying the x coordinate of each vertex by 2. Similarly, the dimensions of the object in other directions can be scaled by multiplying the appropriate coordinates by a suitable scalar. Scaling by various factors in the coordinate directions is accomplished by premultiplying the coordinate matrix by a diagonal matrix having the scale factors as its diagonal entries (see Exercise 21 of Section 1.3). For example, premultiplying the coordinate matrix (13) by the scale matrix

$$\begin{bmatrix} 2 & 0 & 0 \\ 0 & \frac{1}{2} & 0 \\ 0 & 0 & 4 \end{bmatrix}$$

gives the coordinate matrix

$$\begin{bmatrix} -2 & 2 & 2 & -2 & -2 & 2 & 2 & -2 \\ -.5 & -.5 & .5 & 1 & -.5 & -.5 & .5 & 1 \\ 2 & 2 & 2 & 2 & -2 & -2 & -2 & -2 \end{bmatrix}$$

and the corresponding view in Figure 5.11. Similarly, scaling (13) by

$$\begin{bmatrix} \frac{1}{2} & 0 & 0 \\ 0 & 2 & 0 \\ 0 & 0 & 1 \end{bmatrix}$$

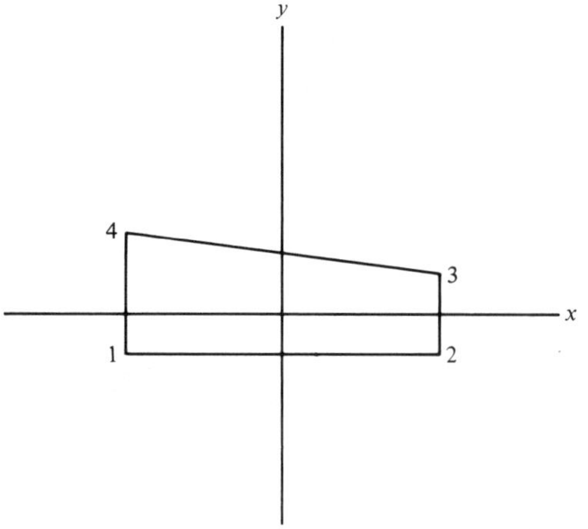

Figure 5.11

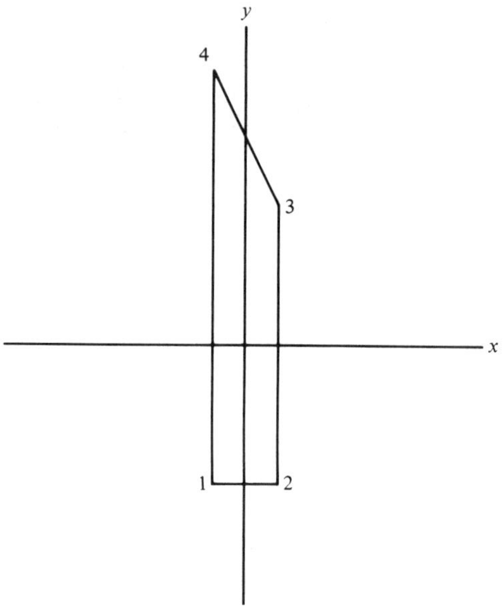

Figure 5.12

results in the view in Figure 5.12. Note that in each case the scale factor in the z direction is irrelevant, due to our point of view.

A common way to view aspects of an object is to rotate it about different axes. In Figure 5.13 rotations about the z, y, and x axes, respectively, are illustrated. In a rotation about the z axis, the z coordinate of no point changes, while the x and y

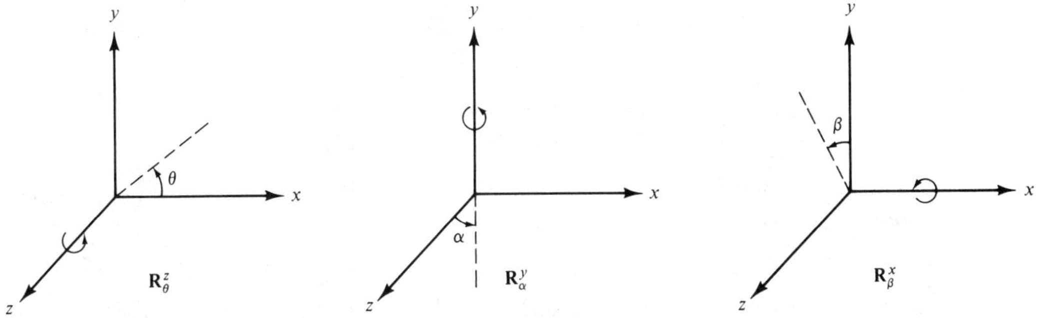

Figure 5.13

coordinates are transformed as in Section 5.3. Therefore, a rotation of θ radians about the z axis, as illustrated in Figure 5.13, is accomplished by the matrix

$$\mathbf{R}^z_\theta = \begin{bmatrix} \cos\theta & -\sin\theta & 0 \\ \sin\theta & \cos\theta & 0 \\ 0 & 0 & 1 \end{bmatrix}.$$

Suppose that $\theta = \pi/4$ and the rotation

$$\mathbf{R}^z_{\pi/4} = \begin{bmatrix} \frac{\sqrt{2}}{2} & \frac{-\sqrt{2}}{2} & 0 \\ \frac{\sqrt{2}}{2} & \frac{\sqrt{2}}{2} & 0 \\ 0 & 0 & 1 \end{bmatrix}$$

is applied to the coordinate matrix (13). The result is the view in Figure 5.14.

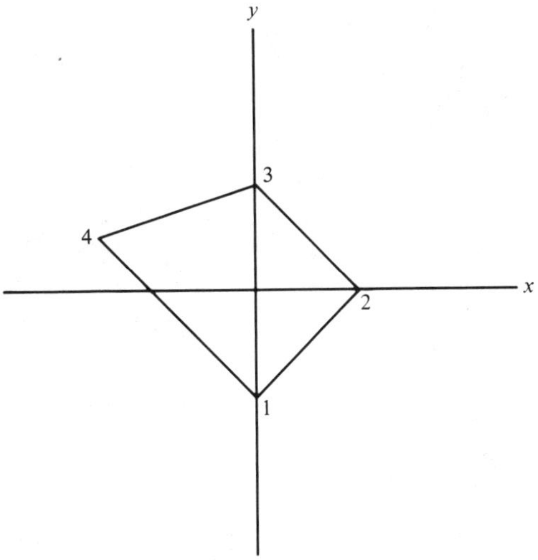

Figure 5.14

The rotation about the x axis as in Figure 5.13 is represented by the matrix

$$\mathbf{R}^x_\beta = \begin{bmatrix} 1 & 0 & 0 \\ 0 & \cos\beta & -\sin\beta \\ 0 & \sin\beta & \cos\beta \end{bmatrix}.$$

For example, the rotation $\mathbf{R}^x_{\pi/4}$ applied to the coordinate matrix (13) results in the new coordinate matrix

$$\begin{bmatrix} -1 & 1 & 1 & -1 & -1 & 1 & 1 & -1 \\ -1.061 & -1.061 & .354 & 1.061 & -.354 & -.354 & 1.061 & 1.768 \\ -.354 & -.354 & 1.061 & 1.768 & -1.061 & -1.061 & .354 & 1.061 \end{bmatrix}, \quad (14)$$

which may be viewed in Figure 5.15.

Finally, a rotation of α radians about the y-axis, as illustrated in Figure 5.13, is represented by the matrix

$$\mathbf{R}^y_\alpha = \begin{bmatrix} \cos\alpha & 0 & \sin\alpha \\ 0 & 1 & 0 \\ -\sin\alpha & 0 & \cos\alpha \end{bmatrix}.$$

If the coordinate matrix (14) is acted on by the rotation $\mathbf{R}^y_{\pi/6}$, the result is a new coordinate matrix which may be viewed in Figure 5.16. If this last coordinate matrix is scaled by

$$\begin{bmatrix} 2 & 0 & 0 \\ 0 & 1.5 & 0 \\ 0 & 0 & 2 \end{bmatrix},$$

the view is Figure 5.17.

By taking combinations (i.e., products) of rotations about the various coordinate axes, it is possible to look upon the object from any point of view.

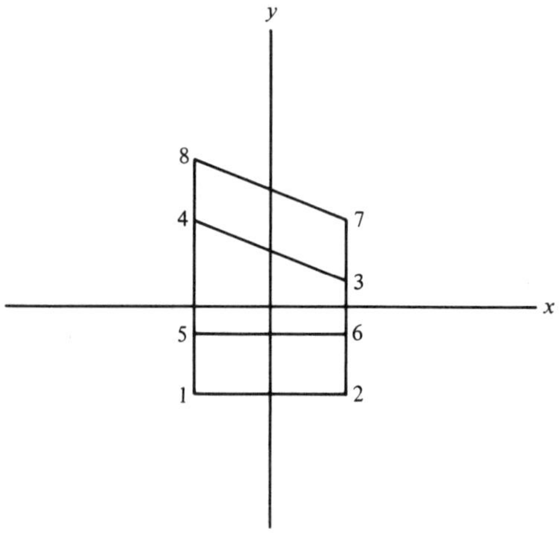

Figure 5.15

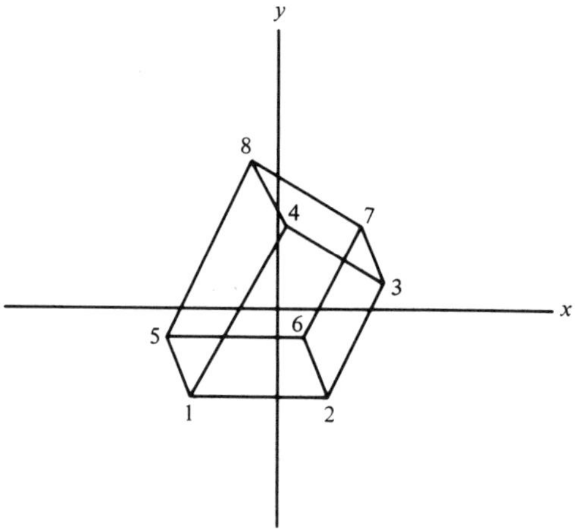

Figure 5.16

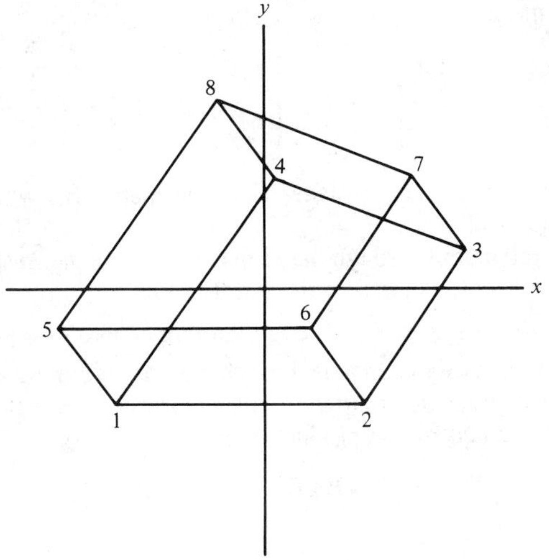

Figure 5.17

EXERCISES 5.6

1. Verify the matrix representations for $\mathbf{R}^y_\alpha$ and $\mathbf{R}^x_\beta$.
2. Do the rotation matrices commute? For example, is $\mathbf{R}^y_\alpha \mathbf{R}^x_\beta = \mathbf{R}^x_\beta \mathbf{R}^y_\alpha$?
3. Does a scaling matrix commute with a rotation matrix?
4. Verify that each of the matrices $\mathbf{R}^z_\theta$, $\mathbf{R}^y_\alpha$, and $\mathbf{R}^x_\beta$ is orthogonal.

Exercises 5–7 refer to the pyramid with base points $(1, 1, -1)$, $(1, -1, -1)$, $(-1, -1, -1)$, $(-1, 1, -1)$ and apex $(0, 0, 1)$. The initial view of the object is specified by the coordinate matrix

$$\begin{bmatrix} 0 & 1 & 1 & -1 & -1 \\ 0 & 1 & -1 & -1 & 1 \\ 1 & -1 & -1 & -1 & -1 \end{bmatrix}.$$

5. Scale the object by

$$\begin{bmatrix} 2 & 0 & 0 \\ 0 & 1 & 0 \\ 0 & 0 & 2 \end{bmatrix}$$

and apply the rotation $\mathbf{R}^x_{\pi/6}$. Sketch.

6. Rotate the result of Exercise 5 by $\mathbf{R}^y_{\pi/3}$. Sketch.
7. Rotate the result of Exercise 6 by $\mathbf{R}^z_{\pi/4}$ and then scale by

$$\begin{bmatrix} 1 & 0 & 0 \\ 0 & 1.5 & 0 \\ 0 & 0 & 1 \end{bmatrix}.$$

Sketch.

8. **(a)** A transformation of the form

$$\begin{bmatrix} x \\ y \\ z \end{bmatrix} \longrightarrow \begin{bmatrix} -x \\ y \\ z \end{bmatrix}$$

is called a reflection through the yz plane. Find the matrix $\mathbf{S}_{yz}$ which produces a reflection throught the yz plane.

(b) Define the notions of reflections through the xz plane and xy plane, respectively. Find the matrices $\mathbf{S}_{xz}$ and $\mathbf{S}_{xy}$ which represent these reflections.

9. Reflect the view resulting from Exercise 5 through the three coordinate planes. Sketch.

***10.** Show that a rotation of ϕ radians about an axis through the origin forming an angle α with the positive z axis and whose projection onto the xy plane forms an angle β with the y axis, as illustrated, may be represented by the product matrix

$$\mathbf{R}^{z}_{-\beta}\mathbf{R}^{x}_{\alpha-\pi/2}\mathbf{R}^{y}_{\phi}\mathbf{R}^{x}_{\pi/2-\alpha}\mathbf{R}^{z}_{\beta}.$$

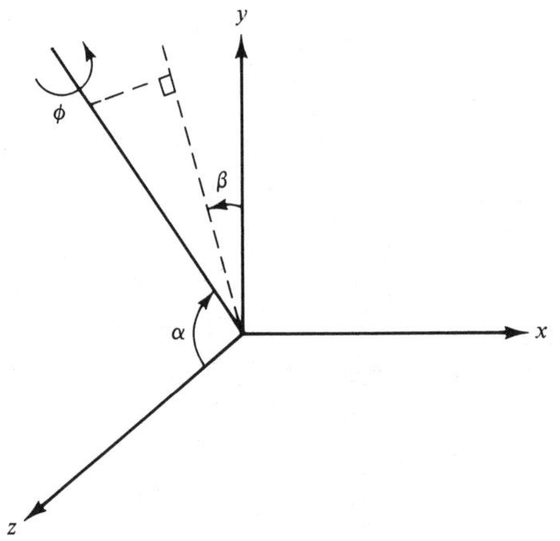

GLOSSARY

adjoint matrix The adjoint of an $m \times n$ matrix $\mathbf{A}$ is the $n \times m$ matrix $\mathbf{A}^*$, where $a^*_{ij} = \overline{a}_{ji}$.

Cauchy–Schwarz inequality $|\langle \mathbf{x}, \mathbf{y} \rangle| \leq \|\mathbf{x}\| \, \|\mathbf{y}\|$.

complex conjugate The conjugate of a complex number $z = a + bi$ is the complex number $\overline{z} = a - bi$.

Gram–Schmidt process A method of converting a set of linearly independent vectors into a set of orthogonal vectors.

inner product A product $\langle \cdot, \cdot \rangle$ defined on vectors which satisfies certain axioms (see p. 136).

modulus The modulus of a complex number $z = a + bi$ is the real number $|z| = \sqrt{a^2 + b^2}$.

norm A generalized "length" $\|\mathbf{x}\|$ associated with a vector $\mathbf{x}$ (see p. 137).

orthogonal matrix A matrix $\mathbf{Q}$ satisfying $\mathbf{Q}^{\mathrm{T}} = \mathbf{Q}^{-1}$.

orthogonal vectors Two vectors $\mathbf{x}$ and $\mathbf{y}$ are orthogonal with respect to the inner product $\langle \cdot, \cdot \rangle$ if $\langle \mathbf{x}, \mathbf{y} \rangle = 0$.

orthonormal A set S of vectors is orthonormal if $\langle \mathbf{x}, \mathbf{y} \rangle = 0$ for $\mathbf{x}, \mathbf{y} \in S (\mathbf{x} \neq \mathbf{y})$ and $\|\mathbf{x}\| = 1$ for $\mathbf{x} \in S$.

QR factorization A matrix $\mathbf{A}$ with linearly independent columns may be factored as $\mathbf{A} = \mathbf{QR}$, where $\mathbf{Q}$ has orthonormal columns and $\mathbf{R}$ is invertible and upper triangular.

self-adjoint matrix A matrix that is equal to its adjoint.

unitary matrix A matrix whose adjoint is its inverse.

NOTES AND COMMENTS

The notions of inner product and orthogonality can be put to important practical use in solving least-squares problems (see Chapter 8). These ideas also play central roles in some areas of mathematical analysis. In particular, inner products of functions are at the heart of the method of Fourier series for solving problems of mathematical physics. The abstract axiomatic development of "spaces" with inner products leads to the theory of Hilbert spaces, an important branch of modern mathematical analysis.

For readers with some background in abstract algebra, the book *Introduction to the Theory of Error-Correcting Codes* by V. Pless (New York: Wiley, 1982) provides an excellent introduction to the subject of digital codes. An account of the use of matrices in computer graphics, complete with Fortran programs, can be found in I. O. Angell, *A Practical Introduction to Computer Graphics* (New York: Wiley, 1981).

6

The Eigenvalue Problem

In this chapter we develop the concepts of an eigenvalue and an eigenvector of a square matrix. Throughout this chapter all matrices are square. We shall see that the idea of an eigenvalue of a nonsquare matrix is meaningless. Our aim is to demonstrate how eigenvalues and eigenvectors can be calculated for small matrices and to interpret their meaning. We introduce these notions by considering an application in population dynamics.

6.1 The Characteristic Equation

Suppose that an ecosystem consists of several, say n, species of animals. We denote by $p_i(t)$ the population of the ith species at time t. Let $\mathbf{p}(t)$ be the n-vector whose ith entry is $p_i(t)$. The animal population of the ecosystem evolves with time and we want to follow this evolution. We suppose for simplicity that there is an $n \times n$ matrix $\mathbf{A}$, called the *transition matrix*, that characterizes the transition of the population from one generation, say at time t, to the population at time $t + 1$. Thus we assume that for each t,

$$\mathbf{p}(t + 1) = \mathbf{A}\mathbf{p}(t).$$

The matrix $\mathbf{A}$ specifies a mathematical model of the ecosystem whose entries take into account the birth and death rates of each species, the competition of some of the species for a common food supply, and the possibility that one or more species may prey on others as a source of food. It is assumed that the entries of $\mathbf{A}$ are known and are independent of time.

Now suppose that the population vector is known at some time, say $t = 0$. The population at some future time $t = m$ is found as follows;

$$\begin{aligned}\mathbf{p}(m) &= \mathbf{A}\mathbf{p}(m - 1)\\ &= \mathbf{A}^2\mathbf{p}(m - 2)\\ &\ \ \vdots\\ &= \mathbf{A}^m\mathbf{p}(0).\end{aligned}$$

Therefore, the evolution of the ecosystem is dependent on powers of the transition matrix and on the initial population.

Of considerable interest to ecologists is the question of a stable population distribution. The population distribution is said to be stable if the population of each species at time $t + 1$ is proportional to its population at time t with the same constant of proportionality for each species. That is, the distribution is stable if

$$\mathbf{p}(t + 1) = \lambda \mathbf{p}(t)$$

for some proportionality constant λ. Recalling that $\mathbf{p}(t + 1) = \mathbf{A}\mathbf{p}(t)$, we see that the stability of the population distribution depends on the existence of a population vector $\mathbf{p}(t)$ and a scalar λ such that

$$\mathbf{A}\mathbf{p}(t) = \lambda \mathbf{p}(t). \tag{1}$$

Assuming that such a $\mathbf{p}(t)$ and λ exist, it is easy to find the population at some future time:

$$\begin{aligned}\mathbf{p}(t + m) &= \mathbf{A}^m\mathbf{p}(t)\\ &= \mathbf{A}^{m-1}\mathbf{A}\mathbf{p}(t) = \mathbf{A}^{m-1}\lambda\mathbf{p}(t)\\ &= \lambda\mathbf{A}^{m-1}\mathbf{p}(t)\end{aligned}$$

and continuing in this way, we find

$$\mathbf{p}(t + m) = \lambda^m\mathbf{p}(t).$$

The total population grows if $\lambda > 1$ and decreases if $0 < \lambda < 1$, but the population distribution is stable when equation (1) is satisfied.

The problem (1) is called the eigenvalue problem for the matrix $\mathbf{A}$. Such problems arise in a variety of applications and their study is the subject of this chapter. We shall examine the eigenvalue problem from a purely mathematical point of view and return to an application in the last section of the chapter.

Let $\mathbf{A}$ be a square matrix. We say that λ is an **eigenvalue** of $\mathbf{A}$ if there exists a vector $\mathbf{x} \neq \boldsymbol{\theta}$ such that

$$\mathbf{A}\mathbf{x} = \lambda\mathbf{x}. \tag{2}$$

The vector $\mathbf{x}$ in (2) is called an **eigenvector** of $\mathbf{A}$ corresponding to the eigenvalue λ. Note that if $\mathbf{x}$ is an n-vector, then the left-hand side of (2) is defined only if $\mathbf{A}$ has n columns and the equality of (2) can hold only if $\mathbf{A}$ has n rows. That is, eigenvalues are defined only for square matrices.

Example 6.1

We show that

$$\mathbf{x} = \begin{bmatrix} 1 \\ 0 \\ -1 \end{bmatrix} \quad \text{is an eigenvector of} \quad \mathbf{A} = \begin{bmatrix} -1 & 2 & 2 \\ 2 & 2 & 2 \\ -3 & -6 & -6 \end{bmatrix}$$

by demonstrating that $\mathbf{A}\mathbf{x}$ is some multiple of $\mathbf{x}$. Direct calculation shows that

$$\mathbf{A}\mathbf{x} = \begin{bmatrix} -1 & 2 & 2 \\ 2 & 2 & 2 \\ -3 & -6 & -6 \end{bmatrix}\begin{bmatrix} 1 \\ 0 \\ -1 \end{bmatrix} = \begin{bmatrix} -3 \\ 0 \\ 3 \end{bmatrix} = (-3)\begin{bmatrix} 1 \\ 0 \\ -1 \end{bmatrix}$$

and thus $\mathbf{Ax} = -3\mathbf{x}$. Therefore, $\mathbf{x}$ is an eigenvector of $\mathbf{A}$ corresponding to the eigenvalue $\lambda = -3$.

■

It should be emphasized that $\mathbf{x} = \boldsymbol{\theta}$ cannot, by definition, be an eigenvector. The reason for disallowing $\mathbf{x} = \boldsymbol{\theta}$ as an eigenvector is simple: If $\mathbf{x} = \boldsymbol{\theta}$ were allowed, then (2) would be satisfied for *every* scalar λ irrespective of the matrix $\mathbf{A}$. However, $\lambda = 0$ can be an eigenvalue. For example, if

$$\mathbf{A} = \begin{bmatrix} -1 & 2 \\ 3 & -6 \end{bmatrix} \quad \text{and} \quad \mathbf{x} = \begin{bmatrix} 2 \\ 1 \end{bmatrix},$$

then $\mathbf{Ax} = \boldsymbol{\theta} = 0\mathbf{x}$ and hence $\lambda = 0$ is an eigenvalue of $\mathbf{A}$. In the next example we show how one can find an eigenvector corresponding to a particular eigenvalue.

Example 6.2

For the matrix $\mathbf{A}$ of Example 6.1 $\lambda = -2$ is an eigenvalue (later we show how this is determined). Therefore, for some $\mathbf{x} \neq \boldsymbol{\theta}$ we must have $\mathbf{Ax} = -2\mathbf{x}$, or equivalently,

$$(\mathbf{A} + 2\mathbf{I})\mathbf{x} = \boldsymbol{\theta}.$$

We solve this homogeneous problem for $\mathbf{x}$ by elimination:

$$\left[\begin{array}{rrr|r} 1 & 2 & 2 & 0 \\ 2 & 4 & 2 & 0 \\ -3 & -6 & -4 & 0 \end{array}\right] \longrightarrow \left[\begin{array}{rrr|r} 1 & 2 & 2 & 0 \\ 0 & 0 & -2 & 0 \\ 0 & 0 & 2 & 0 \end{array}\right]$$

$$\longrightarrow \left[\begin{array}{rrr|r} 1 & 2 & 2 & 0 \\ 0 & 0 & -2 & 0 \\ 0 & 0 & 0 & 0 \end{array}\right].$$

It follows that $x_3 = 0$ and $x_1 + 2x_2 = 0$ (or $x_1 = -2x_2$). Since $\mathbf{x}$ must be nonzero, we find that any vector of the form

$$\mathbf{x} = x_2\begin{bmatrix} -2 \\ 1 \\ 0 \end{bmatrix}, \qquad x_2 \neq 0$$

is an eigenvector corresponding to $\lambda = -2$.

■

Example 6.2 illustrates that eigenvectors corresponding to a given eigenvalue are not unique. Indeed, for each eigenvalue there corresponds infinitely many eigenvectors. To see this in general, suppose that λ is an eigenvalue of $\mathbf{A}$ with corresponding eigenvector $\mathbf{x}$. Thus $\mathbf{Ax} = \lambda\mathbf{x}$ and if α is any nonzero scalar, we then find

$$\mathbf{A}(\alpha\mathbf{x}) = \alpha\mathbf{Ax} = \alpha(\lambda\mathbf{x}) = \lambda(\alpha\mathbf{x}).$$

Hence $\alpha\mathbf{x}$ is an eigenvector corresponding to λ. In other words, any nonzero multiple of an eigenvector is also an eigenvector corresponding to the same eigenvalue.

Another important observation to make about Example 6.2 is the following. The equation

$$\mathbf{Ax} = \lambda \mathbf{x}, \qquad \mathbf{x} \neq \boldsymbol{\theta}$$

is equivalent to the homogeneous matrix equation

$$(\mathbf{A} - \lambda \mathbf{I})\mathbf{x} = \boldsymbol{\theta}, \qquad \mathbf{x} \neq \boldsymbol{\theta}.$$

Thus eigenvectors corresponding to λ are nontrivial solutions of $(\mathbf{A} - \lambda\mathbf{I})\mathbf{x} = \boldsymbol{\theta}$. That is, the eigenvectors corresponding to λ are the nonzero vectors in the nullspace of $\mathbf{A} - \lambda\mathbf{I}$. Since $\mathbf{x} \neq \boldsymbol{\theta}$ it follows that $\mathbf{A} - \lambda\mathbf{I}$ must be singular and hence $\det(\mathbf{A} - \lambda\mathbf{I}) = 0$. Consequently, every eigenvalue λ of $\mathbf{A}$ must satisfy the equation

$$\det(\mathbf{A} - \lambda\mathbf{I}) = 0,$$

which is called the **characteristic equation** for $\mathbf{A}$. We have established the following theorem.

Theorem 6.1. The following are equivalent conditions for λ to be an eigenvalue of $\mathbf{A}$:

(a) For some $\mathbf{x} \neq \boldsymbol{\theta}$, $\mathbf{Ax} = \lambda\mathbf{x}$.
(b) $N(\mathbf{A} - \lambda\mathbf{I}) \neq \{\boldsymbol{\theta}\}$.
(c) $\mathbf{A} - \lambda\mathbf{I}$ is not invertible.
(d) $\det(\mathbf{A} - \lambda\mathbf{I}) = 0$.

■

Note in particular, from (c) in the theorem, that a square matrix $\mathbf{A}$ is invertible if and only if 0 is not an eigenvalue of $\mathbf{A}$. In the next example we find all the eigenvalues by solving the characteristic equation.

Example 6.3

For

$$\mathbf{A} = \begin{bmatrix} 5 & -6 & 0 \\ 1 & -2 & 0 \\ 4 & 6 & -1 \end{bmatrix}$$

the characteristic equation is

$$\begin{vmatrix} 5-\lambda & -6 & 0 \\ 1 & -2-\lambda & 0 \\ 4 & 6 & -1-\lambda \end{vmatrix} = 0.$$

Using the cofactor expansion down the third column gives

$$-(1+\lambda)\begin{vmatrix} 5-\lambda & -6 \\ 1 & -2-\lambda \end{vmatrix} = -(1+\lambda)(\lambda^2 - 3\lambda - 4) = 0.$$

Thus $\det(\mathbf{A} - \lambda\mathbf{I}) = -(1+\lambda)(\lambda-4)(\lambda+1)$ and the eigenvalues are $\lambda = -1, 4, -1$. Note that taking into account the repeated factors, $\mathbf{A}$ has three eigenvalues.

■

The characteristic equation for **A** always results in a polynomial equation in λ,

$$p_{\mathbf{A}}(\lambda) = 0,$$

where $p_{\mathbf{A}}(\lambda) = \det(\mathbf{A} - \lambda\mathbf{I})$ is called the **characteristic polynomial** of **A**. Moreover, if **A** is $n \times n$, then $p_{\mathbf{A}}(\lambda)$ is a polynomial of degree n which has the form

$$p_{\mathbf{A}}(\lambda) = (-1)^n\lambda^n + a_{n-1}\lambda^{n-1} + \cdots + a_1\lambda + a_0,$$

where the coefficients a_i depend on the entries of **A**. The eigenvalues of **A** are the roots of the characteristic equation and these roots can be determined by factoring the characteristic polynomial. Such a factorization may lead to repeated linear factors as in Example 6.3. An eigenvalue that is the root of such a repeated linear factor is listed a corresponding number of times when specifying the eigenvalues. In this way an $n \times n$ matrix has n eigenvalues counting repetitions.

Example 6.4

The characteristic polynomial of

$$\mathbf{A} = \begin{bmatrix} 1 & 0 & 2 \\ 0 & 1 & -1 \\ 2 & -1 & 3 \end{bmatrix}$$

is $p_{\mathbf{A}}(\lambda) = -\lambda^3 + 5\lambda^2 - 2\lambda - 2$. (Check this for yourself.) The eigenvalues of **A** are solutions of the cubic equation

$$\lambda^3 - 5\lambda^2 + 2\lambda + 2 = 0.$$

By trial and error we find that $\lambda = 1$ is a root. Then division by the factor $\lambda - 1$ gives

$$\lambda^3 - 5\lambda^2 + 2\lambda + 2 = (\lambda - 1)(\lambda^2 - 4\lambda - 2),$$

and hence the remaining eigenvalues are found by applying the quadratic formula to the quadratic factor $\lambda^2 - 4\lambda - 2$. This gives $\lambda = 2 \pm \sqrt{6}$ and the eigenvalues of **A** are $\lambda = 1, 2 - \sqrt{6}, 2 + \sqrt{6}$.

■

Example 6.4 was designed so that the cubic polynomial $p_{\mathbf{A}}(\lambda)$ was relatively easy to factor. In general, the problem of factoring a cubic is not so simple, and for polynomials of higher degree this difficulty can only get worse. In fact, for a general polynomial of degree greater than 4, Niels Abel (1802–1829) proved that there is no algebraic formula for the roots. Thus, in general, it is impossible to find the eigenvalues of a matrix by an algebraic formula. Lest we give too pessimistic a view, we should point out that in many applications one is satisfied with a good approximation to one or more of the eigenvalues. There are several good numerical methods (some of which are discussed in Chapter 8) for computing approximate eigenvalues, and the better of these methods do not utilize the characteristic polynomial. A plausible (although somewhat naive) explanation for this is: If **A** is a large matrix, say 20×20, then it would take years to determine the coefficients of the characteristic polynomial by cofactor expansion. Actually, there are more efficient methods for determining the coefficients of $p_{\mathbf{A}}(\lambda)$, one of which is based on the Cayley–Hamilton theorem and is easy to describe (see Krylov's method in Section 6.3). Nonetheless, in actual matrix computations, the approximate eigenvalues are not usually found by use of the characteristic polynomial.

Thus far all our examples have dealt with real eigenvalues. But a real polynomial can have complex roots and hence real matrices can have complex eigenvalues. The next example illustrates this point.

Example 6.5

For

$$\mathbf{A} = \begin{bmatrix} -1 & 4 & 2 \\ 0 & 1 & 0 \\ -4 & 4 & 3 \end{bmatrix}$$

the characteristic polynomial is, by the second-row cofactor expansion,

$$\begin{aligned} p_A(\lambda) &= (1 - \lambda)\begin{vmatrix} -1 - \lambda & 2 \\ -4 & 3 - \lambda \end{vmatrix} \\ &= (1 - \lambda)(\lambda^2 - 2\lambda + 5). \end{aligned}$$

Clearly, $\lambda = 1$ is one eigenvalue and by applying the quadratic formula to $\lambda^2 - 2\lambda + 5$, we find the remaining eigenvalues:

$$\lambda = (2 \pm \sqrt{4 - 20})/2 = 1 \pm 2i.$$

A has eigenvalues $\lambda_1 = 1$, $\lambda_2 = 1 - 2i$, and $\lambda_3 = 1 + 2i$. The characteristic polynomial can be written in factored form as

$$p_A(\lambda) = -(\lambda - 1)(\lambda - 1 + 2i)(\lambda - 1 - 2i).$$

■

The complex eigenvalues of Example 6.5 are conjugates; that is, $\lambda_2 = 1 - 2i = \overline{1 + 2i} = \bar{\lambda}_3$. Complex eigenvalues of a real matrix always occur in conjugate pairs. Because the coefficients of $p_{\mathbf{A}}(\lambda)$ are real, we have

$$p_{\mathbf{A}}(\bar{\lambda}) = \overline{p_{\mathbf{A}}(\lambda)}$$

for any complex number λ. Therefore, if λ is an eigenvalue of $\mathbf{A}$ we find that

$$p_{\mathbf{A}}(\bar{\lambda}) = \overline{p_{\mathbf{A}}(\lambda)} = \bar{0} = 0$$

and $\bar{\lambda}$ is also an eigenvalue of $\mathbf{A}$. When complex eigenvalues of a real matrix occur, the corresponding eigenvectors cannot be real; that is, at least one entry of the corresponding eigenvector is a complex number. For example, it is a simple exercise to verify that eigenvectors corresponding to $\lambda_2 = 1 - 2i$ and $\lambda_3 = 1 + 2i$ from Example 6.5 are

$$\mathbf{x}^{(2)} = \begin{bmatrix} 1 + i \\ 0 \\ 2 \end{bmatrix} \quad \text{and} \quad \mathbf{x}^{(3)} = \begin{bmatrix} 1 - i \\ 0 \\ 2 \end{bmatrix},$$

respectively. We shall be concerned primarily with the real eigenvalues and corresponding real eigenvectors of a matrix.

Given an $n \times n$ matrix $\mathbf{A}$ the **trace** of $\mathbf{A}$, denoted by tr $(\mathbf{A})$, is defined by

$$\text{tr}(\mathbf{A}) = \sum_{i=1}^{n} a_{ii}.$$

That is, the trace of a matrix is the sum of its diagonal entries. It turns out that tr $(\mathbf{A})$ and det $(\mathbf{A})$ are related to the eigenvalues of $\mathbf{A}$. Suppose that the eigenvalues of the $n \times n$ matrix $\mathbf{A}$ are $\lambda_1, \ldots, \lambda_n$. Then $p_{\mathbf{A}}(\lambda)$ can be factored as

$$p_{\mathbf{A}}(\lambda) = (-1)^n(\lambda - \lambda_1)(\lambda - \lambda_2) \cdots (\lambda - \lambda_n).$$

It follows that

$$\begin{aligned}\det(\mathbf{A}) = \det(\mathbf{A} - 0\mathbf{I}) = p_{\mathbf{A}}(0) &= (-1)^n(-\lambda_1)(-\lambda_2) \cdots (-\lambda_n)\\ &= \lambda_1 \cdots \lambda_n\end{aligned}$$

and hence the determinant is equal to the product of the eigenvalues. A somewhat more difficult argument shows that the trace is equal to the sum of the eigenvalues.

Theorem 6.2. Let the $n \times n$ matrix $\mathbf{A}$ have eigenvalues $\lambda_1, \lambda_2, \ldots, \lambda_n$ (not necessarily distinct). Then

$$\det(\mathbf{A}) = \prod_{i=1}^{n} \lambda_i$$

and

$$\text{tr}(\mathbf{A}) = \sum_{i=1}^{n} \lambda_i.$$

■

The determinant and trace can be used as a check on the calculation of eigenvalues.

Example 6.6

For

$$\mathbf{A} = \begin{bmatrix} -1 & 4 & 2 \\ 0 & 1 & 0 \\ -4 & 4 & 3 \end{bmatrix}$$

we found in Example 6.5 that $p_{\mathbf{A}}(\lambda) = (1 - \lambda)(\lambda^2 - 2\lambda + 5) = -\lambda^3 - 3\lambda^2 - 7\lambda + 5$. An easy calculation gives tr $(\mathbf{A}) = a_{11} + a_{22} + a_{33} = 3$ and det $(\mathbf{A}) = p_{\mathbf{A}}(0) = 5$. The eigenvalues, from Example 6.5, are $\lambda_1 = 1$, $\lambda_2 = 1 - 2i$, and $\lambda_3 = 1 + 2i$. Therefore, we find

$$\text{tr}(\mathbf{A}) = 3 = 1 + (1 - 2i) + (1 + 2i) = \sum_{j=1}^{3} \lambda_j$$

and

$$\det(\mathbf{A}) = 5 = (1)(1 - 2i)(1 + 2i) = \prod_{j=1}^{3} \lambda_j,$$

illustrating Theorem 6.2.

■

In our final example of this section we determine all the eigenvalues and corresponding eigenvectors of a matrix.

Example 6.7

The symmetric matrix

$$\mathbf{A} = \begin{bmatrix} 4 & -1 & 0 \\ -1 & 5 & -1 \\ 0 & -1 & 4 \end{bmatrix}$$

has characteristic polynomial $p_{\mathbf{A}}(\lambda) = \det(\mathbf{A} - \lambda\mathbf{I}) = -\lambda^3 + 13\lambda^2 - 54\lambda + 72$. Factoring gives

$$p_{\mathbf{A}}(\lambda) = (4 - \lambda)(\lambda - 6)(\lambda - 3) = 0$$

and the eigenvalues are real and distinct: $\lambda_1 = 3$, $\lambda_2 = 4$, and $\lambda_3 = 6$. To find the eigenvectors we solve the three homogeneous systems:

$$(\mathbf{A} - \lambda_j\mathbf{I})\mathbf{x}^{(j)} = \boldsymbol{\theta}, \qquad 1 \le j \le 3.$$

$(\mathbf{A} - \lambda_1\mathbf{I})\mathbf{x}^{(1)} = \boldsymbol{\theta}$:

$$\left[\begin{array}{rrr|r} 1 & -1 & 0 & 0 \\ -1 & 2 & -1 & 0 \\ 0 & -1 & 1 & 0 \end{array}\right] \longrightarrow \left[\begin{array}{rrr|r} 1 & -1 & 0 & 0 \\ 0 & 1 & -1 & 0 \\ 0 & -1 & 1 & 0 \end{array}\right]$$

$$\longrightarrow \left[\begin{array}{rrr|r} 1 & -1 & 0 & 0 \\ 0 & 1 & -1 & 0 \\ 0 & 0 & 0 & 0 \end{array}\right]$$

We find $x_2^{(1)} = x_3^{(1)}$ and $x_1^{(1)} = x_2^{(1)}$. Thus if $\alpha = x_2^{(1)} \neq 0$, then

$$\mathbf{x}^{(1)} = \alpha\begin{bmatrix} 1 \\ 1 \\ 1 \end{bmatrix}.$$

$(\mathbf{A} - \lambda_2\mathbf{I})\mathbf{x}^{(2)} = \boldsymbol{\theta}$:

$$\left[\begin{array}{rrr|r} 0 & -1 & 0 & 0 \\ -1 & 1 & -1 & 0 \\ 0 & -1 & 0 & 0 \end{array}\right] \longrightarrow \left[\begin{array}{rrr|r} -1 & 1 & -1 & 0 \\ 0 & -1 & 0 & 0 \\ 0 & -1 & 0 & 0 \end{array}\right]$$

$$\longrightarrow \left[\begin{array}{rrr|r} -1 & 1 & -1 & 0 \\ 0 & -1 & 0 & 0 \\ 0 & 0 & 0 & 0 \end{array}\right].$$

It follows that $x_2^{(2)} = 0$ and $x_1^{(2)} = -x_3^{(2)}$. Letting $\beta = x_3^{(2)} \neq 0$, we find

$$\mathbf{x}^{(2)} = \beta\begin{bmatrix} -1 \\ 0 \\ 1 \end{bmatrix}.$$

$(\mathbf{A} - \lambda_3\mathbf{I})\mathbf{x}^{(3)} = \boldsymbol{\theta}$:

$$\left[\begin{array}{rrr|r} -2 & -1 & 0 & 0 \\ -1 & -1 & -1 & 0 \\ 0 & -1 & -2 & 0 \end{array}\right] \longrightarrow \left[\begin{array}{rrr|r} -2 & -1 & 0 & 0 \\ 0 & -\frac{1}{2} & -1 & 0 \\ 0 & -1 & -2 & 0 \end{array}\right]$$

$$\longrightarrow \left[\begin{array}{rrr|r} -2 & -1 & 0 & 0 \\ 0 & -\frac{1}{2} & -1 & 0 \\ 0 & 0 & 0 & 0 \end{array}\right].$$

Here we find $x_2^{(3)} = -2x_3^{(3)}$ and $x_1^{(3)} = -1/2x_2^{(3)} = x_3^{(3)}$. Thus for $\gamma = x_3^{(3)} \neq 0$,

$$\mathbf{x}^{(3)} = \gamma\begin{bmatrix} 1 \\ -2 \\ 1 \end{bmatrix}.$$

To each eigenvalue there corresponds infinitely many eigenvectors, but these eigenvectors are all multiples of a single eigenvector. That is, for each eigenvalue there is only one linearly independent eigenvector.

■

Clearly, the determination of all the eigenvalues and eigenvectors requires a considerable amount of work—much more than is required to solve $\mathbf{Ax} = \mathbf{b}$. Unfortunately, there is no way to avoid these lengthy calculations. For *small* matrices the procedure is clear and straightforward: First determine the eigenvalues by factoring (if possible) the characteristic polynomial, and then successively determine the eigenvectors by solving the equation $(\mathbf{A} - \lambda\mathbf{I})\mathbf{x} = \boldsymbol{\theta}$ for each eigenvalue λ of $\mathbf{A}$.

We conclude this section by giving a geometric interpretation of eigenvalues and eigenvectors. For simplicity we take $\mathbf{A}$ to be a 2×2 matrix. Then an eigenvector $\mathbf{x}$ of $\mathbf{A}$ can be thought of as determining an invariant direction for $\mathbf{A}$. If $\mathbf{Ax} = \lambda\mathbf{x}$, then the direction of $\mathbf{Ax}$ is the same as the direction of $\mathbf{x}$ if λ is positive, while the direction is reversed for λ negative. Moreover, the effect of multiplying $\mathbf{x}$ by $\mathbf{A}$ is to stretch $\mathbf{x}$ if $\lambda > 1$, or to contract $\mathbf{x}$ if $0 < \lambda < 1$, or to leave $\mathbf{x}$ fixed if $\lambda = 1$. This is illustrated graphically in Figure 6.1.

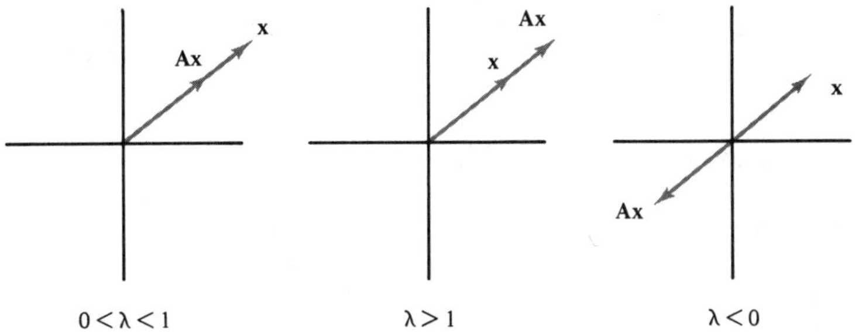

Figure 6.1

For example, $\lambda = 3$ is an eigenvalue of

$$\mathbf{A} = \begin{bmatrix} 4 & -1 \\ 2 & 1 \end{bmatrix}$$

with corresponding eigenvectors $\mathbf{x} = \alpha[1 \quad 1]^T$, $\alpha \neq 0$. The nullspace of $\mathbf{A} - 3\mathbf{I}$ is the line depicted in Figure 6.2. Any nonzero vector on the line is an eigenvector corresponding to $\lambda = 3$.

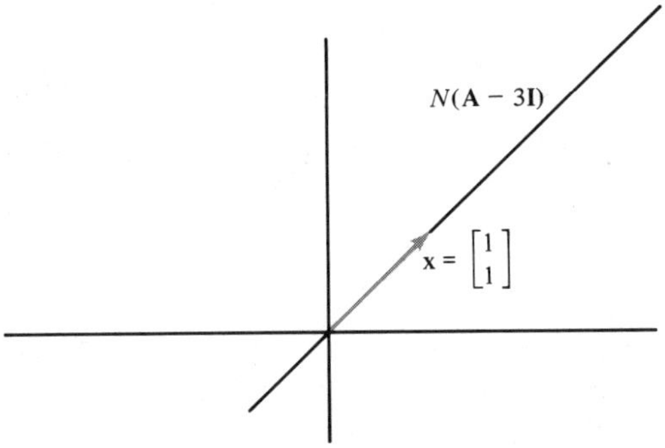

Figure 6.2

There can be a plane of eigenvectors corresponding to a particular eigenvalue. We leave it to the reader to verify that $\lambda = 1$ is an eigenvalue of

$$\mathbf{A} = \begin{bmatrix} 1 & 4 & 4 \\ 0 & 3 & 2 \\ 0 & 2 & 3 \end{bmatrix}.$$

The eigenvectors corresponding to $\lambda = 1$ take the form [verify this by solving $(\mathbf{A} - \lambda\mathbf{I})\mathbf{x} = \boldsymbol{\theta}$]

$$\mathbf{x} = \begin{bmatrix} x_1 \\ -x_3 \\ x_3 \end{bmatrix} \neq \boldsymbol{\theta}.$$

That is, for $\lambda = 1$, the nullspace of $\mathbf{A} - \mathbf{I}$ is a plane P consisting of all vectors $\mathbf{x}$ whose entries satisfy $x_2 = -x_3$. Note that since

$$\begin{bmatrix} x_1 \\ -x_3 \\ x_3 \end{bmatrix} = x_1 \begin{bmatrix} 1 \\ 0 \\ 0 \end{bmatrix} + x_3 \begin{bmatrix} 0 \\ -1 \\ 1 \end{bmatrix},$$

the plane P is the span of $\{\mathbf{u}^{(1)}, \mathbf{u}^{(2)}\}$, where

$$\mathbf{u}^{(1)} = \begin{bmatrix} 1 \\ 0 \\ 0 \end{bmatrix} \quad \text{and} \quad \mathbf{u}^{(2)} = \begin{bmatrix} 0 \\ -1 \\ 1 \end{bmatrix}.$$

Any nonzero vector in the plane P of Figure 6.3 is an eigenvector corresponding to $\lambda = 1$.

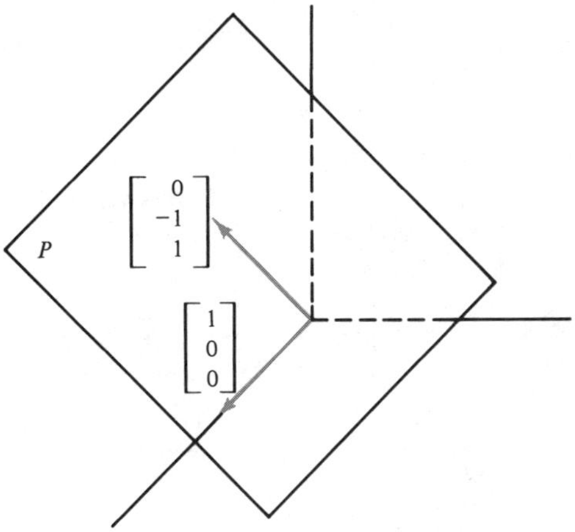

Figure 6.3

EXERCISES 6.1

1. Show that if **A** is an $m \times n$ matrix and **x** is an n-vector such that $\mathbf{Ax} = \lambda\mathbf{x}$ for some scalar λ, then $m = n$.

2. Show that $\mathbf{x} = [1 \quad -4 \quad 5]^{\mathrm{T}}$ is an eigenvector of

$$\mathbf{A} = \begin{bmatrix} 5 & 1 & 0 \\ -2 & -2 & -2 \\ 10 & 5 & 3 \end{bmatrix}.$$

3. An eigenvalue of **A** is $\lambda = -6$. Find a corresponding eigenvector.

$$\mathbf{A} = \begin{bmatrix} -1 & 1 & -3 \\ 1 & -5 & 1 \\ -3 & 1 & -1 \end{bmatrix}$$

In Exercises 4–8, find the characteristic polynomial and the eigenvalues of **A**.

4. $\mathbf{A} = \begin{bmatrix} 2 & 6 \\ 1 & 2 \end{bmatrix}$

5. $\mathbf{A} = \begin{bmatrix} 5 & 4 \\ -6 & -3 \end{bmatrix}$

6. $\mathbf{A} = \begin{bmatrix} 0 & 1 & 0 \\ 0 & 0 & 0 \\ 0 & 2 & 0 \end{bmatrix}$

7. $\mathbf{A} = \begin{bmatrix} 0 & 1 & -1 \\ 1 & 0 & -1 \\ 0 & 1 & 2 \end{bmatrix}$

8. $\mathbf{A} = \begin{bmatrix} 4 & 1 & 7 & 2 \\ -1 & 2 & 5 & -6 \\ 0 & 0 & 3 & 2 \\ 0 & 0 & 4 & -1 \end{bmatrix}$

9. Show that

$$\mathbf{A} = \begin{bmatrix} 1 & -3 \\ 1 & 1 \end{bmatrix}$$

has no real eigenvalues.

10. **(a)** Without calculating any determinants, find the eigenvalues of

$$\mathbf{A} = \begin{bmatrix} 3 & 6 & -1 \\ 0 & -1 & 4 \\ 0 & 0 & 2 \end{bmatrix}.$$

(b) What are the eigenvalues of an upper triangular matrix? Lower triangular matrix?

11. Find the eigenvalues of the transpose of the matrix in Exercise 6.

12. In each of the following matrices, $\lambda = 2$ is a repeated eigenvalue. For each matrix, how many linearly independent eigenvectors correspond to $\lambda = 2$?

$$\mathbf{A} = \begin{bmatrix} 2 & 0 & 0 \\ 0 & 2 & 0 \\ 0 & 0 & 2 \end{bmatrix}, \quad \mathbf{B} = \begin{bmatrix} 2 & 0 & 1 \\ 0 & 2 & 0 \\ 0 & 0 & 2 \end{bmatrix}, \quad \mathbf{C} = \begin{bmatrix} 2 & 1 & 0 \\ 0 & 2 & 1 \\ 0 & 0 & 2 \end{bmatrix}$$

Find a basis for $N(\mathbf{A} - 2\mathbf{I})$, for $N(\mathbf{B} - 2\mathbf{I})$, and for $N(\mathbf{C} - 2\mathbf{I})$.

13. Which vectors $\mathbf{x} \in \mathbf{R}^3$ are eigenvectors for $\mathbf{I}_3$?

14. Verify Theorem 6.2 for the matrix of Exercise 3.

15. Suppose that $\mathbf{A}$ is a 2×2 matrix which satisfies $\det(\mathbf{A}) = 15$ and $\operatorname{tr}(\mathbf{A}) = 8$. Find the eigenvalues of $\mathbf{A}$.

16. Suppose that $\mathbf{A}$ is a 3×3 matrix and $\lambda = -1$ is an eigenvalue of $\mathbf{A}$. Determine the other eigenvalues of $\mathbf{A}$ given that $\det(\mathbf{A}) = -18$ and $\operatorname{tr}(\mathbf{A}) = 8$.

***17.** Suppose that n is an odd integer and $\mathbf{A}$ is a real $n \times n$ matrix. Show that $\mathbf{A}$ has at least one real eigenvalue.

18. If $\mathbf{Ax} = \lambda\mathbf{x}$ with $\mathbf{x} \neq \boldsymbol{\theta}$, show that

$$\lambda = \frac{\langle \mathbf{Ax}, \mathbf{x} \rangle}{\langle \mathbf{x}, \mathbf{x} \rangle}.$$

***19.** Suppose that $\mathbf{x}$ is an eigenvector of $\mathbf{A}$ corresponding to the eigenvalue λ.

(a) Show that $\mathbf{x}$ is an eigenvector of $2\mathbf{A}$. Find the corresponding eigenvalue.

(b) Show that $\lambda + 6$ is an eigenvalue of $\mathbf{A} + 6\mathbf{I}$. Find a corresponding eigenvector.

(c) If $\mathbf{A}$ is invertible, show that $\mathbf{x}$ is an eigenvector of $\mathbf{A}^{-1}$. Find the corresponding eigenvalue.

***20.** Suppose that $\mathbf{Ax} = \lambda\mathbf{x}$ with $\mathbf{x} \neq \boldsymbol{\theta}$. Find an eigenvalue and corresponding eigenvector of $\mathbf{B} = \mathbf{P}^{-1}\mathbf{AP}$.

***21.** If $\mathbf{B} = \mathbf{P}^{-1}\mathbf{AP}$, find a relationship between $p_{\mathbf{A}}(\lambda)$ and $p_{\mathbf{B}}(\lambda)$.

22. A 4×4 Hessenberg matrix has the form

$$\mathbf{A} = \begin{bmatrix} a_{11} & a_{12} & a_{13} & a_{14} \\ a_{21} & a_{22} & a_{23} & a_{24} \\ 0 & a_{32} & a_{33} & a_{34} \\ 0 & 0 & a_{43} & a_{44} \end{bmatrix}$$

with $a_{21}a_{32}a_{43} \neq 0$. Show that $\mathbf{x} = [x_1, x_2, x_3, 0]^T$ cannot be an eigenvector of $\mathbf{A}$. [*Hint:* Solve the homogeneous system $(\mathbf{A} - \lambda\mathbf{I})\mathbf{x} = \boldsymbol{\theta}$.]

6.2 Properties of Eigenvectors and Eigenvalues

Our primary aim in this section is to determine the linearly independent eigenvectors of a given matrix. The best possible situation is for an $n \times n$ matrix $\mathbf{A}$ to have n linearly independent eigenvectors. In this case $\mathbf{R}^n$ has a basis consisting of eigenvectors of $\mathbf{A}$ and the action of $\mathbf{A}$ on any vector $\mathbf{x}$ is completely determined by the action of $\mathbf{A}$ on its eigenvectors. Fortunately, an important class of matrices, the symmetric matrices, have this property.

We begin by deriving some simple properties which are easy consequences of the definition of an eigenvalue. Let $\mathbf{Ax} = \lambda\mathbf{x}$ with $\mathbf{x} \neq \boldsymbol{\theta}$. Then for any scalar α we have

$$(\alpha\mathbf{A})\mathbf{x} = \alpha(\mathbf{Ax}) = \alpha(\lambda\mathbf{x}) = (\alpha\lambda)\mathbf{x}$$

and hence $\alpha\lambda$ is an eigenvalue of $\alpha\mathbf{A}$. Also, it is easy to see that $\lambda + \alpha$ is an eigenvalue of $\mathbf{A} + \alpha\mathbf{I}$:

$$(\mathbf{A} + \alpha\mathbf{I})\mathbf{x} = \lambda\mathbf{x} + \alpha\mathbf{x} = (\lambda + \alpha)\mathbf{x}.$$

If $\mathbf{A}$ is invertible, we have

$$\mathbf{x} = \mathbf{A}^{-1}\mathbf{Ax} = \mathbf{A}^{-1}(\lambda\mathbf{x}) = \lambda\mathbf{A}^{-1}\mathbf{x},$$

and since $\lambda \neq 0$ (See Exercise 18 of Section 6.1), we find that $\mathbf{A}^{-1}\mathbf{x} = \lambda^{-1}\mathbf{x}$. Thus λ^{-1} is an eigenvalue of $\mathbf{A}^{-1}$.

Two other useful properties of eigenvalues are consequences of the characteristic equation. First we show that every eigenvalue of $\mathbf{A}$ is also an eigenvalue of $\mathbf{A}^{\mathrm{T}}$. Recalling (Exercise 14 of Section 3.1) that the determinant of a matrix is equal to the determinant of its transpose, we find

$$0 = \det(\mathbf{A} - \lambda\mathbf{I}) = \det((\mathbf{A} - \lambda\mathbf{I})^{\mathrm{T}}) = \det(\mathbf{A}^{\mathrm{T}} - \lambda\mathbf{I}).$$

Next suppose that $\mathbf{A}$ is either upper or lower triangular. Then so is $\mathbf{A} - \lambda\mathbf{I}$ and hence by properties of the determinant we have

$$p_{\mathbf{A}}(\lambda) = |\mathbf{A} - \lambda\mathbf{I}| = (a_{11} - \lambda)(a_{22} - \lambda) \cdots (a_{nn} - \lambda).$$

It follows that the eigenvalues of a triangular matrix $\mathbf{A}$ equal the diagonal entries of $\mathbf{A}$. Let us summarize these properties of eigenvalues and eigenvectors in the following theorem.

Theorem 6.3. Suppose that $\mathbf{Ax} = \lambda\mathbf{x}$ with $\mathbf{x} \neq \boldsymbol{\theta}$. Then:
(a) $\alpha\lambda$ is an eigenvalue of $\alpha\mathbf{A}$ and $(\alpha\mathbf{A})\mathbf{x} = (\alpha\lambda)\mathbf{x}$.
(b) $\lambda + \alpha$ is an eigenvalue of $(\mathbf{A} + \alpha\mathbf{I})$ and $(\mathbf{A} + \alpha\mathbf{I})\mathbf{x} = (\lambda + \alpha)\mathbf{x}$.
(c) $\mathbf{A}^{-1}\mathbf{x} = \lambda^{-1}\mathbf{x}$ if $\mathbf{A}$ is invertible.
(d) λ is an eigenvalue of $\mathbf{A}^{\mathrm{T}}$.
(e) If $\mathbf{A}$ is triangular, then $\lambda = a_{ii}$ for some i.

■

The following example illustrates many of the results in Theorem 6.3.

Example 6.8

Let

$$\mathbf{A} = \begin{bmatrix} 3 & 0 & 0 \\ -2 & 0 & 0 \\ 1 & -4 & 5 \end{bmatrix} \quad \text{and} \quad \mathbf{B} = \begin{bmatrix} 4 & -1 & 0 \\ -1 & 5 & -1 \\ 0 & -1 & 4 \end{bmatrix}.$$

Then $\mathbf{A}$ is triangular with eigenvalues $\lambda_1 = 3$, $\lambda_2 = 0$, and $\lambda_3 = 5$. Since $\lambda_2 = 0$, $\mathbf{A}$ is not invertible. The eigenvalues of $\mathbf{A} + 6\mathbf{I}$ are $\lambda_1 + 6 = 9$, $\lambda_2 + 6 = 6$, $\lambda_3 + 6 = 11$. In Example 6.7 we found the eigenvalues of $\mathbf{B}$:

$\mu_1 = 3$, $\mu_2 = 4$, $\mu_3 = 6$. Since none of these eigenvalues is zero, $\mathbf{B}$ is invertible and $\mathbf{B}^{-1}$ has eigenvalues $\frac{1}{3}$, $\frac{1}{4}$, and $\frac{1}{6}$.

■

Next we consider the linear independence of the eigenvectors of a matrix. We begin by considering the eigenvectors which correspond to a particular eigenvalue of $\mathbf{A}$. How many linearly independent eigenvectors correspond to λ? We know that for some $\mathbf{x} \neq \boldsymbol{\theta}$,

$$(\mathbf{A} - \lambda\mathbf{I})\mathbf{x} = \boldsymbol{\theta}.$$

Thus $\mathbf{x}$ is in the nullspace of $\mathbf{A} - \lambda\mathbf{I}$. It follows that the number, call it k, of linearly independent eigenvectors corresponding to λ is given by

$$k = n(\mathbf{A} - \lambda\mathbf{I}),$$

or equivalently, if $\mathbf{A}$ is $n \times n$,

$$k = n - r(\mathbf{A} - \lambda\mathbf{I})$$

by the rank-nullity theorem.

Example 6.9

For

$$\mathbf{A} = \begin{bmatrix} 1 & 2 & 6 \\ 0 & 3 & 1 \\ 0 & 2 & 2 \end{bmatrix}$$

the characteristic polynomial is

$$p_{\mathbf{A}}(\lambda) = -(\lambda - 1)(\lambda - 1)(\lambda - 4)$$

and the eigenvalues of $\mathbf{A}$ are $\lambda_1 = \lambda_2 = 1$ and $\lambda_3 = 4$. The number of linearly independent eigenvectors corresponding to the repeated eigenvalue 1 is $k = 3 - r(\mathbf{A} - \mathbf{I})$. Using elimination we find an echelon form for $\mathbf{A} - \mathbf{I}$:

$$\mathbf{A} - \mathbf{I} = \begin{bmatrix} 0 & 2 & 6 \\ 0 & 2 & 1 \\ 0 & 2 & 1 \end{bmatrix} \longrightarrow \begin{bmatrix} 0 & 2 & 6 \\ 0 & 0 & -5 \\ 0 & 0 & -5 \end{bmatrix} \longrightarrow \begin{bmatrix} 0 & 2 & 6 \\ 0 & 0 & -5 \\ 0 & 0 & 0 \end{bmatrix}.$$

Thus the rank of $\mathbf{A} - \mathbf{I}$ is 2 and hence $\mathbf{A}$ has only one linearly independent eigenvector corresponding to $\lambda_1 = \lambda_2 = 1$. We leave it to the reader to verify that $r(\mathbf{A} - 4\mathbf{I}) = 2$ and hence there is only one linearly independent eigenvector corresponding to $\lambda_3 = 4$.

■

We say that μ is an eigenvalue of **multiplicity** m for $\mathbf{A}$ if $(\lambda - \mu)^m$ is a factor of $p_{\mathbf{A}}(\lambda)$ but $(\lambda - \mu)^{m+1}$ is not a factor of $p_{\mathbf{A}}(\lambda)$, that is, μ is a repeated (m times) root of $p_{\mathbf{A}}(\lambda) = 0$. In Example 6.9, $\mu = 1$ is an eigenvalue of multiplicity 2. If $m = 1$, we call μ a **simple eigenvalue** of $\mathbf{A}$. Let $\lambda_1, \ldots, \lambda_j$ denote the distinct eigenvalues of the $n \times n$ matrix $\mathbf{A}$ with multiplicities $m_1, \ldots, m_j$, respectively.

Then, since $p_{\mathbf{A}}(\lambda)$ has n roots, we have

$$n = m_1 + \cdots + m_j.$$

The following theorem, which is proved in more advanced text books, shows that the multiplicity of an eigenvalue is never less than the number of linearly independent eigenvectors corresponding to the eigenvalue.

Theorem 6.4. Let μ be an eigenvalue of multiplicity m for the $n \times n$ matrix $\mathbf{A}$. Then the number, k, of linearly independent eigenvectors corresponding to μ satisfies

$$1 \leq k = n - r(\mathbf{A} - \mu\mathbf{I}) = n(\mathbf{A} - \mu\mathbf{I}) \leq m.$$

■

Clearly, we must have $k \geq 1$ in this theorem since $\mathbf{Ax} = \mu\mathbf{x}$ for some $\mathbf{x} \neq \boldsymbol{\theta}$. That is, $n(\mathbf{A} - \mu\mathbf{I}) \geq 1$. Moreover, if μ is a simple eigenvalue, there corresponds exactly one linearly independent eigenvector.

Example 6.10

We find the number of linearly independent eigenvectors corresponding to each distinct eigenvalue of

$$\mathbf{A} = \begin{bmatrix} 1 & 4 & 4 \\ 0 & 3 & 2 \\ 0 & 2 & 3 \end{bmatrix}.$$

The characteristic polynomial is $p_{\mathbf{A}}(\lambda) = (1 - \lambda)^2(5 - \lambda)$ and the distinct eigenvalues are $\lambda_1 = 1$ and $\lambda_2 = 5$ with multiplicities $m_1 = 2$ and $m_2 = 1$, respectively. From Theorem 6.4 there is one linearly independent eigenvector corresponding to λ_2. For $\lambda_1 = 1$ we calculate $r(\mathbf{A} - \mathbf{I})$:

$$\mathbf{A} - \mathbf{I} = \begin{bmatrix} 0 & 4 & 4 \\ 0 & 2 & 2 \\ 0 & 2 & 2 \end{bmatrix} \longrightarrow \begin{bmatrix} 0 & 4 & 4 \\ 0 & 0 & 0 \\ 0 & 0 & 0 \end{bmatrix}.$$

Clearly, $r(\mathbf{A} - \mathbf{I}) = 1$ and hence there are $k = 3 - r(\mathbf{A} - \mathbf{I}) = 2$ linearly independent eigenvectors corresponding to λ_1. Solving $(\mathbf{A} - \mathbf{I})\mathbf{x} = \boldsymbol{\theta}$ gives the eigenvectors

$$\mathbf{x} = \begin{bmatrix} x_1 \\ -x_3 \\ x_3 \end{bmatrix} = x_1 \begin{bmatrix} 1 \\ 0 \\ 0 \end{bmatrix} + x_3 \begin{bmatrix} 0 \\ -1 \\ 1 \end{bmatrix}.$$

Two linearly independent eigenvectors corresponding to $\lambda_1 = 1$ are

$$\begin{bmatrix} 1 \\ 0 \\ 0 \end{bmatrix} \quad \text{and} \quad \begin{bmatrix} 0 \\ -1 \\ 1 \end{bmatrix}.$$

■

Up to this point we have considered the linearly independent eigenvectors corresponding to a particular eigenvalue. The next theorem guarantees that eigenvectors which correspond to distinct eigenvalues are linearly independent.

Theorem 6.5. Suppose that the distinct eigenvalues of $\mathbf{A}$ are $\lambda_1, \ldots, \lambda_j$ and $\mathbf{x}^{(1)}, \ldots, \mathbf{x}^{(j)}$ are corresponding eigenvectors. Then $\{\mathbf{x}^{(i)}\}_{i=1}^{j}$ is linearly independent.

Proof. For simplicity we prove the theorem for $j = 2$. That is, we suppose that for two nonzero vectors $\mathbf{x}^{(1)}$ and $\mathbf{x}^{(2)}$, $\mathbf{A}\mathbf{x}^{(1)} = \lambda_1\mathbf{x}^{(1)}$ and $\mathbf{A}\mathbf{x}^{(2)} = \lambda_2\mathbf{x}^{(2)}$ with $\lambda_1 \neq \lambda_2$. Now suppose that

$$\alpha_1\mathbf{x}^{(1)} + \alpha_2\mathbf{x}^{(2)} = \boldsymbol{\theta}$$

for some scalars α_1 and α_2. To show that $\mathbf{x}^{(1)}$ and $\mathbf{x}^{(2)}$ are linearly independent, we must demonstrate that $\alpha_1 = \alpha_2 = 0$. Multiplying by $\mathbf{A}$ gives

$$\mathbf{A}(\alpha_1\mathbf{x}^{(1)} + \alpha_2\mathbf{x}^{(2)}) = \alpha_1\lambda_1\mathbf{x}^{(1)} + \alpha_2\lambda_2\mathbf{x}^{(2)} = \mathbf{A}\boldsymbol{\theta} = \boldsymbol{\theta}$$

and multiplying by λ_1 gives

$$\alpha_1\lambda_1\mathbf{x}^{(1)} + \alpha_2\lambda_1\mathbf{x}^{(2)} = \boldsymbol{\theta}.$$

Subtraction of the last two equations gives

$$\alpha_2(\lambda_1 - \lambda_2)\mathbf{x}^{(2)} = \boldsymbol{\theta}.$$

But $\mathbf{x}^{(2)} \neq \boldsymbol{\theta}$ and $\lambda_1 \neq \lambda_2$, so we must have $\alpha_2 = 0$. Therefore,

$$\boldsymbol{\theta} = \alpha_1\mathbf{x}^{(1)} + \alpha_2\mathbf{x}^{(2)} = \alpha_1\mathbf{x}^{(1)} + 0\mathbf{x}^{(2)} = \alpha_1\mathbf{x}^{(1)},$$

and since $\mathbf{x}^{(1)} \neq \boldsymbol{\theta}$ we get $\alpha_1 = 0$.

■

Example 6.11

In Example 6.7 we found that

$$\mathbf{A} = \begin{bmatrix} 4 & -1 & 0 \\ -1 & 5 & -1 \\ 0 & -1 & 4 \end{bmatrix}$$

has eigenvalues $\lambda_1 = 3$, $\lambda_2 = 4$, and $\lambda_3 = 6$. The corresponding eigenvectors are

$$\mathbf{x}^{(1)} = \begin{bmatrix} 1 \\ 1 \\ 1 \end{bmatrix}, \quad \mathbf{x}^{(2)} = \begin{bmatrix} -1 \\ 0 \\ 1 \end{bmatrix}, \quad \mathbf{x}^{(3)} = \begin{bmatrix} 1 \\ -2 \\ 1 \end{bmatrix}.$$

By Theorem 6.5 the set of eigenvectors $\{\mathbf{x}^{(j)}\}_{j=1}^{3}$ is linearly independent since the eigenvalues are distinct.

■

Let us consider the special case of Theorem 6.5 in which the $n \times n$ matrix $\mathbf{A}$ has n distinct real eigenvalues, say $\lambda_1, \ldots, \lambda_n$. Then each eigenvalue is simple and the corresponding set of eigenvectors $\{\mathbf{x}^{(j)}\}_{j=1}^{n}$ is linearly independent. Since any set of n linearly independent vectors in $\mathbf{R}^n$ forms a basis for $\mathbf{R}^n$ (see the discussion following Theorem 4.10), we have:

Theorem 6.6. Suppose that the $n \times n$ matrix $\mathbf{A}$ has distinct real eigenvalues $\lambda_1, \ldots, \lambda_n$ with corresponding eigenvectors $\mathbf{x}^{(1)}, \ldots, \mathbf{x}^{(n)}$. Then $\{\mathbf{x}^{(i)}\}_{i=1}^{n}$ is a basis for $\mathbf{R}^n$.

■

This theorem gives sufficient conditions for $\mathbf{R}^n$ to have a basis consisting of eigenvectors of $\mathbf{A}$. The matrix of Example 6.11 has three distinct eigenvalues and the eigenvectors $\{\mathbf{x}^{(i)}\}_{i=1}^{3}$ constitute a basis for $\mathbf{R}^3$. However, as the next example shows, there are matrices having multiple eigenvalues whose eigenvectors form a basis for $\mathbf{R}^n$.

Example 6.12

The matrix of Example 6.10 has eigenvalues $\lambda_1 = 1$ with multiplicity $m_1 = 2$ and $\lambda_2 = 5$ with $m_2 = 1$. For λ_1 we found two linearly independent eigenvectors:

$$\begin{bmatrix} 1 \\ 0 \\ 0 \end{bmatrix}, \quad \begin{bmatrix} 0 \\ -1 \\ 1 \end{bmatrix}.$$

An easy calculation gives an eigenvector corresponding to $\lambda_2 = 5$:

$$\mathbf{A} - 5\mathbf{I} = \begin{bmatrix} -4 & 4 & 4 \\ 0 & -2 & 2 \\ 0 & 2 & -2 \end{bmatrix} \longrightarrow \begin{bmatrix} -4 & 4 & 4 \\ 0 & -2 & 2 \\ 0 & 0 & 0 \end{bmatrix}.$$

Thus solutions of $(\mathbf{A} - 5\mathbf{I})\mathbf{x} = \boldsymbol{\theta}$ are given by $x_2 = x_3$ and $x_1 = x_2 + x_3 = 2x_3$. The corresponding eigenvectors are

$$\mathbf{x} = x_3 \begin{bmatrix} 2 \\ 1 \\ 1 \end{bmatrix}, \qquad x_3 \neq 0.$$

We claim that the eigenvectors

$$\begin{bmatrix} 1 \\ 0 \\ 0 \end{bmatrix}, \quad \begin{bmatrix} 0 \\ -1 \\ 1 \end{bmatrix}, \quad \begin{bmatrix} 2 \\ 1 \\ 1 \end{bmatrix}$$

form a basis for $\mathbf{R}^3$. To show that we find the rank of the matrix of eigenvectors of $\mathbf{A}$. Thus we reduce to echelon form the matrix $\mathbf{E}$ whose columns are eigenvectors of **A**. We have

$$\mathbf{E} = \begin{bmatrix} 1 & 0 & 2 \\ 0 & -1 & 1 \\ 0 & 1 & 1 \end{bmatrix} \longrightarrow \begin{bmatrix} 1 & 0 & 2 \\ 0 & -1 & 1 \\ 0 & 0 & 2 \end{bmatrix}$$

and hence **E** has rank 3. Thus the eigenvectors form a basis for $\mathbf{R}^3$.

■

An $n \times n$ matrix that does not have n linearly independent eigenvectors is called **defective.** To determine whether a given matrix is defective, it is necessary to calculate all of its eigenvectors. Theorem 6.6 guarantees no difficulty when the eigenvalues are all distinct. When an eigenvalue μ is repeated m times (i.e., its multiplicity is m), the nullity of $\mathbf{A} - \mu\mathbf{I}$ is the key. If $n(\mathbf{A} - \mu\mathbf{I}) = m$, there are enough linearly independent eigenvectors corresponding to μ. Moreover, if each multiple eigenvalue has the same number of linearly independent eigenvectors as its multiplicity, $\mathbf{R}^n$ has a basis consisting of eigenvectors of **A**. To put it another way, **A** is defective if and only if there is an eigenvalue μ of **A** having multiplicity m with $n(\mathbf{A} - \mu\mathbf{I}) < m$. The matrix of Example 6.9 is defective.

Now we examine the eigenvalues and eigenvectors of an important class of matrices—the real symmetric ones. Our first result concerns the eigenvalues of symmetric matrices.

Theorem 6.7. The eigenvalues of a real symmetric matrix are real.

Proof. Suppose that **A** is a real symmetric matrix and $\mathbf{Ax} = \lambda\mathbf{x}$, $\mathbf{x} \neq \boldsymbol{\theta}$. Then, using the complex inner product, we have

$$\lambda\langle\mathbf{x}, \mathbf{x}\rangle = \langle\lambda\mathbf{x}, \mathbf{x}\rangle = \langle\mathbf{Ax}, \mathbf{x}\rangle = \langle\mathbf{x}, \mathbf{Ax}\rangle = \langle\mathbf{x}, \lambda\mathbf{x}\rangle = \bar{\lambda}\langle\mathbf{x}, \mathbf{x}\rangle$$

and hence

$$(\lambda - \bar{\lambda})\langle\mathbf{x}, \mathbf{x}\rangle = 0.$$

But $\langle\mathbf{x}, \mathbf{x}\rangle = \|\mathbf{x}\|^2 \neq 0$, so we must have $\lambda = \bar{\lambda}$; thus λ is real.

■

Recall that eigenvectors corresponding to different eigenvalues are linearly independent. For real symmetric matrices even more is true.

Theorem 6.8. Suppose that **A** is a real symmetric matrix; then eigenvectors of **A** which correspond to distinct eigenvalues are orthogonal.

Proof. Suppose that $\mathbf{Ax} = \lambda\mathbf{x}$ and $\mathbf{Av} = \mu\mathbf{v}$ with $\lambda \neq \mu$. Then

$$\langle\mathbf{Ax}, \mathbf{v}\rangle = \langle\lambda\mathbf{x}, \mathbf{v}\rangle = \lambda\langle\mathbf{x}, \mathbf{v}\rangle$$

and by symmetry,

$$\langle\mathbf{Ax}, \mathbf{v}\rangle = \langle\mathbf{x}, \mathbf{Av}\rangle = \langle\mathbf{x}, \mu\mathbf{v}\rangle = \mu\langle\mathbf{x}, \mathbf{v}\rangle.$$

Therefore, $\lambda\langle \mathbf{x}, \mathbf{v}\rangle = \mu\langle \mathbf{x}, \mathbf{v}\rangle$, or equivalently $(\lambda - \mu)\langle \mathbf{x}, \mathbf{v}\rangle = 0$. Since $\lambda \neq \mu$, we must have $\langle \mathbf{x}, \mathbf{v}\rangle = 0$.

■

Example 6.13

The eigenvalues of

$$\mathbf{A} = \begin{bmatrix} 1 & 2 & -4 \\ 2 & -2 & -2 \\ -4 & -2 & 1 \end{bmatrix}$$

are $\lambda_1 = -3$, $\lambda_2 = -3$, $\lambda_3 = 6$ with corresponding eigenvectors

$$\mathbf{x}^{(1)} = \begin{bmatrix} -1 \\ 2 \\ 0 \end{bmatrix}, \quad \mathbf{x}^{(2)} = \begin{bmatrix} 1 \\ 0 \\ 1 \end{bmatrix}, \quad \text{and} \quad \mathbf{x}^{(3)} = \begin{bmatrix} -2 \\ -1 \\ 2 \end{bmatrix}.$$

Each of the eigenvectors, $\mathbf{x}^{(1)}$ and $\mathbf{x}^{(2)}$, corresponding to the eigenvalue $\lambda = -3$ is orthogonal to $\mathbf{x}^{(3)}$:

$$\langle \mathbf{x}^{(1)}, \mathbf{x}^{(3)}\rangle = (-1)(-2) + (2)(-1) + 0(2) = 0$$

$$\langle \mathbf{x}^{(2)}, \mathbf{x}^{(3)}\rangle = (1)(-2) + 0(-1) + (1)(2) = 0.$$

■

Finally, we state an important theorem for symmetric matrices concerning a basis of eigenvectors. The proof of this theorem may be found in more advanced books.

Theorem 6.9. For each $n \times n$ real symmetric matrix $\mathbf{A}$ there is an orthonormal basis $\{\mathbf{u}^{(i)}\}_{i=1}^{n}$ for $\mathbf{R}^n$ consisting of eigenvectors of $\mathbf{A}$.

■

We shall construct an orthonormal basis of eigenvectors in the next example. First we point out that such an orthonormal basis is easy to find if $\mathbf{A}$ has only simple eigenvalues. Then, by Theorem 6.8, the eigenvectors are orthogonal and it is only necessary to normalize them.

Example 6.14

For the matrix of Example 6.13 there are two linearly independent eigenvectors corresponding to the multiple eigenvalue $\lambda = -3$:

$$\mathbf{x}^{(1)} = \begin{bmatrix} -1 \\ 2 \\ 0 \end{bmatrix} \quad \text{and} \quad \mathbf{x}^{(2)} = \begin{bmatrix} 1 \\ 0 \\ 1 \end{bmatrix}.$$

For the other eigenvalue ($\lambda_3 = 6$) the corresponding eigenvector $\mathbf{x}^{(3)} = [-2 \quad -1 \quad 2]^T$ is normalized to give $\mathbf{u}^{(3)} = \mathbf{x}^{(3)}/\|\mathbf{x}^{(3)}\| = \frac{1}{3}\mathbf{x}^{(3)}$. We apply the

Gram–Schmidt process (see Section 5.3) to $\{\mathbf{x}^{(1)}, \mathbf{x}^{(2)}\}$ in order to obtain an orthonormal basis. First we set

$$\mathbf{u}^{(1)} = \frac{\mathbf{x}^{(1)}}{\|\mathbf{x}^{(1)}\|} = \frac{1}{\sqrt{5}}\mathbf{x}^{(1)}$$

and then calculate

$$\begin{aligned}\mathbf{y}^{(2)} &= \mathbf{x}^{(2)} - \langle \mathbf{x}^{(2)}, \mathbf{u}^{(1)}\rangle \mathbf{u}^{(1)} \\ &= \begin{bmatrix} 1 \\ 0 \\ 1 \end{bmatrix} + \frac{1}{\sqrt{5}}\mathbf{u}^{(1)} = \begin{bmatrix} 1 \\ 0 \\ 1 \end{bmatrix} + \frac{1}{5}\begin{bmatrix} -1 \\ 2 \\ 0 \end{bmatrix} \\ &= \begin{bmatrix} \frac{4}{5} \\ \frac{2}{5} \\ 1 \end{bmatrix}\end{aligned}$$

and finally $\mathbf{u}^{(2)} = \mathbf{y}^{(2)}/\|\mathbf{y}^{(2)}\| = (\sqrt{5}/3)\mathbf{y}^{(2)}$. Thus an orthonormal basis of eigenvectors of $\mathbf{A}$ for $\boldsymbol{R}^3$ is

$$\frac{1}{\sqrt{5}}\begin{bmatrix} -1 \\ 2 \\ 0 \end{bmatrix}, \quad \frac{1}{3\sqrt{5}}\begin{bmatrix} 4 \\ 2 \\ 5 \end{bmatrix}, \quad \frac{1}{3}\begin{bmatrix} -2 \\ -1 \\ 2 \end{bmatrix}.$$

Notice that any linear combination of $\mathbf{x}^{(1)}$ and $\mathbf{x}^{(2)}$ is an eigenvector corresponding to $\lambda = -3$; in particular, $\mathbf{u}^{(1)}$ and $\mathbf{u}^{(2)}$ are eigenvectors corresponding to $\lambda = -3$.

■

The general procedure for producing an orthonormal basis of eigenvectors of a symmetric matrix is as follows. First normalize each eigenvector corresponding to the simple eigenvalues. For an eigenvalue of multiplicity $m > 1$, find m corresponding linearly independent eigenvectors and orthonormalize them by use of the Gram–Schmidt process. Repetition of this for each multiple eigenvalue yields an orthonormal basis.

We conclude this section by presenting a result, due to Gershgorin, which can be used to estimate the eigenvalues of a matrix.

Theorem 6.10. Let $\mathbf{A}$ be an $n \times n$ matrix and define, for $i = 1, 2, \ldots, n$, the off-diagonal row sums

$$r_i = \sum_{\substack{j=1 \\ j \neq i}}^{n} |a_{ij}|.$$

The set $D_i = \{z : |z - a_{ii}| \leq r_i\}$ is a disk in the complex plane of radius r_i centered at a_{ii}. If λ is an eigenvalue of $\mathbf{A}$, then $\lambda \in D_i$ for some i. Moreover, if D_i is disjoint from D_j for all $j \neq i$, D_i contains exactly one eigenvalue of $\mathbf{A}$.

Proof. We prove only the first part of the theorem. Suppose that $\mathbf{A}\mathbf{x} = \lambda\mathbf{x}$ with

$\mathbf{x} \neq \boldsymbol{\theta}$ and let x_i be a component of $\mathbf{x}$ with largest magnitude, that is,

$$|x_j| \leq |x_i| \qquad \text{for } j \neq i.$$

Consider the ith component of the vector equation $\mathbf{Ax} = \lambda\mathbf{x}$:

$$\sum_{j=1}^{n} a_{ij}x_j = \lambda x_i$$

Rewriting, we have

$$(\lambda - a_{ii})x_i = \sum_{\substack{j=1 \\ j\neq i}}^{n} a_{ij}x_j,$$

and hence, by the triangle inequality,

$$|\lambda - a_{ii}||x_i| \leq \sum_{\substack{j=1 \\ j\neq i}}^{n} |a_{ij}||x_j|.$$

But $|x_j| \leq |x_i|$ for all $j \neq i$ and thus

$$|\lambda - a_{ii}||x_i| \leq \sum_{\substack{j=1 \\ j\neq i}}^{n} |a_{ij}||x_i| = |x_i| \sum_{\substack{j=1 \\ j\neq i}}^{n} |a_{ij}|.$$

Dividing by $|x_i| \neq 0$ (why?) gives

$$|\lambda - a_{ii}| \leq \sum_{\substack{j=1 \\ j\neq i}}^{n} |a_{ij}| = r_i$$

and we find that $\lambda \in D_i$.

■

We illustrate Gershgorin's theorem with two examples.

Example 6.15

Consider the matrix

$$\mathbf{A} = \begin{bmatrix} 9 & -2 & -2 & -4 \\ -2 & 11 & 0 & 2 \\ -2 & 0 & 7 & -2 \\ -4 & 2 & -2 & -9 \end{bmatrix}.$$

Then $r_1 = 8$, $r_2 = 4$, $r_3 = 4$, and $r_4 = 8$. Every eigenvalue of $\mathbf{A}$ is in one of the disks:

$$D_1 = \{z : |z - 9| \leq 8\}, \qquad D_2 = \{z : |z - 11| \leq 4\}$$
$$D_3 = \{z : |z - 7| \leq 4\}, \qquad D_4 = \{z : |z + 9| \leq 8\}.$$

But since $\mathbf{A}$ is symmetric, it has real eigenvalues, and hence each eigenvalue is in one of the intervals $D_1 = [1, 17]$, $D_2 = [7, 15]$, $D_3 = [3, 11]$, or $D_4 = [-17, -1]$. Moreover, since D_4 is disjoint from D_1, D_2, and D_3, one of the eigenvalues of $\mathbf{A}$ must satisfy $-17 \le \lambda \le -1$, whereas all of the remaining eigenvalues satisfy the inequality $1 \le \lambda \le 17$. In particular, 0 is not an eigenvalue of $\mathbf{A}$ since $0 \notin D_i$ for $i = 1, 2, 3, 4$, and hence $\mathbf{A}$ is invertible. One eigenvalue of $\mathbf{A}^{-1}$ must be in the interval $[-1, -\frac{1}{17}]$ and the remaining eigenvalues are in $[\frac{1}{17}, 1]$. Thus every eigenvalue μ of $\mathbf{A}^{-1}$ satisfies $0 < |\mu| \le 1$. ■

Example 6.16

The eigenvalues of

$$\mathbf{A} = \begin{bmatrix} -3 & .1 & .2 \\ 0 & 4 & -.3 \\ .4 & -.1 & 6 \end{bmatrix}$$

can be estimated more accurately than in Example 6.15 since its diagonal entries are large relative to the corresponding off-diagonal row sums. The disks are $D_1 = \{z : |z + 3| \le .3\}$, $D_2 = \{z : |z - 4| \le .3\}$, and $D_3 = \{z : |z - 6| \le .5\}$. Moreover, the disks are mutually disjoint (see Figure 6.4) and thus each contains exactly one eigenvalue of $\mathbf{A}$. The eigenvalues of $\mathbf{A}$, possibly complex, are approximately equal to -3, 4, and 6.

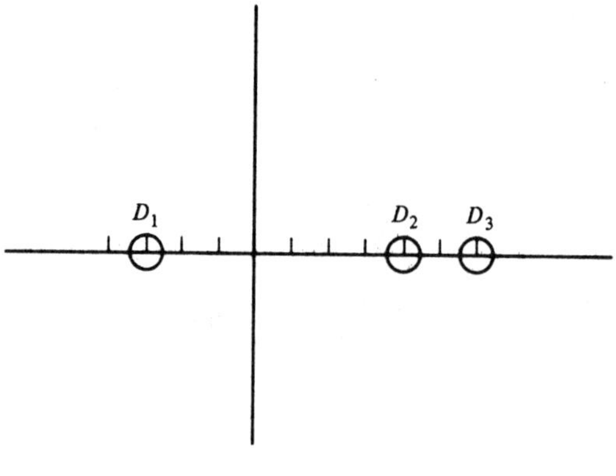

Figure 6.4

■

EXERCISES 6.2

1. The eigenvalues of $\mathbf{A}$ are $\lambda_1 = -2$, $\lambda_2 = 1$, $\lambda_3 = 4$. Find the eigenvalues of
   (a) $4\mathbf{A}$ **(b)** $\mathbf{A} - 4\mathbf{I}$ **(c)** $\mathbf{A}^{-1}$ **(d)** $\mathbf{A}^T$ **(e)** $\mathbf{A}^{-1} + 3\mathbf{I}$

In Exercise 2–5, find the number of linearly independent eigenvectors of $\mathbf{A}$ corresponding to λ.

2. $\mathbf{A} = \begin{bmatrix} -2 & -4 & 2 \\ -4 & -2 & -2 \\ 2 & -2 & -5 \end{bmatrix}$, $\lambda = 3$

3. $\mathbf{A} = \begin{bmatrix} -11 & 6 & -18 \\ -16 & 11 & -18 \\ 4 & -1 & 10 \end{bmatrix}$, $\lambda = 1$

4. $\mathbf{A} = \begin{bmatrix} -4 & 0 & 2 \\ 0 & -4 & 2 \\ 2 & 2 & -3 \end{bmatrix}, \quad \lambda = -4$ **5.** $\mathbf{A} = \begin{bmatrix} 4 & 0 & 0 \\ -1 & 8 & 2 \\ 2 & -4 & 2 \end{bmatrix}, \quad \lambda = 4$

6. Find the number of linearly independent eigenvectors of the matrix in Exercise 3.

7. Find a basis for $\mathbf{R}^3$ consisting of eigenvectors of the matrix

$$\mathbf{A} = \begin{bmatrix} 2 & -4 & 2 \\ -4 & 2 & -2 \\ 2 & -2 & -1 \end{bmatrix}.$$

8. Does $\mathbf{R}^2$ have a basis consisting of eigenvectors of

$$\mathbf{A} = \begin{bmatrix} 0 & 1 \\ 0 & 0 \end{bmatrix}?$$

9. Does $\mathbf{R}^2$ have a basis consisting of eigenvectors of

$$\mathbf{A} = \begin{bmatrix} 2 & -1 \\ 0 & 2 \end{bmatrix}?$$

Justify your answer.

10. Determine which of the following matrices is defective.

(a) $\mathbf{A} = \begin{bmatrix} a & b \\ 0 & a \end{bmatrix}$ **(b)** $\mathbf{A} = \begin{bmatrix} 2 & 2 & 0 \\ 2 & 2 & 1 \\ 0 & 0 & 2 \end{bmatrix}$ **(c)** $\mathbf{A} = \begin{bmatrix} -2 & 2 & 0 & 1 \\ 1 & -1 & 0 & 6 \\ 6 & 2 & -3 & 2 \\ 0 & 0 & 0 & 0 \end{bmatrix}$

In Exercises 11–14, find an orthonormal basis for $\mathbf{R}^n$ consisting of eigenvectors of the $n \times n$ matrix $\mathbf{A}$.

11. $\mathbf{A} = \begin{bmatrix} 1 & 2 \\ 2 & 1 \end{bmatrix}$ **12.** $\mathbf{A} = \begin{bmatrix} 1 & -1 \\ -1 & 1 \end{bmatrix}$ **13.** $\mathbf{A} = \begin{bmatrix} 4 & -4 & 5 \\ -4 & 4 & 5 \\ 5 & 5 & -5 \end{bmatrix}$

14. $\mathbf{A} = \begin{bmatrix} 2 & 2 & 0 & -1 \\ 2 & -1 & 0 & 2 \\ 0 & 0 & 3 & 0 \\ -1 & 2 & 0 & 2 \end{bmatrix}$

15. Draw the Gershgorin disks for $\mathbf{A}$.

$$\mathbf{A} = \begin{bmatrix} 7 & -1 & 1 \\ .2 & 6 & .4 \\ .3 & -.1 & 2 \end{bmatrix}$$

***16.** An $n \times n$ matrix $\mathbf{A}$ is called **row diagonally dominant** if for each $i = 1, 2, \ldots, n$,

$$|a_{ii}| > \sum_{\substack{j=1 \\ j \neq i}}^{n} |a_{ij}|.$$

Show that a row diagonally dominant matrix is invertible. (*Hint:* $\mathbf{A}$ is invertible if 0 is not an eigenvalue.)

17. Show that $\mathbf{A}$ is invertible , where

$$\mathbf{A} = \begin{bmatrix} 4 & 1 & 0 & 0 & 0 \\ 1 & 4 & 1 & 0 & 0 \\ 0 & 1 & 4 & 1 & 0 \\ 0 & 0 & 1 & 4 & 1 \\ 0 & 0 & 0 & 1 & 4 \end{bmatrix}.$$

18. Estimate the eigenvalues of the matrix of Exercise 17 by using the theorem of Gershgorin.

6.3 The Cayley–Hamilton Theorem

In this section we state a famous theorem due to Cayley and Hamilton and examine some of its applications. Before stating the theorem we need to briefly discuss matrix polynomials.

Recall that powers of square matrices are defined inductively by $\mathbf{A}^0 = \mathbf{I}$, $\mathbf{A}^1 = \mathbf{A}$, $\mathbf{A}^2 = \mathbf{AA}$, $\mathbf{A}^3 = \mathbf{A}^2\mathbf{A}$, and so on. Now if $q(t) = a_0 + a_1 t + \cdots + a_n t^n$ is a given polynomial, we define the matrix polynomial $q(\mathbf{A})$ to be the matrix given by

$$q(\mathbf{A}) = a_0\mathbf{I} + a_1\mathbf{A} + \cdots + a_n\mathbf{A}^n.$$

Example 6.17

Let $q(t) = -5 + 3t^2 + t^3$ and

$$\mathbf{A} = \begin{bmatrix} 1 & 1 \\ -1 & 1 \end{bmatrix}.$$

Then $q(\mathbf{A}) = -5\mathbf{I} + 3\mathbf{A}^2 + \mathbf{A}^3$. We find that

$$\mathbf{A}^2 = \begin{bmatrix} 0 & 2 \\ -2 & 0 \end{bmatrix} \quad \text{and} \quad \mathbf{A}^3 = \begin{bmatrix} -2 & 2 \\ -2 & -2 \end{bmatrix}$$

and thus

$$\begin{aligned} q(\mathbf{A}) &= -5\begin{bmatrix} 1 & 0 \\ 0 & 1 \end{bmatrix} + 3\begin{bmatrix} 0 & 2 \\ -2 & 0 \end{bmatrix} + \begin{bmatrix} -2 & 2 \\ -2 & -2 \end{bmatrix} \\ &= \begin{bmatrix} -7 & 8 \\ -8 & -7 \end{bmatrix}. \end{aligned}$$

■

We are now in a position to state the Cayley–Hamilton theorem, which asserts that a given square matrix satisfies its own characteristic equation.

Theorem 6.11. Suppose that $\mathbf{A}$ is a square matrix; then $p_{\mathbf{A}}(\mathbf{A}) = \mathbf{O}$.

■

The proof of this theorem is beyond the scope of this book and we are content to illustrate its validity with an example.

Example 6.18

For

$$\mathbf{A} = \begin{bmatrix} 1 & 0 & 2 \\ 0 & 1 & -1 \\ 2 & -1 & 3 \end{bmatrix}$$

we found, in Example 6.4, that $p_{\mathbf{A}}(\lambda) = -\lambda^3 + 5\lambda^2 - 2\lambda - 2$. Some straightforward calculations give

$$\mathbf{A}^2 = \begin{bmatrix} 5 & -2 & 8 \\ -2 & 2 & -4 \\ 8 & -4 & 14 \end{bmatrix} \quad \text{and} \quad \mathbf{A}^3 = \begin{bmatrix} 21 & -10 & 36 \\ -10 & 6 & -18 \\ 36 & -18 & 62 \end{bmatrix}$$

and thus

$$p_{\mathbf{A}}(\mathbf{A}) = -2\mathbf{I} - 2\begin{bmatrix} 1 & 0 & 2 \\ 0 & 1 & -1 \\ 2 & -1 & 3 \end{bmatrix} + 5\begin{bmatrix} 5 & -2 & 8 \\ -2 & 2 & -4 \\ 8 & -4 & 14 \end{bmatrix}$$
$$- \begin{bmatrix} 21 & -10 & 36 \\ -10 & 6 & -18 \\ 36 & -18 & 62 \end{bmatrix} = \begin{bmatrix} 0 & 0 & 0 \\ 0 & 0 & 0 \\ 0 & 0 & 0 \end{bmatrix}.$$

That is, $p_{\mathbf{A}}(\mathbf{A}) = \mathbf{O}$.

■

One consequence of the Cayley–Hamilton theorem is a new method of calculating $\mathbf{A}^{-1}$. If $\mathbf{A}$ is an $n \times n$ matrix, then $p_{\mathbf{A}}(\lambda)$ takes the form

$$p_{\mathbf{A}}(\lambda) = (-1)^n\lambda^n + a_{n-1}\lambda^{n-1} + \cdots + a_1\lambda + a_0,$$

where $a_0 = p_{\mathbf{A}}(0)$. If the constant term, a_0, is nonzero, then 0 is not an eigenvalue of $\mathbf{A}$ and hence $\mathbf{A}$ is invertible. By the Cayley–Hamilton theorem we have

$$(-1)^n\mathbf{A}^n + a_{n-1}\mathbf{A}^{n-1} + \cdots + a_1\mathbf{A} + a_0\mathbf{I} = \mathbf{O}.$$

Rewriting this gives

$$-(-1)^n\mathbf{A}^n - a_{n-1}\mathbf{A}^{n-1} - \cdots - a_1\mathbf{A} = a_0\mathbf{I},$$

or equivalently, if $a_0 \neq 0$,

$$\frac{-(-1)^n\mathbf{A}^n - a_{n-1}\mathbf{A}^{n-1} - \cdots - a_2\mathbf{A}^2 - a_1\mathbf{A}}{a_0} = \mathbf{I}.$$

From this it follows that

$$\mathbf{B} = \frac{-a_1\mathbf{I} - a_2\mathbf{A} - \cdots - a_{n-1}\mathbf{A}^{n-2} - (-1)^n\mathbf{A}^{n-1}}{a_0}$$

satisfies

$$\mathbf{BA} = \mathbf{AB} = \mathbf{I}$$

and hence $\mathbf{B} = \mathbf{A}^{-1}$. Thus the inverse of $\mathbf{A}$ can be determined as a polynomial of degree $n - 1$ in $\mathbf{A}$.

Example 6.19

For

$$\mathbf{A} = \begin{bmatrix} 1 & 2 \\ 2 & -2 \end{bmatrix}$$

the characteristic polynomial is $p_{\mathbf{A}}(\lambda) = \lambda^2 + \lambda - 6$ and hence $p_{\mathbf{A}}(\mathbf{A}) = \mathbf{A}^2 + \mathbf{A} - 6\mathbf{I} = \mathbf{O}$. Therefore,

$$\mathbf{A}^2 + \mathbf{A} = 6\mathbf{I},$$

or equivalently,

$$\frac{\mathbf{A}^2 + \mathbf{A}}{6} = \frac{\mathbf{A}[\mathbf{I} + \mathbf{A}]}{6} = \mathbf{I}.$$

Thus

$$\mathbf{A}^{-1} = \frac{\mathbf{I} + \mathbf{A}}{6} = \begin{bmatrix} \frac{1}{3} & \frac{1}{3} \\ \frac{1}{3} & -\frac{1}{6} \end{bmatrix}.$$

■

At the begining of the chapter we found that powers of the transition matrix determine the long-term behavior of the population distribution of an ecosystem. Computation of $\mathbf{A}^m$ for large m is tedious and time consuming. However, if the eigenvalues of $\mathbf{A}$ are known, then the Cayley–Hamilton theorem provides a reasonable way of computing $\mathbf{A}^m$. The procedure that we develop for computing $\mathbf{A}^m$ applies to the calculation of any matrix polynomial, say $q(\mathbf{A})$. The essential point is that for an $n \times n$ matrix $\mathbf{A}$, the Cayley–Hamilton theorem reduces the calculation of $q(\mathbf{A})$, no matter how large the degree of q, to the calculation of a matrix polynomial of degree less than n.

Suppose that $\mathbf{A}$ is an $n \times n$ matrix and q is a polynomial of degree $m > n$. Division of $q(t)$ by $p_{\mathbf{A}}(t)$ results in a quotient $Q(t)$ and a remainder $R(t)$, where Q and R are polynomials and degree of $R < n$. That is,

$$\frac{q(t)}{p_{\mathbf{A}}(t)} = Q(t) + \frac{R(t)}{p_{\mathbf{A}}(t)},$$

or equivalently,

$$q(t) = Q(t)p_{\mathbf{A}}(t) + R(t).$$

Now if $q(t) = Q(t)p_{\mathbf{A}}(t) + R(t)$ the Cayley–Hamilton theorem implies that

$$\begin{aligned} q(\mathbf{A}) &= Q(\mathbf{A})p_{\mathbf{A}}(\mathbf{A}) + R(\mathbf{A}) \\ &= R(\mathbf{A}) \end{aligned}$$

since $p_{\mathbf{A}}(\mathbf{A}) = \mathbf{O}$. Therefore, the calculation of $q(\mathbf{A})$ reduces to that of $R(\mathbf{A})$, where the degree of R is less than n.

Example 6.20

Suppose that $q(t) = t^5 - 3t^4 + t^2 - 4$ and

$$\mathbf{A} = \begin{bmatrix} -1 & 2 \\ 1 & -1 \end{bmatrix}.$$

Then $p_{\mathbf{A}}(t) = t^2 + 2t - 1$ and division of $q(t)$ by $p_{\mathbf{A}}(t)$ gives

$$q(t) = \underbrace{(t^3 - 5t^2 + 11t - 26)}_{Q(t)}p_{\mathbf{A}}(t) + \underbrace{63t - 30}_{R(t)}.$$

By the Cayley–Hamilton theorem we then have

$$q(\mathbf{A}) = R(\mathbf{A}) = 63\mathbf{A} - 30\mathbf{I} = \begin{bmatrix} -93 & 126 \\ 63 & -93 \end{bmatrix}.$$

■

This technique may be used in particular to compute large powers of a square matrix such as those which occur in population studies. For example, if $q(t) = t^{500}$ and $\mathbf{A}$ is a 3×3 matrix, the evaluation of $\mathbf{A}^{500} = q(\mathbf{A})$ reduces to the calculation of $R(\mathbf{A})$, where R has degree <3. In theory we could find R by performing the division $t^{500}/p_{\mathbf{A}}(t)$, but there is a much simpler method when the eigenvalues of $\mathbf{A}$ are known.

First, we examine the case where $\mathbf{A}$ has only simple eigenvalues. If $\mathbf{A}$ is an $n \times n$ matrix with distinct eigenvalues $\lambda_1, \ldots, \lambda_n$, then for $m > n$,

$$t^m = Q(t)p_{\mathbf{A}}(t) + R(t),$$

where $R(t) = c_0 + c_1 t + \cdots + c_{n-1}t^{n-1}$ for some coefficients $c_0, c_1, \ldots, c_{n-1}$. Now we use the fact that each eigenvalue of $\mathbf{A}$ satisfies the characteristic equation, that is, $p_{\mathbf{A}}(\lambda_j) = 0$, which yields

$$\lambda_j^m = R(\lambda_j), \qquad 1 \le j \le n.$$

This gives n linear equations which can be solved for the n unknown coefficients of R.

Example 6.21

For

$$\mathbf{A} = \begin{bmatrix} 1 & 2 \\ 2 & -2 \end{bmatrix}$$

we calculate $\mathbf{A}^{500}$. The eigenvalues of $\mathbf{A}$ are $\lambda_1 = -3$, and $\lambda_2 = 2$. Then for some polynomial R of degree 1, say $R(t) = c_0 + c_1 t$,

$$\mathbf{A}^{500} = R(\mathbf{A}) = c_0\mathbf{I} + c_1\mathbf{A}.$$

To find the coefficients of R, we solve the two equations

$$(\lambda_1)^{500} = (-3)^{500} = R(\lambda_1) = c_0 - 3c_1$$

$$(\lambda_2)^{500} = 2^{500} = R(\lambda_2) = c_0 + 2c_1.$$

This yields $c_0 = [2(3)^{500} + 3(2)^{500}]/5$ and $c_1 = [2^{500} - 3^{500}]/5$ and thus $\mathbf{A}^{500} = \{[2(3)^{500} + 3(2)^{500}]/5\}\mathbf{I} + \{[2^{500} - 3^{500}]/5\}\mathbf{A}$.

■

The technique used in the example to determine R must be modified if $\mathbf{A}$ has multiple eigenvalues. Suppose that μ is an eigenvalue of multiplicity $m > 1$ for $\mathbf{A}$. Then $(\lambda - \mu)^m$ is a factor of $p_{\mathbf{A}}(\lambda)$ and it follows that $p_{\mathbf{A}}(\mu) = p'_{\mathbf{A}}(\mu) = \cdots = p_{\mathbf{A}}^{(m-1)}(\mu) = 0$, where $p_{\mathbf{A}}^{(j)}$ denotes the jth derivative of $p_{\mathbf{A}}$. Consequently, if

$$q(t) = Q(t)p_{\mathbf{A}}(t) + R(t),$$

we find by differentiation that

$$q^{(j)}(\mu) = R^{(j)}(\mu), \qquad j = 0, 1, \ldots, m - 1,$$

where $q^{(0)}(\mu) = q(\mu)$. That is, q and its derivatives of order less than or equal to $m - 1$ agree with those of R at μ. Now if the $n \times n$ matrix $\mathbf{A}$ has distinct eigenvalues $\lambda_1, \ldots, \lambda_k$ of multiplicities $m_1, \ldots, m_k$, then to determine R we solve the equations: For $i = 1, 2, \ldots, k$,

$$q^{(j)}(\lambda_i) = R^{(j)}(\lambda_i), \qquad 0 \le j \le m_i - 1.$$

For the ith eigenvalue, λ_i, there are m_i equations and the total number of equations is $m_1 + \cdots + m_k = n$. Thus there are n equations for the n unknown coefficients $c_0, c_1, \ldots, c_{n-1}$ of R.

Example 6.22

The distinct eigenvalues of

$$\mathbf{A} = \begin{bmatrix} 2 & -2 & 1 \\ 2 & -3 & 2 \\ 2 & -4 & 3 \end{bmatrix}$$

are $\lambda_1 = 0$ $(m_1 = 1)$ and $\lambda_2 = 1$ $(m_2 = 2)$. To find $\mathbf{A}^{100}$ we need $R(t) = c_0 + c_1 t + c_2 t^2$, where $q(t) = t^{100}$ and $R(t)$ are related by the equations

$$0 = q(\lambda_1) = R(\lambda_1) = c_0$$

$$1^{100} = q(\lambda_2) = R(\lambda_2) = c_0 + c_1 + c_2$$

$$100(1)^{99} = q'(\lambda_2) = R'(\lambda_2) = c_1 + 2c_2.$$

Solving these equations gives the coefficients: $c_0 = 0$, $c_1 = -98$, and $c_2 = 99$. Therefore, $\mathbf{A}^{100}$ is given by

$$\mathbf{A}^{100} = R(\mathbf{A}) = -98\mathbf{A} + 99\mathbf{A}^2.$$

■

Example 6.23

$$\mathbf{A} = \begin{bmatrix} -1 & 0 & -2 & 2 \\ 3 & 1 & 5 & 3 \\ 0 & 0 & -1 & 0 \\ 0 & 0 & 2 & 1 \end{bmatrix}$$

has eigenvalues $\lambda_1 = -1$ and $\lambda_2 = 1$ with multiplicities $m_1 = m_2 = 2$. To find $q(A) = \mathbf{A}^{50}$ we solve for the coefficients of $R(t) = c_0 + c_1t + c_2t^2 + c_3t^3$:

$$\begin{aligned}(-1)^{50} &= q(\lambda_1) = R(\lambda_1) = c_0 - c_1 + c_2 - c_3\\ 50(-1)^{49} &= q'(\lambda_1) = R'(\lambda_1) = c_1 - 2c_2 + 3c_3\\ (1)^{50} &= q(\lambda_2) = R(\lambda_2) = c_0 + c_1 + c_2 + c_3\\ 50(1)^{49} &= q'(\lambda_2) = R'(\lambda_2) = c_1 + 2c_2 + 3c_3.\end{aligned}$$

The solution of this system is: $c_0 = -24$, $c_1 = 0$, $c_2 = 25$, and $c_3 = 0$. Thus $\mathbf{A}^{50} = -24\mathbf{I} + 25\mathbf{A}^2$.

■

The last application we present is a procedure for calculating the characteristic polynomial without the use of determinants. This procedure, known as Krylov's method, consists in finding a linear system of equations which is satisfied by the unknown coefficients of $p_{\mathbf{A}}(\lambda)$. Suppose that $\mathbf{A}$ is an $n \times n$ matrix whose characteristic polynomial

$$p_{\mathbf{A}}(\lambda) = (-1)^n\lambda^n + a_{n-1}\lambda^{n-1} + \cdots + a_1\lambda + a_0$$

is to be determined. Let $\mathbf{x} \neq \boldsymbol{\theta}$ be an n-vector; then by the Cayley–Hamilton theorem $p_{\mathbf{A}}(\mathbf{A})\mathbf{x} = \mathbf{O}\mathbf{x} = \boldsymbol{\theta}$ and we have

$$(-1)^n\mathbf{A}^n\mathbf{x} + a_{n-1}\mathbf{A}^{n-1}\mathbf{x} + \cdots + a_1\mathbf{A}\mathbf{x} + a_0\mathbf{x} = \boldsymbol{\theta},$$

or equivalently,

$$-(-1)^n\mathbf{A}^n\mathbf{x} = a_{n-1}\mathbf{A}^{n-1}\mathbf{x} + \cdots + a_1\mathbf{A}\mathbf{x} + a_0\mathbf{x}.$$

The last equation is a vector equation in the unknown coefficients $a_0, a_1, \ldots, a_{n-1}$. This vector equation is equivalent to the matrix equation

$$\mathbf{B}\mathbf{a} = -(-1)^n\mathbf{A}^n\mathbf{x}, \tag{3}$$

where $\mathbf{a} = [a_0, \ldots, a_{n-1}]^T$ and the jth column of $\mathbf{B}$ is $\mathbf{B}_j = \mathbf{A}^{j-1}\mathbf{x}$. The columns of $\mathbf{B}$ are determined successively by $\mathbf{B}_1 = \mathbf{x}$, $\mathbf{B}_2 = \mathbf{A}\mathbf{x} = \mathbf{A}\mathbf{B}_1$, $\mathbf{B}_3 = \mathbf{A}\mathbf{B}_2$, and so on. If $\mathbf{x}$ is such that the columns of $\mathbf{B}$ are linearly independent, then $\mathbf{B}$ is invertible and the coefficient vector $\mathbf{a}$ is uniquely determined. We solve for $\mathbf{a}$ by elimination applied to equation (3). The procedure should be clear with the aid of our next example.

Example 6.24

We apply Krylov's method to

$$\mathbf{A} = \begin{bmatrix} 1 & 2 & 4 \\ -1 & 3 & 0 \\ 0 & -1 & 2 \end{bmatrix}.$$

Let

$$\mathbf{x} = \begin{bmatrix} 1 \\ 1 \\ 1 \end{bmatrix}$$

and calculate

$$\mathbf{Ax} = \begin{bmatrix} 7 \\ 2 \\ 1 \end{bmatrix}, \qquad \mathbf{A}^2\mathbf{x} = \mathbf{A}\begin{bmatrix} 7 \\ 2 \\ 1 \end{bmatrix} = \begin{bmatrix} 15 \\ -1 \\ 0 \end{bmatrix},$$

$$\text{and} \quad \mathbf{A}^3\mathbf{x} = \mathbf{A}\begin{bmatrix} 15 \\ -1 \\ 0 \end{bmatrix} = \begin{bmatrix} 15 \\ -18 \\ 1 \end{bmatrix}.$$

It follows that

$$\mathbf{Ba} = \begin{bmatrix} 1 & 7 & 15 \\ 1 & 2 & -1 \\ 1 & 1 & 0 \end{bmatrix}\begin{bmatrix} a_0 \\ a_1 \\ a_2 \end{bmatrix} = -(-1)^3\begin{bmatrix} 13 \\ -18 \\ 1 \end{bmatrix}.$$

Solving gives $a_0 = 14$, $a_1 = -13$, and $a_2 = 6$. Thus we have

$$p_{\mathbf{A}}(\lambda) = (-1)^3\lambda^3 + 6\lambda^2 - 13\lambda - 14.$$

■

We remark that the success of Krylov's method depends upon the choice of $\mathbf{x}$. The important requirement is that $\{\mathbf{x}, \mathbf{Ax}, \ldots, \mathbf{A}^{n-1}\mathbf{x}\}$ be a linearly independent set; that is, $\mathbf{B}$ is invertible. This process is more efficient than calculating det $(\mathbf{A} - \lambda\mathbf{I})$; however, it is not without some difficulties. In addition to the problem of selecting a suitable $\mathbf{x}$, one must exercise caution when using Krylov's method on a digital computer. The reason is that the computed coefficients will not be exact due to round-off error. This is significant because a small error in the coefficients of a polynomial can result in a large perturbation in the roots of the polynomial. For example, real roots can become complex if the coefficients are changed slightly.

EXERCISES 6.3

In Exercises 1–4, verify the Cayley–Hamilton theorem for $\mathbf{A}$.

1. $\mathbf{A} = \begin{bmatrix} 1 & -2 \\ 4 & 3 \end{bmatrix}$ **2.** $\mathbf{A} = \begin{bmatrix} 2 & -1 \\ -4 & 2 \end{bmatrix}$

3. $\mathbf{A} = \begin{bmatrix} 1 & 0 & 1 \\ 0 & 1 & 0 \\ 0 & 0 & 1 \end{bmatrix}$ **4.** $\mathbf{A} = \begin{bmatrix} 3 & -1 & 0 \\ -1 & 2 & 4 \\ 0 & 4 & -1 \end{bmatrix}$

5. Use the Cayley–Hamilton theorem to find $\mathbf{A}^{-1}$, where $\mathbf{A}$ is the matrix of Exercise 1.

6. Show that the constant term a_0 in $p_{\mathbf{A}}(\lambda)$ is equal to the determinant of $\mathbf{A}$.

7. Use the Cayley–Hamilton theorem to find $\mathbf{A}^{-1}$, where

$$\mathbf{A} = \begin{bmatrix} 1 & 1 & 2 \\ -1 & 2 & 4 \\ 0 & 4 & -1 \end{bmatrix}.$$

8. Let $q(t) = t^4 - 4t^2 + t - 6$ and

$$\mathbf{A} = \begin{bmatrix} 5 & -1 \\ 7 & -3 \end{bmatrix}.$$

Perform division to find the quotient and remainder in $q/p_{\mathbf{A}}$. Find $q(\mathbf{A})$.

9. Find $q(\mathbf{A})$ where $q(t) = 2t^5 + 3t^2 + 4$ and $\mathbf{A}$ is the matrix of Exercise 3.

10. Find $\mathbf{A}^{100}$ for

$$\mathbf{A} = \begin{bmatrix} 2 & -1 \\ -4 & 2 \end{bmatrix}.$$

11. Find $\mathbf{A}^m$ for

$$\mathbf{A} = \begin{bmatrix} a & 1-a \\ 1-a & a \end{bmatrix},$$

where $0 < a < 1$.

12. For

$$\mathbf{A} = \begin{bmatrix} 3 & 4 & 2 \\ -1 & -1 & 3 \\ 0 & 0 & 1 \end{bmatrix}$$

find $\mathbf{A}^{10}$.

Use Krylov's method to find the characteristic polynomial of matrix $\mathbf{A}$ in Exercises 13–15.

13. $\mathbf{A} = \begin{bmatrix} -1 & 2 \\ 3 & 6 \end{bmatrix}$ **14.** $\mathbf{A} = \begin{bmatrix} -1 & 4 & 2 \\ 0 & 1 & 0 \\ -4 & 4 & 3 \end{bmatrix}$ **15.** $\mathbf{A} = \begin{bmatrix} 4 & -1 & 0 \\ -1 & 5 & -1 \\ 0 & -1 & 4 \end{bmatrix}$

6.4 Markov Processes

This section requires of the reader some familiarity with, or at least an intuitive understanding of, elementary probability. We shall give a brief introduction to a type of mathematical model known as a Markov process or Markov chain. This type of modeling has been successfully applied to problems arising in biology, business, engineering, and physics.

We consider a process which evolves over fixed time intervals such that at any given time there are only a finite number of possible occurrences called **states.** It is assumed that the probability (or likelihood) of going from one state to another is independent of time. To be more precise we suppose that there are n possible states and we call them state 1, state 2, . . . , and state n. If the process is in state j, then, after the next time interval, it will be in state i with some probability, say p_{ij}. Thus the **transition probabilities** p_{ij} give the likelihood of transitions from one state to another during one time interval. Since each p_{ij} is a probability, these numbers must satisfy

$$0 \le p_{ij} \le 1 \quad \text{for all } i, j = 1, 2, \ldots, n.$$

Moreover, for each fixed j it must be the case that

$$p_{1j} + p_{2j} + \cdots + p_{nj} = 1.$$

This equation states the fact that a process in state j at some time will, with probability 1, be in one of the n possible states after the next time interval.

Using these n^2 transition probabilities, we can form the $n \times n$ matrix **P** whose i, j-entry is p_{ij}. This matrix is called the **stochastic matrix** of the Markov process. Note that **P** has nonnegative entries and each of its columns sums to 1. Some authors refer to stochastic matrices as Markov matrices or transition matrices. To fix ideas, let us consider a specific example.

Example 6.25

A department store has experienced a large increase in the amount of shoplifting. To combat this problem the store manager decides to employ several undercover detectives. The manager instructs the detective assigned to the women's apparel department (Figure 6.5) to monitor the area near each aisle for 15 minutes and then either to move to a neighboring aisle or to remain in the same aisle.

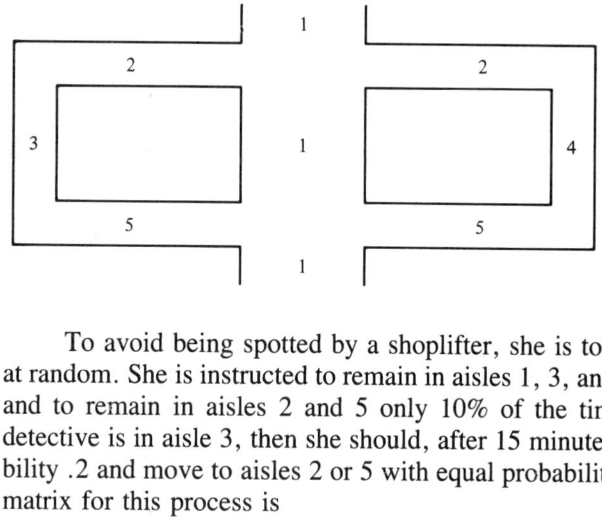

Figure 6.5

To avoid being spotted by a shoplifter, she is told to choose a new aisle at random. She is instructed to remain in aisles 1, 3, and 4 only 20% of the time and to remain in aisles 2 and 5 only 10% of the time. For example, if the detective is in aisle 3, then she should, after 15 minutes, stay there with probability .2 and move to aisles 2 or 5 with equal probabilities of .4. The stochastic matrix for this process is

$$\begin{array}{c} \text{old aisle} \\ \begin{array}{ccccc} 1 & 2 & 3 & 4 & 5 \end{array} \end{array}$$

$$\mathbf{P} = \begin{bmatrix} .2 & .3 & 0 & 0 & .3 \\ .4 & .1 & .4 & .4 & 0 \\ 0 & .3 & .2 & 0 & .3 \\ 0 & .3 & 0 & .2 & .3 \\ .4 & 0 & .4 & .4 & .1 \end{bmatrix} \begin{array}{c} 1 \\ 2 \\ 3 \\ 4 \\ 5 \end{array} \quad \text{new aisle}$$

Note that each column of **P** sums to 1. A zero value in the 3, 1-entry means that the detective cannot move from aisle 1 to aisle 3 because they are not neighboring aisles. ■

Our main objective in this section is to predict the future state of a given system which is modeled as a Markov process. As one would expect, our predictions will be in terms of probabilities. For example, if the store detective begins work at 9:00 A.M. in aisle 1, what is the probability that she will be in aisle 3 at 10:30 A.M? If this probability is low, a smart shoplifter will concentrate his or her efforts near aisle 3 at 10:30 A.M. and a smart store manager will change the monitoring strategy of the detective.

To be precise about our predictions, we introduce some definitions and state some results about stochastic matrices. A probability distribution vector, or more simply a **probability vector,** is a vector with nonnegative entries whose components sum to 1. We denote by $\mathbf{p}^{(k)}$ the probability vector whose ith component, $p_i^{(k)}$, is the probability that the process is in state i during the kth time interval. Such a vector $\mathbf{p}^{(k)}$ is called a **state vector** of the process and in particular $\mathbf{p}^{(0)}$ is the **initial state vector.** Given $\mathbf{P}$ and $\mathbf{p}^{(0)}$, how does one determine $\mathbf{p}^{(k)}$? Using the notion of conditional probability, it can be shown that all future state vectors are determined by $\mathbf{P}$ and the initial state vector.

Theorem 6.12. Suppose that $\mathbf{P}$ is the stochastic matrix of a Markov process, then

$$\mathbf{p}^{(k+1)} = \mathbf{P}\mathbf{p}^{(k)}.$$

■

From this theorem it follows that

$$\begin{aligned}\mathbf{p}^{(1)} &= \mathbf{P}\mathbf{p}^{(0)}\\ \mathbf{p}^{(2)} &= \mathbf{P}\mathbf{p}^{(1)} = \mathbf{P}^2\mathbf{p}^{(0)}\\ &\vdots\\ \mathbf{p}^{(k)} &= \mathbf{P}\mathbf{p}^{(k-1)} = \mathbf{P}^k\mathbf{p}^{(0)}.\end{aligned}$$

That is, the kth state vector equals the initial state vector premultiplied by the kth power of $\mathbf{P}$.

Example 6.26

For the stochastic matrix in Example 6.25 and $\mathbf{p}^{(0)} = [1 \quad 0 \quad 0 \quad 0 \quad 0]^T$, we compute $\mathbf{p}^{(k)}$ for $k = 1, 2, \ldots, 6$.

$$\mathbf{p}^{(1)} = \mathbf{P}\mathbf{p}^{(0)} = \begin{bmatrix} .2 \\ .4 \\ 0 \\ 0 \\ .4 \end{bmatrix}, \quad \mathbf{p}^{(2)} = \mathbf{P}\mathbf{p}^{(1)} = \begin{bmatrix} .28 \\ .12 \\ .24 \\ .24 \\ .12 \end{bmatrix}, \quad \mathbf{p}^{(3)} = \mathbf{P}\mathbf{p}^{(2)} = \begin{bmatrix} .128 \\ .316 \\ .12 \\ .12 \\ .316 \end{bmatrix},$$

$$\mathbf{p}^{(4)} = \mathbf{P}\mathbf{p}^{(3)} = \begin{bmatrix} .2152 \\ .1788 \\ .2136 \\ .2136 \\ .1788 \end{bmatrix}, \quad \mathbf{p}^{(5)} = \mathbf{P}\mathbf{p}^{(4)} = \begin{bmatrix} .15032 \\ .27484 \\ .15 \\ .15 \\ .27484 \end{bmatrix},$$

$$\mathbf{p}^{(6)} = \mathbf{P}\mathbf{p}^{(5)} = \begin{bmatrix} .194968 \\ .207612 \\ .194904 \\ .194904 \\ .207612 \end{bmatrix}$$

The choice of $\mathbf{p}^{(0)}$ corresponds to the detective starting in aisle 1. If time zero is 9:00 A.M., then the probability that the detective is in aisle 3 at 10:30 A.M. is .194904.

■

A probability vector $\mathbf{q}$ is called a **steady-state vector** for a Markov chain with stochastic matrix $\mathbf{P}$ if $\mathbf{Pq} = \mathbf{q}$. That is, $\mathbf{q}$ is a steady-state vector if $\mathbf{q}$ is an eigenvector of $\mathbf{P}$ corresponding to the eigenvalue $\lambda = 1$ and $\mathbf{q}$ is a probability vector. If $\mathbf{P}$ is applied to a steady-state vector, then there is no change in the state probabilities. For this definition to be meaningful we need to know if $\lambda = 1$ is an eigenvalue of $\mathbf{P}$ and moreover if a corresponding eigenvector can be chosen with nonnegative entries. If $\mathbf{Px} = \mathbf{x}$, then, for any $\alpha \neq 0$, $\alpha\mathbf{x}$ is also an eigenvector corresponding to $\lambda = 1$. Thus if $\mathbf{x}$ has nonnegative entries we can choose α so that $\mathbf{q} = \alpha\mathbf{x}$ is a probability vector. The next theroem characterizes the eigenvalues of stochastic matrices.

Theorem 6.13. Suppose that $\mathbf{P}$ is an $n \times n$ stochastic matrix; then $\lambda = 1$ is an eigenvalue of $\mathbf{P}$ and every eigenvalue λ of $\mathbf{P}$ satisfies $|\lambda| \leq 1$.

Proof. First we show that $\lambda = 1$ is an eigenvalue. Consider the matrix $\mathbf{A} = \mathbf{P} - \mathbf{I}$. By adding each of the rows 1 through $n - 1$ of $\mathbf{A}$ to row n of $\mathbf{A}$, we obtain a new matrix, call it $\mathbf{B}$, whose last row entries satisfy

$$b_{nj} = \sum_{i=1}^{n} a_{ij} = \sum_{i=1}^{n} p_{ij} - 1 = 0, \qquad 1 \leq j \leq n$$

because each column sum of $\mathbf{P}$ equals 1. It follows that an echelon form for $\mathbf{A}$ has at least one row of zeros. Hence $r(\mathbf{A}) = r(\mathbf{P} - \mathbf{I}) < n$ and $\mathbf{P} - \mathbf{I}$ is not invertible. This implies by Theorem 6.1 that $\lambda = 1$ is an eigenvalue of $\mathbf{P}$.

To show that $|\lambda| \leq 1$ for every eigenvalue we apply Gershgorin's theorem to $\mathbf{P}^T$ (recall that $\mathbf{P}$ and $\mathbf{P}^T$ have the same eigenvalues). Every eigenvalue λ of $\mathbf{P}^T$ satisfies, for some i,

$$|\lambda - p_{ii}| \leq \sum_{\substack{j=1 \\ j\neq i}}^{n} |p_{ji}| = \sum_{\substack{j=1 \\ j\neq i}}^{n} p_{ji} = 1 - p_{ii}$$

since each $p_{ij} \geq 0$ and the column sums of $\mathbf{P}$ equal 1. But

$$|\lambda - p_{ii}| \geq |\lambda| - |p_{ii}| = |\lambda| - p_{ii}$$

and hence $|\lambda| - p_{ii} \leq 1 - p_{ii}$, which gives $|\lambda| \leq 1$.

■

Every stochastic matrix $\mathbf{P}$ has $\lambda = 1$ as an eigenvalue but the question remains:

Does $\mathbf{P}$ have a steady-state vector? For certain types of stochastic matrices the answer is yes. We say that $\mathbf{P}$ is a **regular stochastic matrix** if, for some $k \geq 1$, $\mathbf{P}^k$ has no zero entries. Thus $\mathbf{P}$ is regular if some power of $\mathbf{P}$ has only positive entries.

Theorem 6.14. Suppose that $\mathbf{P}$ is a regular stochastic matrix; then there is a unique probability vector $\mathbf{q}$ satisfying $\mathbf{Pq} = \mathbf{q}$. That is, $\mathbf{q}$ is the unique steady-state vector.

■

Let us consider a simple 2×2 example in complete detail.

Example 6.27

A survey is taken of homeowners in which each homeowner is asked the question: Do you prefer living in a contemporary or a traditional house? Based on the survey it was determined that the probability that an owner of a contemporary house will either keep his house or buy another contemporary house is .95. The probability that the owner of a traditional house will sell his home and buy a contemporary house is .18. Thus the stochastic matrix

$$\begin{array}{c} \\ T \\ C \end{array}\!\!\begin{array}{c} \begin{array}{cc} T & \;\;C \end{array} \\ \begin{bmatrix} .82 & .05 \\ .18 & .95 \end{bmatrix} \end{array}$$

determines the likelihood of home-ownership style for future time. If we begin with a sample u of traditional homeowners and v owners of contemporary houses, then after one year

$$.82u + .05v \text{ will own traditional houses}$$

$$.18u + .95v \text{ will own contemporary houses}$$

on the average. That is,

$$\mathbf{P}\begin{bmatrix} u \\ v \end{bmatrix} = \begin{bmatrix} .82 & .05 \\ .18 & .95 \end{bmatrix}\begin{bmatrix} u \\ v \end{bmatrix}$$

gives the home-style distribution after 1 year. Starting with an initial sample $\mathbf{x}^{(0)} = [u \quad v]^{\mathrm{T}}$, the central question is whether we can predict any trends for the future. After k years the distribution of homeowners is governed by $\mathbf{P}^k$, where $\mathbf{P}$ is the regular stochastic matrix given above.

■

Rather than analyze this particular stochastic matrix, we examine the general 2×2 case where

$$\mathbf{P} = \begin{bmatrix} a & b \\ 1 - a & 1 - b \end{bmatrix}$$

with $0 < a < 1$ and $0 < b < 1$. It is not difficult to see that $p_{\mathbf{P}}(\lambda) = \lambda^2 - (a + 1 - b)\lambda + a - b$. To find $\mathbf{P}^k$ we use the method of section 3, wherein

$\mathbf{P}^k = R(\mathbf{P})$ with $R(t) = c_0 + c_1 t$. The coefficients of R are determined by

$$(\lambda_1)^k = R(\lambda_1)$$
$$(\lambda_2)^k = R(\lambda_2),$$

where the eigenvalues are $\lambda_1 = 1$ and $\lambda_2 = a - b$ (check this). This gives the equations

$$c_0 + c_1 = 1$$
$$c_0 + (a - b)c_1 = (a - b)^k$$

which have the unique solution

$$c_0 = (a - b)\frac{1 - (a - b)^{k-1}}{a - b - 1}$$
$$c_1 = \frac{(a - b)^k - 1}{a - b - 1}.$$

Therefore, we get

$$\mathbf{P}^k = \frac{(a - b)[(a - b)^{k-1} - 1]\mathbf{I} + [1 - (a - b)^k]\mathbf{P}}{1 - a + b}$$

and the long-term behavior of the Markov chain is determined by considering $\mathbf{P}^k$ as $k \to \infty$. Since $0 < a < 1$ and $0 < b < 1$ we have $|a - b| < 1$ and hence $(a - b)^k \to 0$ as $k \to \infty$. Therefore, as $k \to \infty$,

$$\mathbf{P}^k \longrightarrow \mathbf{A} = \frac{b - a}{1 - a + b}\mathbf{I} + \frac{1}{1 - a + b}\mathbf{P} = \frac{1}{1 - a + b}\begin{bmatrix} b & b \\ 1 - a & 1 - a \end{bmatrix}.$$

Moreover, if $\mathbf{x}^{(0)}$ is any vector, then as $k \to \infty$,

$$\mathbf{P}^k\mathbf{x}^{(0)} \longrightarrow \frac{1}{1 - a + b}\begin{bmatrix} b & b \\ 1 - a & 1 - a \end{bmatrix}\mathbf{x}^{(0)}.$$

In particular, if $\mathbf{x}^{(0)}$ is any probability vector, that is, $x_1^{(0)} + x_2^{(0)} = 1$, then

$$\frac{1}{(1 - a + b)}\begin{bmatrix} b & b \\ 1 - a & 1 - a \end{bmatrix}\mathbf{x}^{(0)} = \frac{1}{1 - a + b}\begin{bmatrix} b(x_1^{(0)} + x_2^{(0)}) \\ (1 - a)(x_1^{(0)} + x_2^{(0)}) \end{bmatrix}$$
$$= \frac{1}{1 - a + b}\begin{bmatrix} b \\ 1 - a \end{bmatrix}.$$

Also, it is easy to see that the vector $\mathbf{q} = 1/(1 - a + b)[b \quad 1 - a]^{\mathrm{T}}$ satisfies $\mathbf{P}\mathbf{q} = \mathbf{q}$. The unique steady-state vector for $\mathbf{P}$ is $\mathbf{q}$ and $\mathbf{P}^k \to \mathbf{A}$, as $k \to \infty$, where each column of $\mathbf{A}$ equals $\mathbf{q}$.

Let us return to the homeowners' survey example.

Example 6.28

For the data of Example 6.27 suppose that 1000 homeowners are sampled, of which 900 own traditional homes. After 10 years, what will be the distribution of traditional and contemporary houses? To answer the question we compute

$$\mathbf{P}^{10}\begin{bmatrix} 900 \\ 100 \end{bmatrix}.$$

From our previous analysis with $a = .82$ and $b = .05$, we find

$$\mathbf{P}^{10} = \frac{.77[(.77)^9 - 1]\mathbf{I} + [1 - (.77)^{10}]\mathbf{P}}{.23}$$

$$= \begin{bmatrix} .2747303 & .2014637 \\ .7252696 & .7985362 \end{bmatrix}$$

and hence

$$\mathbf{P}^{10}\begin{bmatrix} 900 \\ 100 \end{bmatrix} = \begin{bmatrix} 267 \\ 733 \end{bmatrix}$$

when rounded off to the nearest integer. This shows a significant trend toward contemporary home ownership and as $k \to \infty$,

$$\mathbf{P}^k\begin{bmatrix} 900 \\ 100 \end{bmatrix} \longrightarrow \begin{bmatrix} 217 \\ 783 \end{bmatrix}.$$

■

The following theorem summarizes the most important features of regular Markov processes.

Theorem 6.15. Suppose that $\mathbf{P}$ is an $n \times n$ regular stochastic matrix with steady-state vector $\mathbf{q}$. Then

(a) As $k \to \infty$, $\mathbf{P}^k \to \mathbf{A}$, where $\mathbf{A}_j$, the jth column of $\mathbf{A}$, satisfies $\mathbf{A} = \mathbf{q}$ for $j = 1, 2, \ldots, n$.

(b) For any initial probability vector $\mathbf{p}^{(0)}$,

$$\mathbf{P}^k\mathbf{p}^{(0)} \longrightarrow \mathbf{q}, \qquad \text{as } k \longrightarrow \infty.$$

■

We do not give a complete proof of this theorem but instead show only how part (b) follows from part (a). To see this we note that part (a) implies that

$$\mathbf{P}^k\mathbf{p}^{(0)} \longrightarrow \mathbf{A}\mathbf{p}^{(0)}$$

for any probability vector $\mathbf{p}^{(0)}$. By definition of matrix multiplication we have

$$\begin{aligned} \mathbf{A}\mathbf{p}^{(0)} &= p_1^{(0)}\mathbf{A}_1 + p_2^{(0)}\mathbf{A}_2 + \cdots + p_n^{(0)}\mathbf{A}_n \\ &= p_1^{(0)}\mathbf{q} + p_2^{(0)}\mathbf{q} + \cdots + p_n^{(0)}\mathbf{q} \\ &= \left(\sum_{i=1}^{n} p_i^{(0)}\right)\mathbf{q} = (1)\mathbf{q} = \mathbf{q} \end{aligned}$$

Therefore, $\mathbf{A}$ transforms any probability vector $\mathbf{p}^{(0)}$ into the steady-state vector $\mathbf{q}$.

Example 6.29

For the store detective problem of Example 6.25, the stochastic matrix is regular since

$$\mathbf{P}^2 = \begin{bmatrix} .28 & .09 & .24 & .24 & .09 \\ .12 & .37 & .12 & .12 & .36 \\ .24 & .09 & .28 & .24 & .09 \\ .24 & .09 & .24 & .28 & .09 \\ .12 & .36 & .12 & .12 & .37 \end{bmatrix}$$

has no zero entries. By solving $(\mathbf{P} - \mathbf{I})\mathbf{x} = \boldsymbol{\theta}$, we find that eigenvectors corresponding to $\lambda = 1$ take the form

$$\mathbf{x} = \alpha \begin{bmatrix} 3 \\ 4 \\ 3 \\ 3 \\ 4 \end{bmatrix}.$$

Choosing $\alpha = \frac{1}{17}$ gives the steady-state vector $\mathbf{q}$. Thus the long-term probabilities of finding the detective in aisles 1 through 5 of the women's apparel department are given by the entries of

$$\mathbf{q} = \begin{bmatrix} 3/17 \\ 4/17 \\ 3/17 \\ 3/17 \\ 4/17 \end{bmatrix}.$$

■

We conclude by summarizing the analysis of a Markov process:

1. Determine the stochastic matrix $\mathbf{P}$ and initial state vector $\mathbf{p}^{(0)}$.
2. The kth state vector $\mathbf{p}^{(k)}$ can be determined by
   (a) successively computing $\mathbf{p}^{(1)} = \mathbf{P}\mathbf{p}^{(0)}$, $\mathbf{p}^{(2)} = \mathbf{P}\mathbf{p}^{(1)}, \ldots, \mathbf{p}^{(k)} = \mathbf{P}\mathbf{p}^{(k-1)}$ or if convenient by
   (b) $\mathbf{p}^{(k)} = \mathbf{P}^k\mathbf{p}^{(0)}$, where $\mathbf{P}^k$ is computed by the method of Section 6.3 based on the Cayley–Hamilton theorem
3. For regular $\mathbf{P}$ the steady-state vector $\mathbf{q}$ is found by solving $(\mathbf{P} - \mathbf{I})\mathbf{x} = \boldsymbol{\theta}$ using elimination and then choosing α so that $\mathbf{q} = \alpha\mathbf{x}$ is a probability vector.

If $\mathbf{P}$ is a large matrix, then $\mathbf{q}$ can be determined approximately by calculation of $\mathbf{p}^{(k)} = \mathbf{P}^k\mathbf{p}^{(0)}$ for sufficiently large k starting with any initial state vector $\mathbf{p}^{(0)}$. For example, if four-digit accuracy is desired, then compute until the entries of $\mathbf{p}^{(k)}$ and $\mathbf{p}^{(k-1)}$ agree to four digits.

EXERCISES 6.4

1. A man plays two slot machines. The first pays off with probability .2 and the second with probability .25. If he loses he plays the same machine again, and when he wins he switches to the other slot machine. Let the states be: state i—playing the ith slot machine $i = 1, 2$. Find the transition matrix. After 10 plays what is the probability that he will be playing the second machine if he started with the first?

2. In Memphis they never have two snowy days in a row. The day after a snowy day they are likely to have a rainy day or a nice day with equal probability. If the weather is rainy or nice then it stays the same with probability .4. Given that the weather changes from rainy or nice, it changes to snowy one-tenth of the time. Let the states be: 1—nice weather, 2—rainy, and 3—snowy. Find the stochastic matrix. If today's weather is rainy, what type of weather is most likely to occur 4 days from now?

3. A diplomat engaged in shuttle diplomacy always goes from Geneva to Beirut and always from Beirut to Cairo. However, from Cairo she goes with probability $\frac{1}{2}$ to Geneva and with probability $\frac{1}{2}$ to Beirut. Form the stochastic matrix for her travels and determine a steady-state vector.

4. Consider a Markov process that has two states, "success," S, and "failure," F. Suppose that S follows S with probability p and S follows F with probability q. Find the stochastic matrix that represents the transitions of the process.

***5.** Suppose that the $n \times n$ stochastic matrix $\mathbf{P}$ is given with each of its row sums equal to 1. Show that the steady-state vector $\boldsymbol{q}$ has entries $\mathbf{q} = 1/n$ for $i = 1, 2, \ldots, n$.

6. Determine if

$$\mathbf{P} = \begin{bmatrix} 1 & a \\ 0 & 1-a \end{bmatrix},$$

where $0 < a < 1$, is regular. For $a = \frac{1}{3}$, find $\mathbf{P}^m$. Estimate the entries of $\mathbf{P}^m$ for large m.

7. Determine which of the following stochastic matrices are regular.

$$\text{(a) } \mathbf{P} = \begin{bmatrix} \frac{1}{2} & \frac{1}{3} \\ \frac{1}{2} & \frac{2}{3} \end{bmatrix} \quad \text{(b) } \mathbf{P} = \begin{bmatrix} \frac{1}{3} & 0 \\ \frac{2}{3} & 1 \end{bmatrix} \quad \text{(c) } \mathbf{P} = \begin{bmatrix} \frac{1}{4} & \frac{1}{2} & \frac{1}{3} \\ \frac{1}{4} & \frac{1}{2} & 0 \\ \frac{1}{2} & 0 & \frac{2}{3} \end{bmatrix}$$

8. Find the steady-state vectors for the following regular stochastic matrices.

$$\text{(a) } \mathbf{P} = \begin{bmatrix} \frac{2}{3} & \frac{1}{2} \\ \frac{1}{3} & \frac{1}{2} \end{bmatrix} \quad \text{(b) } \mathbf{P} = \begin{bmatrix} \frac{1}{2} & 0 & \frac{1}{2} \\ \frac{1}{4} & \frac{1}{4} & \frac{1}{2} \\ \frac{1}{4} & \frac{3}{4} & 0 \end{bmatrix} \quad \text{(c) } \mathbf{P} = \begin{bmatrix} .3 & .9 \\ .7 & .1 \end{bmatrix}$$

9. In a certain geographic region the population is categorized in three demographic groups: city dwellers, suburbanites, or rural dwellers. A statistical study shows that city dwellers remain in the cities with probability .76 and move to the suburbs with probability .23. Those who live in the suburbs stay there with probability .88 and move to the cities with probability .1. Those who live in the rural areas move to the cities with probability .08 and to the suburbs with probability .04. Set up the stochastic matrix and determine the long-term probability distribution of the region's population.

10. Show that the following matrix $\mathbf{P}$ is not regular:

$$\mathbf{P} = \begin{bmatrix} 1 & \frac{1}{4} & 0 \\ 0 & \frac{1}{2} & 0 \\ 0 & \frac{1}{4} & 1 \end{bmatrix}.$$

If

$$\mathbf{p}^{(0)} = \begin{bmatrix} \frac{1}{4} \\ \frac{1}{2} \\ \frac{1}{4} \end{bmatrix},$$

estimate $\mathbf{P}^m\mathbf{p}^{(0)}$ for large m.

GLOSSARY

Cayley–Hamilton theorem Every square matrix satisfies its characteristic equation.

characteristic equation For a square matrix $\mathbf{A}$ the characteristic equation is $\det(\mathbf{A} - \lambda\mathbf{I}) = 0$.

characteristic polynomial For a square matrix $\mathbf{A}$ the characteristic polynomial is $p_{\mathbf{A}}(\lambda) = \det(\mathbf{A} - \lambda\mathbf{I})$.

defective matrix An $n \times n$ matrix that has fewer than n linearly independent eigenvectors.

eigenvalue of a matrix A A scalar λ which satisfies, for some nonzero vector $\mathbf{x}$, $\mathbf{A}\mathbf{x} = \lambda\mathbf{x}$.

eigenvector of a matrix A A nonzero vector, $\mathbf{x}$, which satisfies $\mathbf{A}\mathbf{x} = \lambda\mathbf{x}$ for some scalar λ.

Krylov's method A procedure for determining the characteristic polynomial without using determinants.

multiplicity An eigenvalue μ of $\mathbf{A}$ has multiplicity m if $(\lambda - \mu)^m$ is a factor of $p_{\mathbf{A}}(\lambda)$, but $(\lambda - \mu)^{m+1}$ is not a factor of $p_{\mathbf{A}}(\lambda)$.

probability vector An n-vector $\mathbf{p}$ whose entries satisfy $0 \le p_i \le 1$ and $\sum_{i=1}^{n} p_i = 1$.

steady-state vector A probability vector which is an eigenvector of a stochastic matrix corresponding to the eigenvalue 1.

stochastic matrix A square matrix with nonnegative entries each of whose columns sums to 1.

trace of a matrix A The sum of the diagonal entries of the square matrix $\mathbf{A}$.

NOTES AND COMMENTS

The theory of Markov processes has numerous applications to the physical and social sciences. A well-written introductory treatment is given in Chapter 10 of the book by D. R. Hunkins and L. R. Mugridge; *Applied Finite Mathematics*, 2nd edition (Boston: Prindle, Weber & Schmidt, 1985).

The method of Krylov for the determination of the characteristic polynomial of a matrix is conceptually simple but is not recommended as a means of determining eigenvalues. Unfortunately, the linear systems of equations that arise in this method are typically ill-conditioned (see Section 8.1). The best methods for determining eigenvalues do not make use of the characteristic polynomial. Rather, one uses a similarity transformation to produce a similar (see Section 7.1) matrix which has a special simple structure and whose eigenvalues are readily determined by some iterative method.

7

Coordinate Transformations

In this chapter we study a type of coordinate transformation, or "change of variables," which is frequently useful in solving problems such as difference or differential equations.

For a square matrix **A** consider the difference equation

$$\mathbf{u}^{(k+1)} = \mathbf{A}\mathbf{u}^{(k)},$$

where $\mathbf{u}^{(1)}, \mathbf{u}^{(2)}, \ldots$ is a sequence of vectors to be determined given the initial vector, $\mathbf{u}^{(0)}$. A particular equation of this type arises in the study of Markov processes (see Section 6.4). Suppose that a change of variables $\mathbf{u} = \mathbf{P}\mathbf{v}$ is introduced; then in terms of the new unknown **v** we have

$$\mathbf{P}\mathbf{v}^{(k+1)} = \mathbf{A}\mathbf{P}\mathbf{v}^{(k)}.$$

Now if **P** is invertible, we obtain

$$\mathbf{v}^{(k+1)} = \mathbf{P}^{-1}\mathbf{A}\mathbf{P}\mathbf{v}^{(k)}.$$

This gives a new difference equation involving the matrix $\mathbf{B} = \mathbf{P}^{-1}\mathbf{A}\mathbf{P}$. If **P** can be chosen in such a way that **B** has a simple form, say diagonal, then the equation $\mathbf{v}^{(k+1)} = \mathbf{B}\mathbf{v}^{(k)}$ has a solution which is easy to compute: Let

$$\mathbf{v}^{(0)} = \mathbf{P}^{-1}\mathbf{u}^{(0)};$$

then

$$\mathbf{v}^{(1)} = \mathbf{B}\mathbf{v}^{(0)}$$

$$\mathbf{v}^{(2)} = \mathbf{B}\mathbf{v}^{(1)} = \mathbf{B}^2\mathbf{v}^{(0)}$$

and continuing in this way we find $\mathbf{v}^{(k)} = \mathbf{B}^k\mathbf{v}^{(0)}$. But for a diagonal matrix **B** the determination of the *k*th power, $\mathbf{B}^k$, is an easy calculation.

A central theme in this chapter is the following. Given a problem involving a square matrix **A** and an unknown vector **x**, we introduce a change of variables $\mathbf{x} = \mathbf{P}\mathbf{y}$, with **P** invertible, so that $\mathbf{P}^{-1}\mathbf{A}\mathbf{P}$ takes some simple form. In this context **P** is called a **coordinate transformation matrix.** This is a standard device in mathematics whereby a change of variables is introduced with the aim of producing a new problem whose solution is more readily obtained than that of the original problem.

7.1 Similarity

Suppose that **A** and **B** are $n \times n$ matrices; then we say that **A** is **similar** to **B** if there exists an invertible matrix **P** such that

$$\mathbf{A} = \mathbf{P}\mathbf{B}\mathbf{P}^{-1}. \tag{1}$$

If **A** is similar to **B** we write $\mathbf{A} \sim \mathbf{B}$. That is, $\mathbf{A} \sim \mathbf{B}$ if equation (1) is true for some invertible matrix **P**.

Example 7.1

Let

$$\mathbf{A} = \begin{bmatrix} 1 & -3 \\ 2 & 0 \end{bmatrix} \quad \text{and} \quad \mathbf{B} = \begin{bmatrix} 0 & 2 \\ -3 & 1 \end{bmatrix}.$$

Then $\mathbf{A} \sim \mathbf{B}$ because

$$\mathbf{A} = \begin{bmatrix} 0 & 1 \\ 1 & 0 \end{bmatrix}\begin{bmatrix} 0 & 2 \\ -3 & 1 \end{bmatrix}\begin{bmatrix} 0 & 1 \\ 1 & 0 \end{bmatrix} = \mathbf{P}\mathbf{B}\mathbf{P}^{-1}$$

with

$$\mathbf{P} = \mathbf{P}^{-1} = \begin{bmatrix} 0 & 1 \\ 1 & 0 \end{bmatrix}.$$

■

Observe that equation (1) can be rewritten as

$$\mathbf{B} = \mathbf{P}^{-1}\mathbf{A}\mathbf{P},$$

and hence if we let $\mathbf{Q} = \mathbf{P}^{-1}$, then

$$\mathbf{B} = \mathbf{Q}\mathbf{A}\mathbf{Q}^{-1}.$$

Therefore, **B** is similar to **A** whenever **A** is similar to **B**. Henceforth we simply say that **A** and **B** are similar matrices if $\mathbf{A} \sim \mathbf{B}$ or equivalently, $\mathbf{B} \sim \mathbf{A}$.

Next suppose that $\mathbf{A} \sim \mathbf{B}$ and $\mathbf{B} \sim \mathbf{C}$; that is, for some invertible matrices **P** and **Q**,

$$\mathbf{A} = \mathbf{P}\mathbf{B}\mathbf{P}^{-1}$$

and

$$\mathbf{B} = \mathbf{Q}\mathbf{C}\mathbf{Q}^{-1}.$$

It then follows that

$$\begin{aligned} \mathbf{A} &= \mathbf{P}(\mathbf{Q}\mathbf{C}\mathbf{Q}^{-1})\mathbf{P}^{-1} \\ &= (\mathbf{P}\mathbf{Q})\mathbf{C}(\mathbf{P}\mathbf{Q})^{-1}. \end{aligned}$$

Thus if $\mathbf{S} = \mathbf{P}\mathbf{Q}$, then $\mathbf{A} = \mathbf{S}\mathbf{C}\mathbf{S}^{-1}$ and hence $\mathbf{A} \sim \mathbf{C}$. Therefore, with Exercise 1 we have established the following properties of similarity.

Suppose that **A**, **B**, **C** are $n \times n$ matrices; then:

1. $\mathbf{A} \sim \mathbf{A}$.
2. $\mathbf{A} \sim \mathbf{B}$ if and only if $\mathbf{B} \sim \mathbf{A}$.
3. If $\mathbf{A} \sim \mathbf{B}$ and $\mathbf{B} \sim \mathbf{C}$, then $\mathbf{A} \sim \mathbf{C}$.

This gives the basic properties of similarity; however, for our purposes, the most important relationship between similar matrices involves their eigenvalues. To discover this relationship suppose that $\mathbf{A} \sim \mathbf{B}$; that is,

$$\mathbf{A} = \mathbf{PBP}^{-1}$$

for some invertible matrix **P**. It follows that

$$\begin{aligned}\mathbf{A} - \lambda\mathbf{I} &= \mathbf{PBP}^{-1} - \lambda\mathbf{I} \\ &= \mathbf{PBP}^{-1} - \lambda\mathbf{PP}^{-1} = \mathbf{PBP}^{-1} - \mathbf{P}(\lambda\mathbf{I})\mathbf{P}^{-1} \\ &= \mathbf{P}(\mathbf{B} - \lambda\mathbf{I})\mathbf{P}^{-1}\end{aligned}$$

and hence

$$p_{\mathbf{A}}(\lambda) = \det(\mathbf{A} - \lambda\mathbf{I}) = \det(\mathbf{P})\det(\mathbf{B} - \lambda\mathbf{I})\det(\mathbf{P}^{-1}).$$

But $\det(\mathbf{P})\det(\mathbf{P}^{-1}) = \det(\mathbf{PP}^{-1}) = \det(\mathbf{I}) = 1$ and we find that

$$p_{\mathbf{A}}(\lambda) = p_{\mathbf{B}}(\lambda).$$

Thus similiar matrices have the same characteristic polynomial and hence they have exactly the same eigenvalues, including multiplicities.

Theorem 7.1. Suppose that **A** and **B** are similar matrices; then $p_{\mathbf{A}}(\lambda) = p_{\mathbf{B}}(\lambda)$.

We caution the reader that two matrices can have the same characteristic polynomial and yet might not be similar.

Example 7.2

Consider the matrices

$$\mathbf{A} = \begin{bmatrix} 1 & 0 \\ 0 & 1 \end{bmatrix} \text{ and } \mathbf{B} = \begin{bmatrix} 1 & 1 \\ 0 & 1 \end{bmatrix},$$

whose characteristic polynomials are $p_{\mathbf{A}}(\lambda) = p_{\mathbf{B}}(\lambda) = (1 - \lambda)^2$. For **A** and **B** to be similar there must be an invertible matrix **P** such that $\mathbf{A} = \mathbf{PBP}^{-1}$, or equivalently,

$$\mathbf{AP} = \mathbf{PB}.$$

Let

$$\mathbf{P} = \begin{bmatrix} a & b \\ c & d \end{bmatrix},$$

then we find

$$\mathbf{AP} = \mathbf{P} = \begin{bmatrix} a & b \\ c & d \end{bmatrix}, \qquad \mathbf{PB} = \begin{bmatrix} a & a+b \\ c & c+d \end{bmatrix}.$$

Therefore, $\mathbf{AP} = \mathbf{PB}$ implies that $a = c = 0$, which, in turn, implies that $\mathbf{P}$ is singular. This is a contradiction and hence $\mathbf{A}$ and $\mathbf{B}$ are not similar.

■

The eigenvectors of similar matrices, although generally not the same, are related by a coordinate transformation. Suppose that $\mathbf{A}$ and $\mathbf{B}$ are similar with $\mathbf{A} = \mathbf{PBP}^{-1}$ and let λ be an eigenvalue of $\mathbf{A}$ (and hence of $\mathbf{B}$). If $\mathbf{Ax} = \lambda\mathbf{x}$, then it follows that

$$\mathbf{PBP}^{-1}\mathbf{x} = \lambda\mathbf{x},$$

or equivalently,

$$\mathbf{BP}^{-1}\mathbf{x} = \lambda\mathbf{P}^{-1}\mathbf{x}.$$

Thus $\mathbf{y} = \mathbf{P}^{-1}\mathbf{x}$ is an eigenvector of $\mathbf{B}$ corresponding to the eigenvalue λ. That is, an eigenvector $\mathbf{x}$ of $\mathbf{A}$ is related to an eigenvector $\mathbf{y}$ of $\mathbf{B}$ (corresponding to the same eigenvalue λ) by the change of variables $\mathbf{x} = \mathbf{Py}$ and $\mathbf{P}$ is a coordinate transformation matrix.

EXERCISES 7.1

1. Show that $\mathbf{A} \sim \mathbf{A}$ for any square matrix A.

2. Which 2×2 matrices are similar to $\mathbf{I}$?

3. Let

$$\mathbf{A} = \begin{bmatrix} 1 & 1 & 0 \\ 0 & 2 & 1 \\ 0 & 0 & 3 \end{bmatrix} \quad \text{and} \quad \mathbf{P} = \begin{bmatrix} 1 & 1 & 1 \\ 0 & 1 & 2 \\ 0 & 0 & 2 \end{bmatrix}.$$

Verify that $\mathbf{AP} = \mathbf{PD}$, where $\mathbf{D} = \mathbf{diag}\,(1, 2, 3)$. Hence $\mathbf{A}$ is similar to a diagonal matrix.

4. $\mathbf{A}$ is said to be **congruent** to $\mathbf{B}$ if for some invertible matrix $\mathbf{Q}$, $\mathbf{A} = \mathbf{Q}^{\mathrm{T}}\mathbf{BQ}$. If $\mathbf{A}$ is congruent to $\mathbf{B}$, does it follow that $\mathbf{B}$ is congruent to $\mathbf{A}$?

5. Suppose that $\mathbf{A}$ is congruent to $\mathbf{B}$ and $\mathbf{B}$ is congruent to $\mathbf{C}$. Is $\mathbf{A}$ congruent to $\mathbf{C}$?

6. If $\mathbf{A}$ is symmetric and congruent to $\mathbf{B}$, does it follow that $\mathbf{B}$ is symmetric?

7. Show by example that congruent matrices do not necessarily have the same eigenvalues.

8. Suppose $\mathbf{A} \sim \mathbf{B}$. Show that $\mathbf{A}^{\mathrm{T}} \sim \mathbf{B}^{\mathrm{T}}$.

9. Suppose $\mathbf{A} \sim \mathbf{B}$. Show that $\mathbf{A}$ is nonsingular if and only if $\mathbf{B}$ is nonsingular.

10. Let

$$\mathbf{A} = \begin{bmatrix} 3 & 0 & 0 \\ -5 & 1 & 4 \\ -1 & 0 & 2 \end{bmatrix}$$

and consider $\mathbf{P}^{-1}\mathbf{AP}$, where $\mathbf{P}$ is a permutation matrix. Recall that a permutation matrix is the product of elementary permutation matrices. Find such a permutation matrix $\mathbf{P}$ so that $\mathbf{P}^{-1}\mathbf{AP}$ is lower triangular.

11. Suppose that $\mathbf{A} = \mathbf{PDP}^{-1}$ with $\mathbf{D} = \mathbf{diag}\,(\lambda_1, \ldots, \lambda_n)$ and $\lambda_i \neq 0$ for $i = 1, 2, \ldots, n$. Show that $\mathbf{A}$ is invertible. Find a formula for $\mathbf{A}^{-1}$ in terms of $\mathbf{P}$ and $\mathbf{D}$.
12. Suppose that

$$\mathbf{A} = \begin{bmatrix} 4 & -2 \\ 0 & -3 \end{bmatrix}, \qquad \mathbf{P} = \frac{1}{\sqrt{2}}\begin{bmatrix} 1 & 1 \\ 1 & -1 \end{bmatrix}, \qquad \text{and} \qquad \mathbf{A} = \mathbf{P}^{-1}\mathbf{BP}.$$

Without calculating $\mathbf{B}$, find **(a)** the eigenvalues of $\mathbf{B}$ and **(b)** the corresponding eigenvectors of $\mathbf{B}$.

7.2 Diagonalizable Matrices

For a given matrix $\mathbf{A}$, a particularly advantageous type of coordinate transformation is one which can be used to produce a diagonal matrix that is similar to $\mathbf{A}$. Of course, not every matrix $\mathbf{A}$ is similar to a diagonal matrix. We say that $\mathbf{A}$ is **diagonalizable** if $\mathbf{A}$ is similar to a diagonal matrix.

For a diagonalizable matrix $\mathbf{A}$ the solution of the difference equation

$$\mathbf{u}^{(k)} = \mathbf{A}\mathbf{u}^{(k-1)}$$

is greatly simplified. Suppose that $\mathbf{A}$ is similar to the diagonal matrix $\mathbf{D}$, say $\mathbf{A} = \mathbf{PDP}^{-1}$. Then, given the initial vector $\mathbf{u}^{(0)}$, the solution for any $k > 0$ is given by

$$\mathbf{u}^{(k)} = \mathbf{A}^k\mathbf{u}^{(0)},$$

where $\mathbf{A}^k = \mathbf{PD}^k\mathbf{P}^{-1}$. Thus the calculation of $\mathbf{A}^k = \mathbf{PD}^k\mathbf{P}^{-1}$ is relatively easy once $\mathbf{P}$, $\mathbf{P}^{-1}$, and $\mathbf{D}$ are known.

In what follows we shall determine necessary and sufficient conditions for a matrix to be diagonalizable. The diagonalizability of an $n \times n$ matrix $\mathbf{A}$ depends on the number of linearly independent eigenvectors of $\mathbf{A}$. More precisely, we shall see that $\mathbf{A}$ is diagonalizable if $\mathbf{A}$ is not defective. Suppose that $\mathbf{A}$ is diagonalizable with $\mathbf{A} = \mathbf{P\Lambda P}^{-1}$ and $\mathbf{\Lambda} = \mathbf{diag}\,(\lambda_1, \ldots, \lambda_n)$. Then

$$\mathbf{AP} = \mathbf{P\Lambda}$$

and by comparing the jth columns of both sides, we find

$$j\text{th column of } (\mathbf{AP}) = \mathbf{AP}_j$$

and

$$j\text{th column of } (\mathbf{P\Lambda}) = \lambda_j\mathbf{P}_j.$$

Thus we conclude that

$$\mathbf{AP}_j = \lambda_j\mathbf{P}_j, \qquad j = 1, \ldots, n.$$

That is, the columns of $\mathbf{P}$ are eigenvectors of $\mathbf{A}$ and the diagonal entries of $\mathbf{\Lambda}$ are the corresponding eigenvalues of **A**. Moreover, since **P** is invertible, the eigenvectors of **A** (the columns of **P**) must be linearly independent. In summary we have found that a necessary condition for an $n \times n$ matrix **A** to be diagonalizable is that **A** has n linearly independent eigenvectors. In other words, if **A** is diagonalizable, then **A** is not defective.

Conversely, suppose that **A** has n linearly independent eigenvectors, $\{\mathbf{u}^{(j)}\}_{j=1}^{n}$, with corresponding eigenvalues $\lambda_1, \ldots, \lambda_n$ (not necessarily distinct). Then we define $\mathbf{\Lambda} = \mathbf{diag}\,(\lambda_1, \ldots, \lambda_n)$ and the matrix **P** by

$$j\text{th column of } \mathbf{P} = \mathbf{u}^{(j)}.$$

From $\mathbf{A}\mathbf{u}^{(j)} = \lambda_j \mathbf{u}^{(j)}$ it follows that $\mathbf{AP} = \mathbf{P\Lambda}$. Moreover, since the columns of **P** are linearly independent, $\mathbf{P}^{-1}$ exists and hence $\mathbf{A} = \mathbf{P\Lambda P}^{-1}$. That is, **A** is diagonalizable.

Theorem 7.2. An $n \times n$ matrix **A** is diagonalizable if and only if **A** has n linearly independent eigenvectors.

■

This theorem says that defective matrices are precisely those square matrices which are not diagonalizable. In Chapter 6 we found some sufficient (but not necessary) conditions which guarantee that a given square matrix is not defective. If an $n \times n$ matrix **A** has n distinct real eigenvalues, then Theorem 6.6 guarantees that **A** is diagonalizable. Recall also Theorem 6.9, which says that real symmetric matrices are diagonalizable.

Example 7.3

The matrix

$$\mathbf{A} = \begin{bmatrix} -17 & 30 \\ -10 & 18 \end{bmatrix}$$

has eigenvalues $\lambda_1 = -2$, $\lambda_2 = 3$. The corresponding eigenvectors are

$$\mathbf{u}^{(1)} = \begin{bmatrix} 2 \\ 1 \end{bmatrix}, \qquad \mathbf{u}^{(2)} = \begin{bmatrix} 3 \\ 2 \end{bmatrix}.$$

A has simple eigenvalues and therefore is diagonalizable. A coordinate transformation matrix which diagonalizes **A** is given by the matrix of eigenvectors

$$\mathbf{P} = \begin{bmatrix} 2 & 3 \\ 1 & 2 \end{bmatrix}.$$

Then $\mathbf{P}^{-1}\mathbf{AP}$ is a diagonal matrix:

$$\mathbf{P}^{-1}\mathbf{AP} = \mathbf{\Lambda} = \begin{bmatrix} -2 & 0 \\ 0 & 3 \end{bmatrix}.$$

■

In the next example we encounter a diagonalizable matrix that has multiple eigenvalues.

Example 7.4

For the matrix

$$\mathbf{A} = \begin{bmatrix} 2 & 3 & -3 \\ -2 & -3 & 2 \\ 2 & 2 & -3 \end{bmatrix}$$

the eigenvalues are $\lambda_1 = \lambda_2 = -1$, and $\lambda_3 = -2$. Corresponding to the multiple eigenvalue -1 we find eigenvectors by solving $(\mathbf{A} + \mathbf{I})\mathbf{x} = \boldsymbol{\theta}$:

$$\left[\begin{array}{ccc|c} 3 & 3 & -3 & 0 \\ -2 & -2 & 2 & 0 \\ 2 & 2 & -2 & 0 \end{array}\right] \longrightarrow \left[\begin{array}{ccc|c} 3 & 3 & -3 & 0 \\ 0 & 0 & 0 & 0 \\ 0 & 0 & 0 & 0 \end{array}\right].$$

Thus a corresponding eigenvector $\mathbf{x}$ satisfies $x_1 = -x_2 + x_3$ and hence the eigenvectors take the form

$$\mathbf{x} = x_2\begin{bmatrix} -1 \\ 1 \\ 0 \end{bmatrix} + x_3\begin{bmatrix} 1 \\ 0 \\ 1 \end{bmatrix}.$$

It follows that two linearly independent eigenvectors corresponding to the eigenvalue -1 are

$$\mathbf{u}^{(1)} = \begin{bmatrix} -1 \\ 1 \\ 0 \end{bmatrix} \quad \text{and} \quad \mathbf{u}^{(2)} = \begin{bmatrix} 1 \\ 0 \\ 1 \end{bmatrix}.$$

An eigenvector corresponding to λ_3 is

$$\mathbf{u}^{(3)} = \begin{bmatrix} 3 \\ -2 \\ 2 \end{bmatrix}.$$

We form the matrix of eigenvectors

$$\mathbf{P} = \begin{bmatrix} -1 & 1 & 3 \\ 1 & 0 & -2 \\ 0 & 1 & 2 \end{bmatrix}$$

and find that

$$\mathbf{P}^{-1}\mathbf{AP} = \boldsymbol{\Lambda} = \begin{bmatrix} -1 & 0 & 0 \\ 0 & -1 & 0 \\ 0 & 0 & -2 \end{bmatrix}.$$

■

For a real symmetric matrix there is an orthonormal basis consisting of eigenvectors (see Theorem 6.9). In this case the matrix $\mathbf{P}$ of eigenvectors can be chosen to be orthogonal and with this choice the calculation of $\mathbf{P}^{-1}$ is very simple: $\mathbf{P}^{-1} = \mathbf{P}^{T}$.

Example 7.5

In Example 6.14 we found that the symmetric matrix

$$\mathbf{A} = \begin{bmatrix} 1 & 2 & -4 \\ 2 & -2 & -2 \\ -4 & -2 & 1 \end{bmatrix}$$

has eigenvalues $\lambda_1 = \lambda_2 = -3$, and $\lambda_3 = 6$. Using the Gram–Schmidt process we found a corresponding orthonormal set of eigenvectors:

$$\mathbf{u}^{(1)} = \frac{1}{\sqrt{5}}\begin{bmatrix} -1 \\ 2 \\ 0 \end{bmatrix}, \qquad \mathbf{u}^{(2)} = \frac{1}{3\sqrt{5}}\begin{bmatrix} 4 \\ 2 \\ 5 \end{bmatrix}, \qquad \mathbf{u}^{(3)} = \frac{1}{3}\begin{bmatrix} 2 \\ -1 \\ 2 \end{bmatrix}.$$

We define the matrix of eigenvectors by

$$\mathbf{P} = [\mathbf{u}^{(1)} \quad \mathbf{u}^{(2)} \quad \mathbf{u}^{(3)}] = \begin{bmatrix} \frac{-1}{\sqrt{5}} & \frac{4}{3\sqrt{5}} & \frac{2}{3} \\ \frac{2}{\sqrt{5}} & \frac{2}{3\sqrt{5}} & -\frac{1}{3} \\ 0 & \frac{5}{3\sqrt{5}} & \frac{2}{3} \end{bmatrix};$$

then $\mathbf{P}$ is orthogonal and

$$\mathbf{P}^{-1}\mathbf{A}\mathbf{P} = \mathbf{P}^{\mathrm{T}}\mathbf{A}\mathbf{P} = \mathbf{\Lambda} = \begin{bmatrix} -3 & 0 & 0 \\ 0 & -3 & 0 \\ 0 & 0 & 6 \end{bmatrix}.$$

■

The diagonalization of symmetric matrices is of particular importance for the applications that we present later in the chapter.

Theorem 7.3. Every real symmetric matrix $\mathbf{A}$ is diagonalizable. The matrix $\mathbf{P}$, for which $\mathbf{P}^{-1}\mathbf{A}\mathbf{P}$ is a diagonal matrix, can be chosen to be orthogonal.

■

In view of this theorem we say that a real symmetric matrix is **orthogonally similar** to a diagonal matrix.

EXERCISES 7.2

1. If $\mathbf{A} = \mathbf{P}\mathbf{D}\mathbf{P}^{-1}$ with

$$\mathbf{P} = \frac{1}{\sqrt{2}}\begin{bmatrix} 1 & 1 \\ 1 & -1 \end{bmatrix} \quad \text{and} \quad \mathbf{D} = \begin{bmatrix} 1 & 0 \\ 0 & .5 \end{bmatrix},$$

then find $\mathbf{A}$. If $\mathbf{u}^{(0)} = [.5,\ .5]^{\mathrm{T}}$ and $\mathbf{u}^{(k)} = \mathbf{A}\mathbf{u}^{(k-1)}$ for $k > 0$, find $\mathbf{u}^{(k)}$.

2. Determine whether the following matrices are diagonalizable.

(a) $\mathbf{A} = \begin{bmatrix} -1 & 2 \\ -2 & 3 \end{bmatrix}$ (b) $\mathbf{A} = \begin{bmatrix} 4 & 5 \\ 2 & 1 \end{bmatrix}$

(c) $\mathbf{A} = \begin{bmatrix} 1 & 1 & 1 \\ -2 & 2 & 3 \\ 3 & 1 & -1 \end{bmatrix}$ (d) $\mathbf{A} = \begin{bmatrix} 0 & 0 & 3 \\ 0 & 1 & 0 \\ 1 & 0 & 0 \end{bmatrix}$

(e) $\mathbf{A} = \begin{bmatrix} 1 & 1 & 2 \\ 0 & 1 & 0 \\ 0 & 0 & -1 \end{bmatrix}$ (f) $\mathbf{A} = \begin{bmatrix} 3 & 1 & 2 & 2 \\ -4 & 1 & 2 & 0 \\ 0 & -4 & -5 & -8 \\ 2 & 3 & 3 & 7 \end{bmatrix}$

[*Hint:* $p_{\mathbf{A}}(\lambda) = (\lambda^2 - 1)(\lambda - 3)^2$.]

3. Show that

$$\mathbf{A} = \begin{bmatrix} 1 & 2 \\ 0 & 1 \end{bmatrix}$$

is not diagonalizable.

4. For each of the following matrices $\mathbf{A}$, find a matrix $\mathbf{P}$ such that $\mathbf{P}^{-1}\mathbf{AP}$ is a diagonal matrix.

(a) $\mathbf{A} = \begin{bmatrix} 2 & 2 \\ 2 & -1 \end{bmatrix}$ (b) $\mathbf{A} = \begin{bmatrix} 1 & 0 & 0 \\ -1 & 2 & 0 \\ 2 & 1 & 3 \end{bmatrix}$

(c) $\mathbf{A} = \begin{bmatrix} -1 & 2 & 2 \\ 2 & -1 & 2 \\ 2 & 2 & -1 \end{bmatrix}$ (d) $\mathbf{A} = \begin{bmatrix} -8 & 4 & 5 \\ 4 & 0 & 1 \\ 5 & 1 & 3 \end{bmatrix}$

5. Let

$$\mathbf{A} = \begin{bmatrix} 3 & 2 \\ -4 & -3 \end{bmatrix}.$$

Compute $\mathbf{A}^{101}$. (*Hint:* diagonalize $\mathbf{A}$).

6. Suppose that $\mathbf{A}$ is diagonalizable and that $\mathbf{A} \sim \mathbf{B}$. Is $\mathbf{B}$ diagonalizable?

7. (a) If $\mathbf{D}$ is a diagonal matrix and $\mathbf{D}^2 = \mathbf{O}$, show that $\mathbf{D} = \mathbf{O}$.
   (b) Suppose that $\mathbf{A} = \mathbf{PDP}^{-1}$ and $\mathbf{A}^2 = \mathbf{O}$; what can you conclude about $\mathbf{A}$?

8. Find an orthogonal matrix $\mathbf{P}$ such that $\mathbf{P}^{-1}\mathbf{AP}$ is a diagonal matrix, where

$$\mathbf{A} = \begin{bmatrix} 1 & 6 & 0 \\ 6 & -2 & 6 \\ 0 & 6 & -5 \end{bmatrix}.$$

9. Find a matrix $\mathbf{A}$ that is not a diagonal matrix and whose characteristic polynomial is $p_{\mathbf{A}}(\lambda) = -(\lambda + 1)(\lambda - 1)^2$.

10. Suppose that $\mathbf{A} = \mathbf{QDQ}^{-1}$, where $\mathbf{D} = \mathbf{diag}\,(\lambda_1, \ldots, \lambda_n)$. Show that for any polynomial p, $p(\mathbf{A}) = \mathbf{Q}p(\mathbf{D})\mathbf{Q}^{-1}$, where $p(\mathbf{D}) = \mathbf{diag}\,(p(\lambda_1), \ldots, p(\lambda_n))$.

*11. Suppose that $\mathbf{A}$ is an $n \times n$ diagonalizable matrix, say $\mathbf{A} = \mathbf{PDP}^{-1}$ and consider the difference equation $\mathbf{x}^{(k+1)} = \mathbf{Ax}^{(k)}$, where $\mathbf{x}^{(0)}$ is given.
   (a) Let $\mathbf{c} = \mathbf{P}^{-1}\mathbf{x}^{(0)}$ and show that the solution of the difference equation can be written as

$$\mathbf{x}^{(k)} = \sum_{j=1}^{n} c_j \lambda_j^k \mathbf{u}^{(j)},$$

where $\mathbf{Au}^{(j)} = \lambda_j \mathbf{u}^{(j)}$ for $j = 1, 2, \ldots, n$.

(b) Suppose that $|\lambda_j| < 1$ for $j = 1, 2, \ldots, n$. Describe the behavior of the solution as $k \to \infty$.

(c) Suppose that for some j, $|\lambda_j| > 1$ and $c_j \neq 0$. Describe the behavior of the solution as $k \to \infty$.

12. Suppose that $\mathbf{A} = \mathbf{PDP}^{-1}$ and $\mathbf{B} = \mathbf{QD'Q}^{-1}$, where $\mathbf{D}$ is diagonal and $\mathbf{D'}$ is a diagonal matrix obtained from $\mathbf{D}$ by permuting its diagonal entries. Show that $\mathbf{A} \sim \mathbf{B}$.

13. For each of the following symmetric matrices, find an orthogonal matrix $\mathbf{Q}$ and a diagonal $\mathbf{D}$ with $\mathbf{Q}^{\mathrm{T}}\mathbf{AQ} = \mathbf{D}$.

(a) $\mathbf{A} = \begin{bmatrix} 1 & 3 \\ 3 & 1 \end{bmatrix}$ **(b)** $\mathbf{A} = \begin{bmatrix} 5 & 1 & -1 \\ 1 & 5 & 1 \\ -1 & 1 & 5 \end{bmatrix}$

(c) $\mathbf{A} = \begin{bmatrix} -2 & 0 & -36 \\ 0 & -3 & 0 \\ -36 & 0 & -23 \end{bmatrix}$

14. Use diagonalization to solve the difference equation $\mathbf{u}^{(k+1)} = \mathbf{Au}^{(k)}$, $\mathbf{u}^{(0)} = [1 \quad 1 \quad 1]^{\mathrm{T}}$, where

$$\mathbf{A} = \begin{bmatrix} -1 & 1 & 0 \\ 0 & -2 & 1 \\ 0 & 0 & -3 \end{bmatrix}.$$

7.3 Quadratic Forms and Conic Sections

In this section we show how the diagonalization of symmetric matrices can be applied to some problems in analytic geometry which involve planar curves called conic sections. The coordinate transformation matrix used in diagonalization has the geometric effect of rotation of the axes and simplifies the equation of the conic section.

Consider a general second-degree polynomial equation in two variables:

$$ax^2 + 2bxy + cy^2 + dx + ey = f, \tag{2}$$

where a, b, c, d, e, and f are given constants. The expression $ax^2 + 2bxy + cy^2$ is called the **quadratic form** corresponding to equation (2). We assume that at least one of the coefficients a, b, or c is nonzero.

The graph of equation (2) is the set of all points (x, y) in R^2 which satisfy the equation. The graph is either a parabola, a hyperbola, or an ellipse (a circle is considered as a special case of an ellipse) except for some degenerate cases. The degenerate cases are treated in the exercises and consist of those situations when the graph is the empty set, a point, a line, or two lines.

The standard forms for the equations of a parabola, hyperbola, and ellipse are as follows:

Parabola: $y = kx^2$.
Hyperbola: $x^2/A^2 - y^2/B^2 = 1$.
Ellipse: $x^2/A^2 + y^2/B^2 = 1$.

In each case an axis of symmetry is along one of the coordinate axes. Also, for the hyperbola and ellipse the center is at the origin. The graphs of these equations are given in Figure 7.1. If an axis is not along one of the coordinate axes or the center is not at the origin, the equation of the conic section does not take a simple standard form.

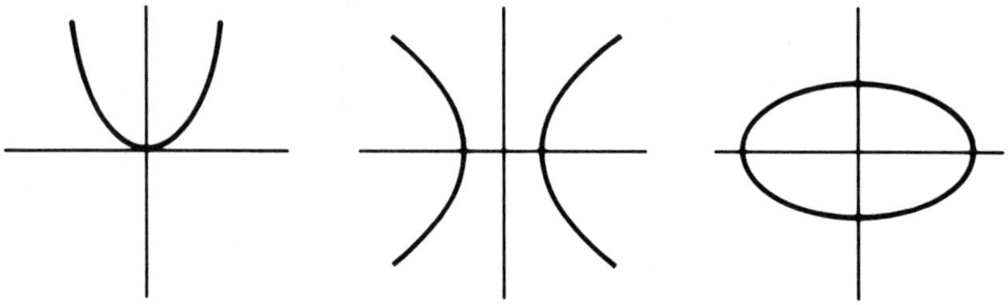

$y = kx^2$ $\frac{x^2}{A^2} - \frac{y^2}{B^2} = 1$ $\frac{x^2}{A^2} + \frac{y^2}{B^2} = 1$

Parabola Hyperbola Ellipse

Figure 7.1

Example 7.6

The graph of the quadratic equation

$$5x^2 + 6xy + 5y^2 - 16x - 16y = -8$$

is the ellipse of Figure 7.2. Note that the ellipse has its center at (1, 1) and has its axes rotated 45 degrees or equivalently, $\pi/4$ radians.

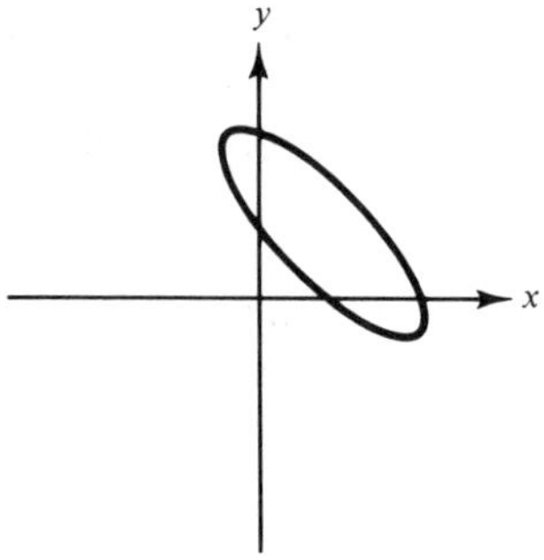

Figure 7.2 ■

We shall develop a technique using rotations and translations that will enable us to recognize the type of conic section and sketch its graph. This technique depends on a 2×2 symmetric matrix associated with the quadratic form. Associated with (2) we define the symmetric matrix

$$\mathbf{A} = \begin{bmatrix} a & b \\ b & c \end{bmatrix} \tag{3}$$

and the vector $\mathbf{w} = [x \quad y]^T$. Then the quadratic form corresponding to (3), that is, $ax^2 + 2bxy + cy^2$, can be written as

$$ax^2 + 2bxy + cy^2 = \langle \mathbf{w}, \mathbf{Aw} \rangle.$$

To see this, we expand the inner product:

$$\begin{aligned} \langle \mathbf{w}, \mathbf{Aw} \rangle &= [x \quad y] \begin{bmatrix} a & b \\ b & c \end{bmatrix} \begin{bmatrix} x \\ y \end{bmatrix} \\ &= [x \quad y] \begin{bmatrix} ax + by \\ bx + cy \end{bmatrix} = ax^2 + 2bxy + cy^2. \end{aligned}$$

Example 7.7

The quadratic form

$$x^2 + 6xy - 4y^2$$

has associated symmetric matrix

$$\mathbf{A} = \begin{bmatrix} 1 & 3 \\ 3 & -4 \end{bmatrix}$$

and if $\mathbf{w} = [x \quad y]^T$, we find that $\langle \mathbf{w}, \mathbf{Aw} \rangle = x^2 + 6xy - 4y^2$.

■

Since the 2×2 matrix $\mathbf{A}$ associated with (2) is symmetric, there is an orthogonal matrix $\mathbf{Q}$ which diagonalizes $\mathbf{A}$, say $\mathbf{Q}^{-1}\mathbf{AQ} = \mathbf{\Lambda} = \mathbf{diag}\,(\lambda_1, \lambda_2)$. Now introduce the coordinate transformation

$$\mathbf{w} = \begin{bmatrix} x \\ y \end{bmatrix} = \mathbf{Qz}, \qquad \text{where } \mathbf{z} = \begin{bmatrix} u \\ v \end{bmatrix}.$$

Then we find

$$\langle \mathbf{w}, \mathbf{Aw} \rangle = \langle \mathbf{Qz}, \mathbf{AQz} \rangle = \langle \mathbf{z}, \mathbf{Q}^T\mathbf{AQz} \rangle,$$

and since $\mathbf{Q}$ is orthogonal, that is, $\mathbf{Q}^T = \mathbf{Q}^{-1}$, it follows that

$$\langle \mathbf{w}, \mathbf{Aw} \rangle = \langle \mathbf{z}, \mathbf{Q}^{-1}\mathbf{AQz} \rangle = \langle \mathbf{z}, \mathbf{\Lambda z} \rangle = \lambda_1 u^2 + \lambda_2 v^2.$$

In terms of the new variables, equation (2) becomes

$$\lambda_1 u^2 + \lambda_2 v^2 + d'u + e'v = f, \tag{4}$$

where $d' = (dq_{11} + eq_{21})$ and $e' = (dq_{12} + eq_{22})$.

The coordinate transformation $\mathbf{w} = \mathbf{Qz}$ has the *algebraic* effect of eliminating the cross-product term, $2bxy$, from the quadratic equation (2). *Geometrically,* the coordinate transformation results in a rotation of the coordinate axes. To see this we need to examine the coordinates of a point with respect to a coordinate system that has been rotated through some angle ψ in the counterclockwise direction.

Rotation of Axes

Consider the point $P(x_1, y_1)$ in the plane with xy coordinate axes. Now suppose that the axes are rotated counterclockwise through ψ radians to form the uv coordinate axes (see Figure 7.3). What are the coordinates of P in the uv coordinate system? Let r be the distance from the origin to P, that is,

$$r = \sqrt{x_1^2 + y_1^2}.$$

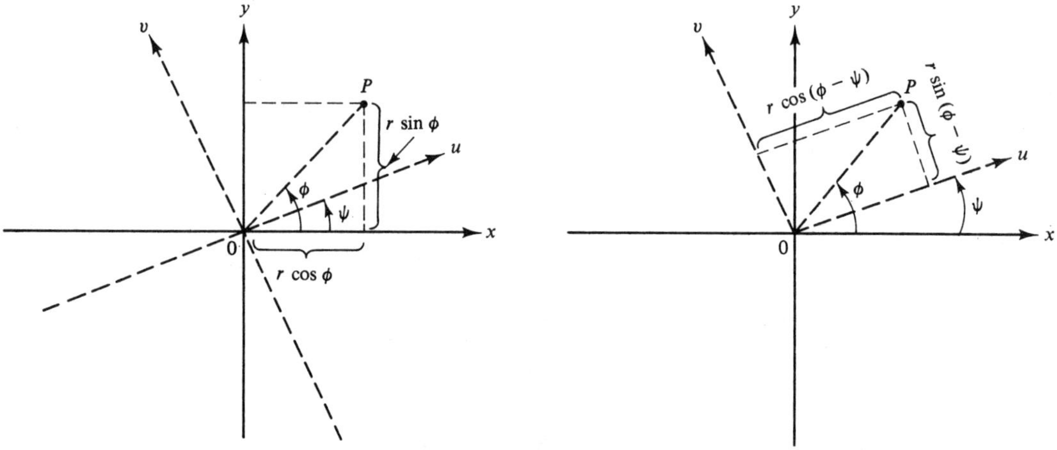

xy coordinates *uv* coordinates

Figure 7.3

Then the coordinates of P in the xy coordinate system are

$$x_1 = r \cos \phi \quad \text{and} \quad y_1 = r \sin \phi,$$

where ϕ is the angle between OP and the x axis. The coordinates of P in the uv coordinate system are given by

$$u_1 = r \cos (\phi - \psi), \qquad v_1 = r \sin (\phi - \psi).$$

Using trigonometric identities for the difference of two angles yields

$$\begin{aligned} u_1 &= r\{\cos \phi \cos \psi + \sin \phi \sin \psi\} \\ &= \{r \cos \phi\} \cos \psi + \{r \sin \phi\} \sin \psi \\ &= x_1 \cos \psi + y_1 \sin \psi \end{aligned}$$

and

$$\begin{aligned} v_1 &= r\{\sin \phi \cos \phi - \cos \phi \sin \psi\} \\ &= \{r \sin \phi\} \cos \psi - \{r \cos \phi\} \sin \psi \\ &= y_1 \cos \psi - x_1 \sin \psi. \end{aligned}$$

Therefore, we obtain

$$\begin{bmatrix} u_1 \\ v_1 \end{bmatrix} = \begin{bmatrix} \cos\psi & \sin\psi \\ -\sin\psi & \cos\psi \end{bmatrix} \begin{bmatrix} x_1 \\ y_1 \end{bmatrix}.$$

Recall from Section 5.3 the rotation matrix

$$R_\psi = \begin{bmatrix} \cos\psi & -\sin\psi \\ \sin\psi & \cos\psi \end{bmatrix}.$$

It follows that the relationship between the coordinates of the point P in Figure 7.4 is given by

$$\begin{bmatrix} u_1 \\ v_1 \end{bmatrix} = R_{-\psi} \begin{bmatrix} x_1 \\ y_1 \end{bmatrix},$$

where (x_1, y_1) are the coordinates of P in the xy coordinate system and (u_1, v_1) are the coordinates of P in the uv coordinate system.

We note that a rotation matrix, R_ψ or $R_{-\psi}$, is orthogonal. Moreover, there is a simple characterization of **plane rotation** matrices.

Theorem 7.4. A 2×2 orthogonal matrix $\mathbf{Q}$ represents a plane rotation if and only if $\det(\mathbf{Q}) = 1$.

■

Now if $\mathbf{Q}$ is the orthogonal matrix of eigenvectors for the 2×2 matrix $\mathbf{A}$ associated with the quadratic equation (2), then from Chapter 5 we know that $\det(\mathbf{Q}) = \pm 1$. If $\det(\mathbf{Q}) = 1$, then $\mathbf{Q}$ represents a plane rotation, whereas if $\det(\mathbf{Q}) = -1$, we need only interchange the columns of $\mathbf{Q}$ to form a plane rotation matrix. This reordering of the columns of $\mathbf{Q}$ simply corresponds to labeling the eigenvalues of $\mathbf{A}$ in reverse order and has no effect on the resultant equation of the conic section.

Example 7.8

Consider the conic section with equation

$$xy = 1.$$

The associated symmetric matrix for the quadratic form xy is

$$\mathbf{A} = \begin{bmatrix} 0 & \frac{1}{2} \\ \frac{1}{2} & 0 \end{bmatrix}.$$

Indeed, if $\mathbf{w} = [x \quad y]^T$, then

$$\langle \mathbf{w}, \mathbf{A}\mathbf{w} \rangle = [x \quad y] \begin{bmatrix} 0 & \frac{1}{2} \\ \frac{1}{2} & 0 \end{bmatrix} \begin{bmatrix} x \\ y \end{bmatrix} = \frac{xy}{2} + \frac{yx}{2} = xy.$$

By doing an eigenanalysis of $\mathbf{A}$ we find that $\mathbf{Q}^T\mathbf{A}\mathbf{Q} = \mathbf{\Lambda}$, where

$$\mathbf{\Lambda} = \begin{bmatrix} -\frac{1}{2} & 0 \\ 0 & \frac{1}{2} \end{bmatrix} \quad \text{and} \quad \mathbf{Q} = \frac{1}{\sqrt{2}} \begin{bmatrix} 1 & 1 \\ -1 & 1 \end{bmatrix}.$$

Therefore, if we change variables by $\mathbf{w} = \mathbf{Qz}$, where $\mathbf{z} = [u \quad v]^T$, then

$$\langle \mathbf{w}, \mathbf{Aw} \rangle = \langle \mathbf{Qz}, \mathbf{AQz} \rangle = \langle \mathbf{z}, \mathbf{Q}^T\mathbf{AQz} \rangle = \langle \mathbf{z}, \mathbf{\Lambda z} \rangle$$
$$= -\tfrac{1}{2} u^2 + \tfrac{1}{2} v^2.$$

The equation $xy = 1$ is equivalent to the equation

$$\frac{-u^2}{2} + \frac{v^2}{2} = 1,$$

which represents a hyperbola in the uv coordinate system. Note that $\mathbf{Q} = \mathbf{R}_{-\pi/4}$, and hence

$$\begin{bmatrix} x \\ y \end{bmatrix} = \mathbf{Q}\begin{bmatrix} u \\ v \end{bmatrix} = \mathbf{R}_{-\pi/4}\begin{bmatrix} u \\ v \end{bmatrix}.$$

That is,

$$\begin{bmatrix} u \\ v \end{bmatrix} = \mathbf{R}^{-1}_{-\pi/4}\begin{bmatrix} x \\ y \end{bmatrix} = \mathbf{R}_{\pi/4}\begin{bmatrix} x \\ y \end{bmatrix}.$$

Therefore, the uv coordinate axes result from a rotation of the xy coordinate axes through $-\pi/4$ radians as is sketched in Figure 7.4.

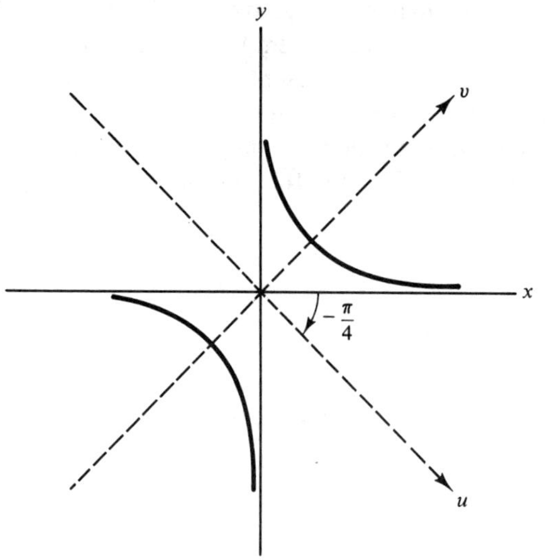

Figure 7.4 ■

Example 7.9

We identify the conic section whose equation is

$$x^2 - 2xy + y^2 + 3x + y = 0.$$

We perform a rotation to eliminate the cross-product term. The associated symmetric matrix is

$$\mathbf{A} = \begin{bmatrix} 1 & -1 \\ -1 & 1 \end{bmatrix},$$

which has eigenvalues $\lambda_1 = 0$ and $\lambda_2 = 2$. The corresponding orthonormal eigenvectors are

$$\mathbf{u}^{(1)} = \frac{1}{\sqrt{2}}\begin{bmatrix} 1 \\ 1 \end{bmatrix}, \qquad \mathbf{u}^{(2)} = \frac{1}{\sqrt{2}}\begin{bmatrix} 1 \\ -1 \end{bmatrix}.$$

But the matrix of eigenvectors has determinant

$$\begin{vmatrix} \frac{1}{\sqrt{2}} & \frac{1}{\sqrt{2}} \\ \frac{1}{\sqrt{2}} & -\frac{1}{\sqrt{2}} \end{vmatrix} = -1,$$

and hence we interchange columns $\mathbf{u}^{(1)}$ and $\mathbf{u}^{(2)}$ so that

$$\mathbf{Q} = [\mathbf{u}^{(2)} \quad \mathbf{u}^{(1)}] = \frac{1}{\sqrt{2}}\begin{bmatrix} 1 & 1 \\ -1 & 1 \end{bmatrix} = \begin{bmatrix} \cos\frac{-\pi}{4} & -\sin\frac{-\pi}{4} \\ \sin\frac{-\pi}{4} & \cos\frac{-\pi}{4} \end{bmatrix}.$$

Therefore, $\mathbf{Q}$ is the rotation matrix $\mathbf{R}_{-\pi/4}$, that is

$$\mathbf{Q} = \mathbf{R}_{-\pi/4}.$$

Performing the coordinate transformation

$$\begin{bmatrix} x \\ y \end{bmatrix} = \mathbf{Q}\begin{bmatrix} u \\ v \end{bmatrix}$$

on the equation

$$x^2 - 2xy + y^2 + 3x + y = 0$$

gives

$$\lambda_2 u^2 + \lambda_1 v^2 + \sqrt{2}\,u + 2\sqrt{2}\,v = 0.$$

Since $\lambda_2 = 2$ and $\lambda_1 = 0$ we get the equation

$$2u^2 + \sqrt{2}\,u = -2\sqrt{2}\,v,$$

or equivalently,

$$v = -\frac{u^2 + \frac{u}{\sqrt{2}}}{\sqrt{2}}.$$

To translate the axes we complete the square:

$$v = -\left(\frac{u + \frac{1}{2}\sqrt{2}}{\sqrt{2}}\right)^2 + \frac{1}{(8\sqrt{2})}.$$

Thus we find

$$v - \frac{1}{8\sqrt{2}} = -\frac{\left(u + \dfrac{1}{2\sqrt{2}}\right)^2}{\sqrt{2}}$$

and if we let $z = v - 1/(8\sqrt{2})$, $w = u + 1/(2\sqrt{2})$, there results the equation

$$z = \frac{-w^2}{\sqrt{2}},$$

which is the equation of a parabola in standard form. The graph of the parabola is given in Figure 7.5 together with the xy, uv, and wz coordinate axes.

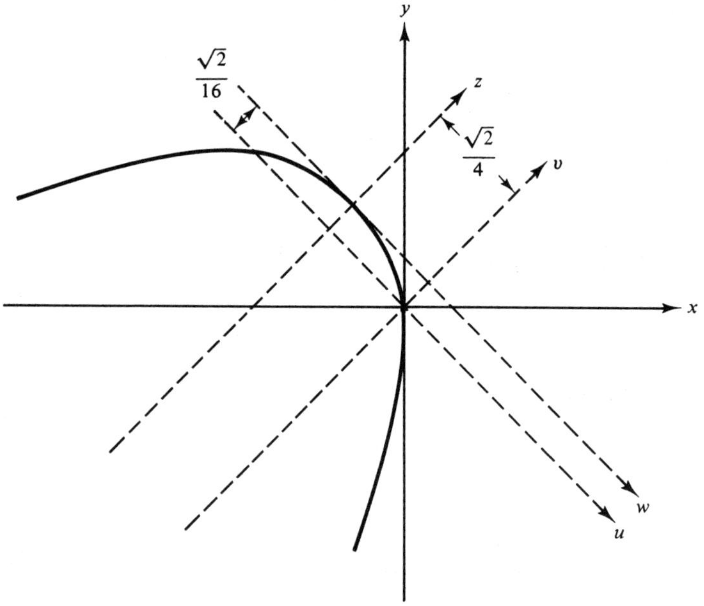

Figure 7.5

■

In Example 7.9 we made a translation of axes of the form

$$z = v + \alpha, \qquad w = u + \beta$$

and the resultant equation was recognized as that of a parabola. In general, a rotation is applied to equation (2) to eliminate the cross term, $2bxy$. The resulting equation [see (4)] takes the form

$$Au^2 + Bv^2 + Du + Ev = F.$$

If A and B are nonzero, we complete the square (if necessary) in both u and v:

$$A\left(\frac{u + D}{2A}\right)^2 + B\left(\frac{v + E}{2B}\right)^2 = F + \frac{D^2/A + E^2/B}{4}.$$

The translation $w = u + D/2A$, $z = v + E/2B$ then gives

$$Aw^2 + Bz^2 = G,$$

or equivalently, if $G \neq 0$,

$$\frac{w^2}{G/A} + \frac{z^2}{G/B} = 1,$$

where $G = F + (D^2/A + E^2/B)/4$. We recognize this as the equation of an ellipse or hyperbola, depending on the signs of A, B, and G. If G is negative and A, B are positive, the equation has no solution and the graph is the empty set. The procedure of completing the square is used to determine an appropriate translation of axes.

Example 7.10

The quadratic equation

$$5x^2 + 4xy + 2y^2 - x + 2y = 1$$

has associated symmetric matrix

$$\mathbf{A} = \begin{bmatrix} 5 & 2 \\ 2 & 2 \end{bmatrix}$$

with eigenvalues $\lambda_1 = 6$, $\lambda_2 = 1$ and corresponding orthonormal eigenvectors:

$$\mathbf{u}^{(1)} = \frac{1}{\sqrt{5}}\begin{bmatrix} 2 \\ 1 \end{bmatrix}, \qquad \mathbf{u}^{(2)} = \frac{1}{\sqrt{5}}\begin{bmatrix} -1 \\ 2 \end{bmatrix}.$$

Then

$$\mathbf{Q} = \frac{1}{\sqrt{5}}\begin{bmatrix} 2 & -1 \\ 1 & 2 \end{bmatrix}$$

is a rotation matrix since det $(\mathbf{Q}) = 1$. The rotation of axes is accomplished by the coordinate transformation

$$\begin{bmatrix} x \\ y \end{bmatrix} = \frac{1}{\sqrt{5}}\begin{bmatrix} 2 & -1 \\ 1 & 2 \end{bmatrix}\begin{bmatrix} u \\ v \end{bmatrix}$$

and results in the equation

$$6u^2 + v^2 + \sqrt{5}\, v = 1.$$

We complete the square in v to obtain

$$6u^2 + \left(\frac{v + \sqrt{5}}{2}\right)^2 = \frac{9}{4}.$$

Finally, we translate axes by using

$$w = u, \qquad z = v + \frac{\sqrt{5}}{2}$$

and obtain the equation of an ellipse in standard form:

$$\frac{w^2}{(\sqrt{6}/4)^2} + \frac{z^2}{(3/2)^2} = 1.$$

To sketch the graph (see Figure 7.6) we recall that $\mathbf{Q} = \mathbf{R}_\psi$ for some ψ, that is,

$$\mathbf{Q} = \frac{1}{\sqrt{5}}\begin{bmatrix} 2 & -1 \\ 1 & 2 \end{bmatrix} = \begin{bmatrix} \cos\psi & -\sin\psi \\ \sin\psi & \cos\psi \end{bmatrix} = \mathbf{R}_\psi.$$

Thus $\cos\psi = 2/\sqrt{5}$ and $\sin\psi = 1/\sqrt{5}$. It follows that $0 < \psi < \pi/2$ and $\tan\psi = \frac{1}{2}$. That is, the u-axis has a slope of $\frac{1}{2}$ with respect to the x-axis.

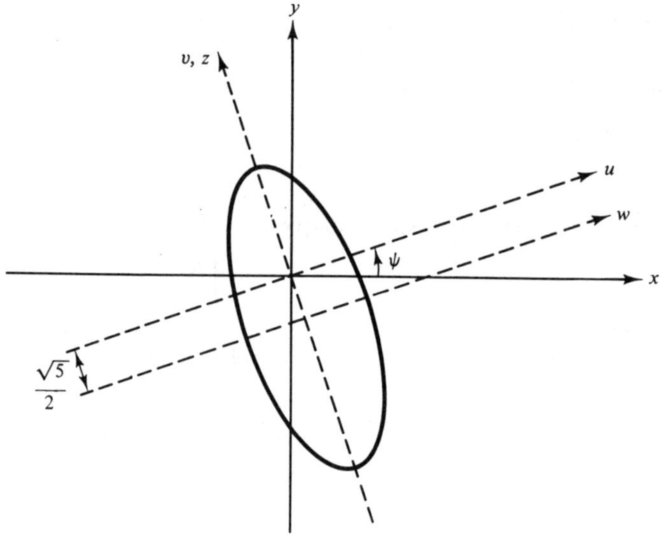

Figure 7.6

■

In the next example we point out the connection between the geometry of an ellipse and the eigenvectors of the matrix associated with the quadratic equation. We show how the graph of the ellipse of Example 7.6 is obtained.

Example 7.11

The quadratic equation we consider is

$$5x^2 + 6xy + 5y^2 - 16x - 16y = -8,$$

which has associated symmetric matrix

$$\mathbf{A} = \begin{bmatrix} 5 & 3 \\ 3 & 5 \end{bmatrix}.$$

$\mathbf{A}$ has characteristic polynomial $p_{\mathbf{A}}(\lambda) = (5 - \lambda)^2 - 9 = 16 - 10\lambda + \lambda^2$ and hence the eigenvalues are $\lambda_1 = 2$ and $\lambda_2 = 8$. The corresponding orthonormal eigenvectors are

$$\mathbf{u}^{(1)} = \frac{1}{\sqrt{2}}\begin{bmatrix} 1 \\ -1 \end{bmatrix}, \qquad \mathbf{u}^{(2)} = \frac{1}{\sqrt{2}}\begin{bmatrix} 1 \\ 1 \end{bmatrix}.$$

The matrix of eigenvectors $\mathbf{Q} = [\mathbf{u}^{(1)} \ \ \mathbf{u}^{(2)}]$ is a rotation matrix since $\det(\mathbf{Q}) = 1$. We introduce the rotation

$$\begin{bmatrix} x \\ y \end{bmatrix} = \mathbf{Q}\begin{bmatrix} u \\ v \end{bmatrix} \qquad \text{where } \mathbf{Q} = \mathbf{R}_{-\pi/4}$$

and obtain

$$2u^2 + 8v^2 - 8\sqrt{2}(u + v) - 8\sqrt{2}(-u + v) = -8.$$

This can be written as

$$2u^2 + 8(v^2 - 2\sqrt{2}\, v) = -8,$$

and hence by completing the square in v we get

$$2u^2 + 8(v - \sqrt{2})^2 = 8,$$

or equivalently,

$$\frac{u^2}{4} + (v - \sqrt{2})^2 = 1.$$

This is an ellipse that is centered at $u = 0$ and $v = \sqrt{2}$ and whose axes are in the same directions as the eigenvectors of $\mathbf{A}$. By introducing the translation $w = u$ and $z = v - \sqrt{2}$, we get the standard form equation of the ellipse: $w^2/4 + z^2 = 1$. For the sketch, see Figure 7.2.

■

EXERCISES 7.3

1. Find the quadratic form corresponding to each of the following equations.
   (a) $-3x^2 + 4xy - 8y^2 + y = 10$
   (b) $5uv + 6v^2 - u + 3v = 8$
   (c) $uv = 1$
   (d) $8x_1^2 + 2\sqrt{2}\, x_1x_2 - 4x_2^2 - x_1 + x_2 = 0$
2. Determine the symmetric matrices associated with the equations in Exercise 1.
3. For each of the following matrices, expand the quadratic form $\langle \mathbf{x}, \mathbf{Ax} \rangle$.
   (a) $\mathbf{A} = \begin{bmatrix} 3 & 1 \\ 1 & 2 \end{bmatrix}$ **(b)** $\mathbf{A} = \begin{bmatrix} 3 & -2 \\ -2 & 0 \end{bmatrix}$ **(c)** $\mathbf{A} = \begin{bmatrix} 0 & -1 \\ -1 & 0 \end{bmatrix}$
   (d) $\mathbf{A} = \begin{bmatrix} 1 & 3 & -1 \\ 3 & 2 & 4 \\ -1 & 4 & 6 \end{bmatrix}$

4. Determine a coordinate transformation $\mathbf{u} = \mathbf{Qx}$ to eliminate the cross term in $\langle \mathbf{x}, \mathbf{Ax} \rangle$.

(a) $\mathbf{A} = \begin{bmatrix} 5 & 3 \\ 3 & 5 \end{bmatrix}$ **(b)** $\mathbf{A} = \begin{bmatrix} 7 & 1 \\ 1 & 7 \end{bmatrix}$ **(c)** $\mathbf{A} = \begin{bmatrix} 3 & -2 \\ -2 & 0 \end{bmatrix}$

(d) $\mathbf{A} = \begin{bmatrix} 0 & -1 \\ -1 & 0 \end{bmatrix}$

5. For each matrix in Exercise 4, determine a rotation matrix $\mathbf{R}$ whose columns are eigenvectors of $\mathbf{A}$.

6. For each of the following equations, identify the graph as a line, two lines, a point, or the empty set.

(a) $x^2 + 2xy + y^2 = 1$
(b) $4x^2 - 4xy + y^2 + 2y - 4x = -1$
(c) $x^2 + y^2 - 2x + 4y = -6$
(d) $x^2 + y^2 + 2x - 4y = -5$

***7.** Consider the general quadratic equation (2). Show that the discriminant $D = b^2 - ac$ is invariant under any rotation. That is, if (2) is transformed to

$$Au^2 + 2Buv + Cv^2 + Du + Ev = F$$

by rotation $\mathbf{R}$, then $B^2 - AC = b^2 - ac$.

8. Suppose that $\mathbf{Q}$ is an orthogonal 2×2 matrix with $\det(\mathbf{Q}) = -1$. Show that $\mathbf{Q} = \mathbf{RS}$, where $\mathbf{R}$ is a rotation matrix and $\mathbf{S}$ is the matrix

$$\mathbf{S} = \begin{bmatrix} 0 & 1 \\ 1 & 0 \end{bmatrix}.$$

9. In each case identify the conic section as an ellipse, a parabola, or a hyperbola.

(a) $xy = 4$
(b) $2xy + 4\sqrt{2}\, y = 6$
(c) $13x^2 - 10xy + 13y^2 - 36x + 36y = 0$
(d) $x^2 - 2xy + y^2 - 2x + 2y = 8$
(e) $5x^2 + 4xy + 2y^2 - 6 = 0$
(f) $3x^2 - 6xy - 5y^2 + 3 = 0$

For each conic section sketch the graph and identify any rotation of axes or translation.

10. Show that the discriminant of (2), $b^2 - ac$, is invariant under the translation $x = u - \alpha$, $y = v - \beta$.

7.4 Simultaneous Diagonalization and Positive Definite Matrices

In this section we examine a different type of change of variables in which a congruence (see Exercise 4 of Section 7.1) transformation is used to diagonalize matrices. Under appropriate conditions we show how a congruence transformation can be used to diagonalize two matrices simultaneously. First we take up the case of an $n \times n$ real symmetric matrix.

The notion of quadratic form and associated symmetric matrix can be extended to n dimensions. Let $\mathbf{x}$ be an n-vector and consider the general quadratic in the variables $x_1, \ldots, x_n$ given by

$$\sum_{i,j=1}^{n} a_{ij}x_i x_j,$$

where $a_{ij} = a_{ji}$. As in the two-dimensional case, the quadratic form can be written as

$$\langle \mathbf{x}, \mathbf{Ax} \rangle = \sum_{i,j=1}^{n} a_{ij}x_i x_j.$$

We say that a real symmetric matrix **A** is **positive definite** if its quadratic form is positive for all nonzero vectors **x**. That is, **A** is positive definite if

$$\langle \mathbf{x}, \mathbf{Ax} \rangle > 0 \qquad \text{for all } \mathbf{x} \neq \boldsymbol{\theta}.$$

We shall see that **A** is positive definite if and only if its quadratic form can be written, after a change of variables, as a particular sum of squares. Let **Q** be the orthogonal matrix of eigenvectors of **A**; then

$$\mathbf{A} = \mathbf{Q\Lambda Q}^{\mathrm{T}},$$

where $\mathbf{\Lambda} = \mathbf{diag}\,(\lambda_1, \ldots, \lambda_n)$ and $\lambda_1, \ldots, \lambda_n$ are the eigenvalues of **A**. That is, **A** is congruent to the diagonal matrix **Λ** and in this context **Q** is called a **congruence transformation matrix**. Note also that **Q** is orthogonal ($\mathbf{Q}^{\mathrm{T}} = \mathbf{Q}^{-1}$), and hence **Q** is also a coordinate transformation matrix. If we introduce the coordinate transformation $\mathbf{x} = \mathbf{Qu}$, then the quadratic form for **A** becomes

$$\langle \mathbf{x}, \mathbf{Ax} \rangle = \langle \mathbf{Qu}, \mathbf{AQu} \rangle = \langle \mathbf{u}, \mathbf{Q}^{\mathrm{T}}\mathbf{AQu} \rangle = \langle \mathbf{u}, \mathbf{\Lambda u} \rangle$$

$$= \sum_{i=1}^{n} \lambda_i u_i^2.$$

From this it follows that $\langle \mathbf{u}, \mathbf{\Lambda u} \rangle$ is positive for all nonzero vectors **u** provided that all of the eigenvalues of **A** are positive. Moreover, since **Q** is invertible, **u** is nonzero if and only if **x** is nonzero. Thus if all of the eigenvalues of **A** are positive, then **A** is positive definite. Conversely, suppose that **A** is positive definite and let **x** be an eigenvector corresponding to some eigenvalue, say λ, of **A**. Then $\mathbf{x} \neq \boldsymbol{\theta}$ and since **A** is positive definite we must have

$$\langle \mathbf{x}, \mathbf{Ax} \rangle = \lambda \langle \mathbf{x}, \mathbf{x} \rangle = \lambda \|\mathbf{x}\|^2 > 0$$

and hence

$$\lambda = \frac{\langle \mathbf{x}, \mathbf{Ax} \rangle}{\|\mathbf{x}\|^2} > 0.$$

Therefore, if **A** is positive definite, then every eigenvalue of **A** must be positive. We have established the following fundamental result.

Theorem 7.5. A real symmetric matrix, **A**, is positive definite if and only if every eigenvalue of **A** is positive. In this case the quadratic form for **A**, after a change of variables $\mathbf{x} = \mathbf{Qu}$, can be written in the form

$$\langle \mathbf{x}, \mathbf{Ax} \rangle = \lambda_1 u_1^2 + \cdots + \lambda_n u_n^2.$$

■

Example 7.12

We determine if $\mathbf{A}$ is a positive definite matrix, where

$$\mathbf{A} = \begin{bmatrix} 7 & 2 & 1 \\ 2 & 7 & -1 \\ 1 & -1 & 4 \end{bmatrix}.$$

The characteristic polynomial of $\mathbf{A}$ is $p_{\mathbf{A}}(\lambda) = \det(\mathbf{A} - \lambda\mathbf{I}) = -\lambda^3 + 18\lambda^2 - 99\lambda + 162$. Factoring gives $p_{\mathbf{A}}(\lambda) = -(\lambda - 3)(\lambda - 6)(\lambda - 9)$ and hence the eigenvalues of $\mathbf{A}$ are all positive: $\lambda_1 = 3$, $\lambda_2 = 6$, and $\lambda_3 = 9$. $\mathbf{A}$ is positive definite. If $\mathbf{Q}$ is the orthogonal matrix of eigenvectors of $\mathbf{A}$ and the change of variables $\mathbf{x} = \mathbf{Qu}$ is introduced, then

$$\langle \mathbf{x}, \mathbf{Ax} \rangle = 3u_1^2 + 6u_2^2 + 9u_3^2.$$

■

Now we show how to *diagonalize* two matrices *simultaneously* by use of a particular congruence transformation matrix. We shall make use of this technique in the next section when solving systems of differential equations that arise in mechanics and electrical networks.

Suppose that $\mathbf{M}$ is a positive definite matrix and $\mathbf{Q}$ is the orthogonal matrix whose columns are the normalized eigenvectors of $\mathbf{M}$. Then

$$\mathbf{M} = \mathbf{QDQ}^{\mathrm{T}},$$

where $\mathbf{D} = \mathbf{diag}\,(d_1, \ldots, d_n)$. Now let $\mathbf{S} = \mathbf{diag}\,(\sqrt{d_1}, \ldots, \sqrt{d_n})$; then

$$\mathbf{M} = \mathbf{QSSQ}^{\mathrm{T}} = (\mathbf{QS})(\mathbf{QS})^{\mathrm{T}}.$$

If $\mathbf{R} = (\mathbf{SQ}^{\mathrm{T}})^{-1} = \mathbf{QS}^{-1}$, then $\mathbf{R}$ is invertible, $\mathbf{R}^{\mathrm{T}} = \mathbf{S}^{-1}\mathbf{Q}^{\mathrm{T}} = (\mathbf{QS})^{-1}$, and hence

$$\begin{aligned} \mathbf{R}^{\mathrm{T}}\mathbf{MR} &= (\mathbf{QS})^{-1}\mathbf{MQS}^{-1} \\ &= (\mathbf{QS})^{-1}(\mathbf{QS})(\mathbf{QS})^{\mathrm{T}}\mathbf{QS}^{-1} \\ &= \mathbf{SQ}^{\mathrm{T}}\mathbf{QS}^{-1} = \mathbf{I}. \end{aligned} \tag{5}$$

Next suppose that $\mathbf{K}$ is a real symmetric matrix of the same size as $\mathbf{M}$ and consider the matrix $\mathbf{C} = \mathbf{R}^{\mathrm{T}}\mathbf{KR}$. Then $\mathbf{C}$ is symmetric since

$$\mathbf{C}^{\mathrm{T}} = (\mathbf{R}^{\mathrm{T}}\mathbf{KR})^{\mathrm{T}} = \mathbf{R}^{\mathrm{T}}\mathbf{K}^{\mathrm{T}}\mathbf{R} = \mathbf{R}^{\mathrm{T}}\mathbf{KR} = \mathbf{C}$$

and there is an orthogonal matrix $\mathbf{U}$ such that

$$\mathbf{U}^{\mathrm{T}}\mathbf{CU} = \mathbf{U}^{\mathrm{T}}\mathbf{R}^{\mathrm{T}}\mathbf{KRU} = \mathbf{\Lambda}$$

with $\mathbf{\Lambda}$ a diagonal matrix. We also have, from (5), that

$$\mathbf{U}^{\mathrm{T}}\mathbf{R}^{\mathrm{T}}\mathbf{MRU} = \mathbf{U}^{\mathrm{T}}\mathbf{IU} = \mathbf{I},$$

and hence if we let $\mathbf{P} = \mathbf{RU}$, there follows

$$\mathbf{P}^{\mathrm{T}}\mathbf{MP} = \mathbf{I}, \qquad \mathbf{P}^{\mathrm{T}}\mathbf{KP} = \mathbf{\Lambda}.$$

Therefore, we have established the following result on simultaneous diagonalization.

Theorem 7.6. Suppose that $\mathbf{M}$ and $\mathbf{K}$ are $n \times n$ real symmetric matrices and $\mathbf{M}$ is positive definite. Then there is an invertible matrix $\mathbf{P}$ such that $\mathbf{P}^T\mathbf{M}\mathbf{P} = \mathbf{I}$ and $\mathbf{P}^T\mathbf{K}\mathbf{P} = \mathbf{\Lambda} = \mathbf{diag}\,(\lambda_1, \ldots, \lambda_n)$.

■

This theorem gives sufficient conditions for $\mathbf{M}$ and $\mathbf{K}$ to be simultaneously congruent to diagonal matrices. In fact, the theorem shows more: The positive definite matrix is congruent to the identity.

What we need now is a procedure for determining the diagonal entries, $\lambda_1, \ldots, \lambda_n$, of $\mathbf{\Lambda}$ and the columns of $\mathbf{P}$. Consider the equation

$$(\mathbf{\Lambda} - \lambda\mathbf{I})\mathbf{y} = \boldsymbol{\theta}.$$

Substituting $\mathbf{\Lambda} = \mathbf{P}^T\mathbf{K}\mathbf{P}$ and $\mathbf{I} = \mathbf{P}^T\mathbf{M}\mathbf{P}$, we obtain

$$(\mathbf{P}^T\mathbf{K}\mathbf{P} - \lambda\mathbf{P}^T\mathbf{M}\mathbf{P}) = \boldsymbol{\theta},$$

or equivalently,

$$\mathbf{P}^T(\mathbf{K} - \lambda\mathbf{M})\mathbf{P}\mathbf{y} = \boldsymbol{\theta}.$$

Thus, if we let $\mathbf{x} = \mathbf{P}\mathbf{y}$, we obtain

$$\mathbf{P}^T(\mathbf{K} - \lambda\mathbf{M})\mathbf{x} = \boldsymbol{\theta},$$

and since $\mathbf{P}^T$ is invertible, we find

$$(\mathbf{K} - \lambda\mathbf{M})\mathbf{x} = \boldsymbol{\theta}.$$

That is, $(\mathbf{\Lambda} - \lambda\mathbf{I})\mathbf{y} = \boldsymbol{\theta}$ if and only if $\mathbf{x} = \mathbf{P}\mathbf{y}$ satisfies $(\mathbf{K} - \lambda\mathbf{M})\mathbf{x} = \boldsymbol{\theta}$. To find the λ_i's we solve for λ in the equation

$$\det\,(\mathbf{K} - \lambda\mathbf{M}) = 0.$$

For $\lambda = \lambda_j$ we find that the solution of $(\mathbf{\Lambda} - \lambda\mathbf{I})\mathbf{y} = \boldsymbol{\theta}$ satisfies (to within a constant multiple) $\mathbf{y} = \mathbf{e}^{(j)} = j$th column of $\mathbf{I}$. It follows that the solution $\mathbf{x}$ of $(\mathbf{K} - \lambda_j\mathbf{M})\mathbf{x} = \boldsymbol{\theta}$ is given by $\mathbf{x} = \mathbf{P}\mathbf{y} = \mathbf{P}\mathbf{e}^{(j)} = j$th column of $\mathbf{P}$. Therefore, if $(\mathbf{K} - \lambda_j\mathbf{M})\mathbf{x} = \boldsymbol{\theta}$, then $\mathbf{x} = j$th column of $\mathbf{P}$. Actually, solutions of $(\mathbf{K} - \lambda\mathbf{M})\mathbf{x} = \boldsymbol{\theta}$ are not unique since any multiple of $\mathbf{x}$ is also a solution, that is,

$$(\mathbf{K} - \lambda\mathbf{M})(\alpha\mathbf{x}) = \alpha(\mathbf{K} - \lambda\mathbf{M})\mathbf{x} = \boldsymbol{\theta}.$$

The procedure outlined above determines the columns of $\mathbf{P}$ to within constant multiples. To determine $\mathbf{P}$, we note that

$$\mathbf{I} = \mathbf{P}^T\mathbf{M}\mathbf{P},$$

and hence it follows that the jth column of $\mathbf{P}$ is $\alpha_j\mathbf{x}$, where $(\mathbf{K} - \lambda_j\mathbf{M})\mathbf{x} = \boldsymbol{\theta}$ and the constant α_j is chosen so that

$$\alpha_j^2\langle\mathbf{x}, \mathbf{M}\mathbf{x}\rangle = \langle\alpha_j\mathbf{x}, \mathbf{M}(\alpha_j\mathbf{x})\rangle = 1.$$

To summarize, we find the λ_i's by solving the equation

$$\det\,(\mathbf{K} - \lambda\mathbf{M}) = 0.$$

The jth column of **P**, up to a constant multiple, is found by solving the equation for **x**:

$$(\mathbf{K} - \lambda_j \mathbf{M})\mathbf{x} = \boldsymbol{\theta},$$

where λ_j is the jth diagonal entry of $\boldsymbol{\Lambda}$.

Example 7.13

We simultaneously diagonalize the matrices

$$\mathbf{M} = \begin{bmatrix} 3 & 0 \\ 0 & 3/2 \end{bmatrix} \qquad \mathbf{K} = \begin{bmatrix} -3 & 1 \\ 1 & -1 \end{bmatrix}.$$

Note that **M** is positive definite and **K** is real and symmetric. First, we find the diagonal entries of $\boldsymbol{\Lambda}$. We have

$$\det(\mathbf{K} - \lambda\mathbf{M}) = \begin{vmatrix} -3 - 3\lambda & 1 \\ 1 & -1 - \dfrac{3}{2}\lambda \end{vmatrix}$$

$$= \frac{9}{2}\lambda^2 + \frac{15}{2}\lambda + 2.$$

Therefore, the roots of $0 = \det(\mathbf{K} - \lambda\mathbf{M})$ are $\lambda_1 = -\frac{4}{3}$ and $\lambda_2 = -\frac{1}{3}$. To find the columns of **P** we solve $(\mathbf{K} - \lambda\mathbf{M})\mathbf{x} = \boldsymbol{\theta}$:

$$(\mathbf{K} - \lambda_1 \mathbf{M})\mathbf{x}^{(1)} = \boldsymbol{\theta} = \begin{bmatrix} 1 & 1 \\ 1 & 1 \end{bmatrix}\begin{bmatrix} x_1^{(1)} \\ x_2^{(1)} \end{bmatrix}.$$

Hence we may take $\mathbf{x}^{(1)} = [1 \quad -1]^{\mathrm{T}}$. The first column of **P** will then be $\alpha_1\mathbf{x}^{(1)}$, where the constant α_1 is chosen so that

$$\alpha_1^2 \langle \mathbf{x}^{(1)}, \mathbf{M}\mathbf{x}^{(1)} \rangle = 1.$$

We find that $\alpha_1 = \sqrt{2}/3$. Also,

$$(\mathbf{K} - \lambda_2 \mathbf{M})\mathbf{x}^{(2)} = \boldsymbol{\theta} = \begin{bmatrix} -2 & 1 \\ 1 & -1/2 \end{bmatrix}\begin{bmatrix} x_1^{(2)} \\ x_2^{(2)} \end{bmatrix},$$

and hence we may take $\mathbf{x}^{(2)} = [1 \quad 2]^{\mathrm{T}}$. The second column of **P** is then $\alpha_2\mathbf{x}^{(2)}$, where

$$\alpha_2^2 \langle \mathbf{x}^{(2)}, \mathbf{M}\mathbf{x}^{(2)} \rangle = 1,$$

so that $\alpha_2 = \frac{1}{3}$. Therefore, we find that

$$\mathbf{P} = \frac{1}{3}\begin{bmatrix} \sqrt{2} & 1 \\ -\sqrt{2} & 2 \end{bmatrix}$$

satisfies

$$\mathbf{P}^{\mathrm{T}}\mathbf{K}\mathbf{P} = \begin{bmatrix} -\frac{4}{3} & 0 \\ 0 & -\frac{1}{3} \end{bmatrix} \text{ and } \mathbf{P}^{\mathrm{T}}\mathbf{M}\mathbf{P} = \mathbf{I}.$$

■

EXERCISES 7.4

1. Determine which of the following matrices is positive definite.

(a) $\mathbf{A} = \begin{bmatrix} 2 & 1 & 1 \\ 1 & 2 & 1 \\ 1 & 1 & 2 \end{bmatrix}$ **(b)** $\mathbf{A} = \begin{bmatrix} -1 & -6 & 0 \\ -6 & 2 & -6 \\ 0 & -6 & 5 \end{bmatrix}$

(c) $\mathbf{A} = \begin{bmatrix} 1 & 2 & 2 \\ 2 & 1 & -2 \\ 2 & -2 & 1 \end{bmatrix}$ **(d)** $\mathbf{A} = \begin{bmatrix} 3 & 2 \\ 2 & 4 \end{bmatrix}$

2. For the matrix in part (c) of Exercise 1, express $\langle \mathbf{x}, \mathbf{Ax} \rangle$ as a sum of the form $\lambda_1 u_1^2 + \lambda_2 u_2^2 + \lambda_3 u_3^2$.

***3.** Suppose that $\mathbf{B}$ is an invertible matrix and let $\mathbf{A} = \mathbf{B}^T\mathbf{B}$. Show that $\mathbf{A}$ is positive definite.

***4.** Suppose that $\mathbf{A}$ is a 2×2 real symmetric matrix whose trace and determinant are both positive. Show that $\mathbf{A}$ is positive definite.

5. Let

$$\mathbf{M} = \begin{bmatrix} 2 & -3 \\ -3 & 5 \end{bmatrix} \quad \text{and} \quad \mathbf{K} = \begin{bmatrix} 3 & 1 \\ 1 & 21 \end{bmatrix}.$$

Find the roots of the equation $\det(\mathbf{K} - \lambda\mathbf{M}) = 0$, and for each root λ solve the equation $(\mathbf{K} - \lambda\mathbf{M})\mathbf{x} = \boldsymbol{\theta}$ for $\mathbf{x}$. Determine the congruence transformation matrix $\mathbf{P}$ which simultaneously diagonalizes the matrices $\mathbf{M}$ and $\mathbf{K}$.

6. Simultaneously diagonalize the following matrices.

$$\mathbf{M} = \begin{bmatrix} 2 & -1 \\ -1 & 5 \end{bmatrix}, \quad \mathbf{K} = \begin{bmatrix} -1 & 5 \\ 5 & -7 \end{bmatrix}$$

7. Use simultaneous diagonalization to solve the difference equation $\mathbf{Mu}^{(k)} = \mathbf{Ku}^{(k-1)}$, $\mathbf{u}^{(0)} = [1 \quad 1]^T$, where $\mathbf{M}$ and $\mathbf{K}$ are given in Exercise 6. (*Hint:* Find the congruence transformation matrix $\mathbf{P}$ which simultaneously diagonalizes $\mathbf{M}$ and $\mathbf{K}$ and show that $\mathbf{u}^{(k)} = \mathbf{P\Lambda P}^{-1}\mathbf{u}^{(k-1)}$.)

***8.** Suppose that $\mathbf{A}$ is a positive definite $n \times n$ matrix. Verify that $\|\mathbf{x}\|_{\mathbf{A}} = \sqrt{\langle \mathbf{Ax}, \mathbf{x} \rangle}$ is a norm.

9. Show that every positive definite matrix is invertible.

7.5 Systems of Ordinary Differential Equations

An ordinary differential equation is an equation involving an unknown function, say $x(t)$, and one or more of its derivatives. Many physical problems are modeled by a differential equation or a system of differential equations. We begin by describing a simple, yet very important differential equation which arises in the dating of fossils.

A radioactive isotope of carbon, carbon 14 (^{14}C), is present in living organisms, and when an organism dies, the concentration of ^{14}C decreases through radioactive decay. It has been determined that the rate of radioactive decay of a substance is directly proportional to the amount of the substance present. Thus if $x(t)$ denotes the concentration of ^{14}C present in a fossil at time t, then $x'(t) = dx(t)/dt$ is the concentration that decays per unit time and is proportional to $x(t)$. That is, for some proportionality constant k,

$$x'(t) = -kx(t).$$

The positive constant k is the decay constant of the substance. For ^{14}C it is known that $k = .1245 \times 10^{-3}$ when t is measured in years.

If the concentration $x^{(0)}$ of ^{14}C in a fossil is known at some time t_0, then the concentration $x(t)$ at any time t is given by solution of the initial-value problem

$$x'(t) = -kx(t)$$
$$x(t_0) = x^{(0)}.$$

The chemist Walter Libby received a Nobel Prize in 1960 for his discovery (in 1949) of a procedure whereby this initial-value problem can be used to determine the age of certain fossils.

It turns out that the differential equation

$$x'(t) = ax(t) \tag{6}$$

occurs in a variety of applications. Moreover, the solution of this equation describes the local behavior of solutions to much more complicated differential equations. Our first task is to solve this very basic problem.

In calculus it is shown that

$$\frac{d(e^{at})}{dt} = ae^{at}$$

and hence $u(t) = e^{at}$ is a solution of (6). In fact, as we now demonstrate, all solutions of (6) are exponential functions. To see this we multiply (6) by e^{-at} to obtain

$$e^{-at}x'(t) = ae^{-at}x(t),$$

or equivalently,

$$e^{-at}x'(t) - ae^{-at}x(t) = 0.$$

But by the product rule for differentiation, we know that

$$\frac{d(e^{-at}x(t))}{dt} = e^{-at}x'(t) - ae^{-at}x(t).$$

Therefore, we obtain

$$\frac{d(e^{-at}x(t))}{dt} = 0$$

and consequently, for some constant C,

$$e^{-at}x(t) = C.$$

From this it follows that $x(t) = Ce^{at}$; that is, every solution of (6) has the form

$$x(t) = Ce^{at}$$

for some constant C. The constant C is determined by the initial condition $x(t_0) = x^{(0)}$:

$$x(t_0) = x^{(0)} = Ce^{at_0}$$

and hence $C = x^{(0)}e^{-at_0}$.

The unique solution of $x'(t) = ax(t)$, $x(t_0) = x^{(0)}$ is given by

$$x(t) = x^{(0)}e^{a(t-t_0)}.$$

Example 7.14

Consider the initial-value problem

$$\frac{dx}{dt} = 4x, \qquad x(1) = -3.$$

According to our previous work, the unique solution is

$$x(t) = -3e^{4(t-1)}.$$

■

One of our aims in this section is to solve, using coordinate transformations, a system of differential equations that is analogous to the scalar equation (6). In many physical problems there are several interactive processes, and mathematical models for such problems involve several differential equations with several unknown functions. Typically, the rate of change of each unknown $x_i(t)$, that is, its derivative $dx_i(t)/dt$, depends on $x_i(t)$ as well as some or all of the other unknowns. To be more precise, let the unknown functions be $x_1(t), \ldots, x_n(t)$ for some n. Denote by $\mathbf{x}(t)$ the n-vector whose ith entry is $x_i(t)$. The derivative of $\mathbf{x}(t)$ is found by differentiation of each entry of $\mathbf{x}(t)$, that is,

$$\frac{d\mathbf{x}(t)}{dt} = \begin{bmatrix} \dfrac{dx_1(t)}{dt} \\ \dfrac{dx_2(t)}{dt} \\ \vdots \\ \dfrac{dx_n(t)}{dt} \end{bmatrix}.$$

A first-order system of differential equations has the form

$$\frac{d\mathbf{x}(t)}{dt} = \begin{bmatrix} f_1(t, \mathbf{x}(t)) \\ f_2(t, \mathbf{x}(t)) \\ \vdots \\ f_n(t, \mathbf{x}(t)) \end{bmatrix}, \tag{7}$$

where $f_1, \ldots, f_n$ are given functions of t and $\mathbf{x}$. Equation (7) is a first-order system since it involves only the first derivative of the unknown functions $x_1(t), \ldots, x_n(t)$. Equation (7) describes a very general nonlinear system of ordinary differential equations which is beyond the scope of this book. We shall be concerned with a special case of (7), where each of the functions $f_i(t, \mathbf{x})$ depends only on $\mathbf{x}$ and in fact is a linear function of $\mathbf{x}$. Thus we consider only those functions $f_i(\mathbf{x})$ of the form

$$f_i(\mathbf{x}) = a_{i1}x_1 + a_{i2}x_2 + \cdots + a_{in}x_n,$$

where $a_{i1}, a_{i2}, \ldots, a_{in}$ are given constants. In this case the system (7) takes the form

$$\frac{d\mathbf{x}(t)}{dt} = \begin{bmatrix} x_1'(t) \\ x_2'(t) \\ \vdots \\ x_n'(t) \end{bmatrix} = \begin{bmatrix} a_{11}x_1(t) + \cdots + a_{1n}x_n(t) \\ a_{21}x_1(t) + \cdots + a_{2n}x_n(t) \\ \vdots \qquad\qquad \vdots \\ a_{n1}x_1(t) + \cdots + a_{nn}x_n(t) \end{bmatrix}.$$

We recognize the vector on the right-hand side of the preceding equation as the matrix product $\mathbf{Ax}(t)$, where $\mathbf{A}$ is the coefficient matrix whose i, j-entry is a_{ij}. It follows that this system can be written in the more compact form

$$\mathbf{x}'(t) = \mathbf{Ax}(t). \tag{8}$$

The reader should recognize that (8) is the direct analogue in n dimensions of the scalar equation (6). In addition, if initial conditions are specified for the unknown vector at some time t_0, say $\mathbf{x}(t_0) = \mathbf{x}^{(0)}$, then the initial-value problem

$$\mathbf{x}'(t) = \mathbf{Ax}(t)$$
$$\mathbf{x}(t_0) = \mathbf{x}^{(0)}$$

results.

Example 7.15

The analysis of electric circuits, such as that in Figure 7.7, is based on Kirchhoff's laws and the current–voltage laws for each circuit element. Kirchhoff's laws are as follows: (1) the sum of the voltage drops around each closed loop of the circuit is zero, and (2) the algebraic sum of the currents through each node is zero. The physical laws that govern the current i, in amperes, through a circuit element and the voltage drop v, in volts, across each element are as follows:

$v = Ri$ $\quad$ R is the resistance measured in ohms

$C\dfrac{dv}{dt} = i$ $\quad$ C is the capacitance measured in farads

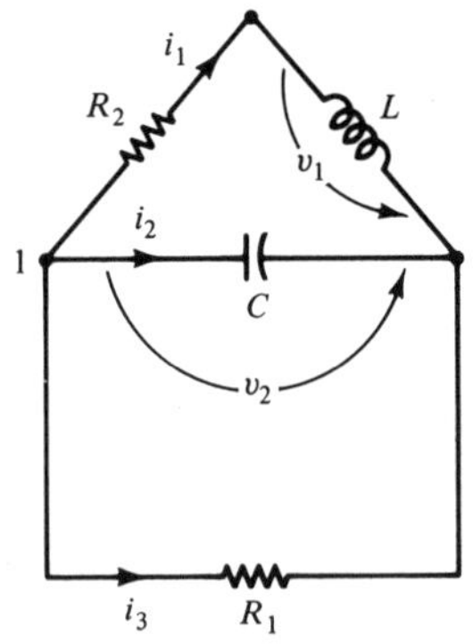

Figure 7.7

$$L\frac{di}{dt} = v \qquad L \text{ is the inductance measured in henrys.}$$

In Figure 7.7 we have (arbitrarily) chosen directions, indicated by arrows, in which the current and voltage drops are taken to be positive. At node 1, Kirchhoff's current law gives the equation

$$i_1 + i_2 + i_3 = 0.$$

For the upper and lower loops, respectively, Kirchhoff's voltage law gives the equations.

$$-v_2 + i_1 R_2 + v_1 = 0$$
$$v_2 - i_3 R_1 = 0.$$

But $v_1 = L\, di_1/dt$ and $i_2 = C\, dv_2/dt$, so that

$$L\frac{di_1}{dt} = v_1 = v_2 - i_1 R_2$$

$$C\frac{dv_2}{dt} = i_2 = -i_1 - i_3 = -\frac{v_2}{R_1} - i_1.$$

Thus the current through the inductor and the voltage across the capacitor satisfy the system of differential equations:

$$\frac{di_1}{dt} = -\frac{R_2}{L}i_1 + \frac{1}{L}v_2$$

$$\frac{dv_2}{dt} = -\frac{1}{C}i_1 - \frac{1}{R_1 C}v_2.$$

In matrix form this system may be written $d\mathbf{x}/dt = \mathbf{A}\mathbf{x}$, where

$$\mathbf{x} = \begin{bmatrix} i_1 \\ v_2 \end{bmatrix} \quad \text{and} \quad \mathbf{A} = \begin{bmatrix} -\dfrac{R_2}{L} & \dfrac{1}{L} \\ -\dfrac{1}{C} & -\dfrac{1}{R_1 C} \end{bmatrix}.$$

■

Example 7.16

Consider a body attached to an elastic spring which is immersed in a fluid that impedes its motion, as in Figure 7.8. Suppose that the mass of the body is m and that the spring is initially stretched to position $x^{(0)}$ and has an initial velocity of $v^{(0)}$. Newton's second law of motion asserts that the force acting on an object is equal to the product of its mass and acceleration. Thus the position, $x(t)$, of the body at time t satisfies $m\, d^2x(t)/dt^2 = F$. The force F acting on the body consists of two components: that due to the elongation or compression of the spring and that exerted by the fluid. An elastic spring obeys Hooke's law, which states that the force required to stretch or compress the spring a distance d (which

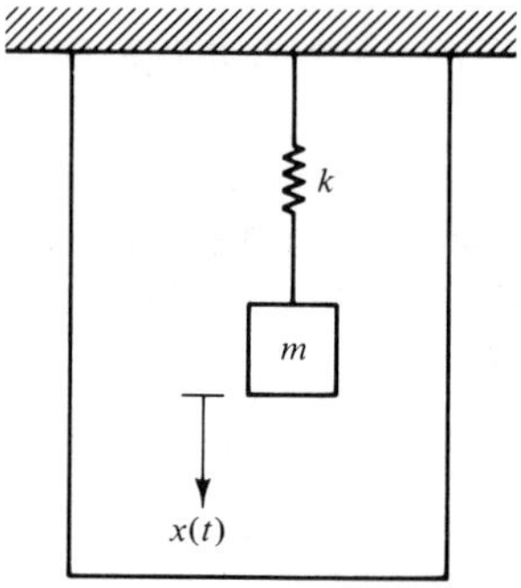

Figure 7.8

is small in comparison to its natural length) is kd. The constant k is a measure of the stiffness of the spring and is called the *spring constant*. The damping (or drag) force which the fluid exerts on the body is directly proportional to the velocity of the body and acts in a direction opposite the direction of motion. The constant of proportionality p is called the *drag coefficient*.

It follows that the position $x(t)$ of the body at time t satisfies the second-order differential equation

$$m\frac{d^2x(t)}{dt^2} = -kx(t) - p\frac{dx(t)}{dt}$$

and the initial conditions

$$\frac{dx(0)}{dt} = v^{(0)}, \qquad x(0) = x^{(0)}.$$

The constants k and p are nonnegative.

This second-order initial-value problem can be rewritten as an initial-value problem for a first-order system of differential equations. We accomplish this as follows. Let $v(t) = dx(t)/dt$; then

$$\frac{dx(t)}{dt} = v(t)$$

$$\frac{dv(t)}{dt} = -\left(\frac{k}{m}\right)x(t) - \left(\frac{p}{m}\right)v(t)$$

and $v(0) = v^{(0)}$, $x(0) = x^{(0)}$. In matrix form we have

$$\begin{bmatrix} x'(t) \\ v'(t) \end{bmatrix} = \begin{bmatrix} 0 & 1 \\ -\dfrac{k}{m} & -\dfrac{p}{m} \end{bmatrix} \begin{bmatrix} x(t) \\ v(t) \end{bmatrix}$$

and

$$\begin{bmatrix} x(0) \\ v(0) \end{bmatrix} = \begin{bmatrix} x^{(0)} \\ v^{(0)} \end{bmatrix}.$$

■

We show how to solve the system (8) using a coordinate transformation when the coefficient matrix is diagonalizable. Thus we assume that

$$\mathbf{A} = \mathbf{P\Lambda P}^{-1}$$

for some invertible matrix $\mathbf{P}$ and diagonal matrix $\mathbf{\Lambda}$. Recall that the diagonal entries of $\mathbf{\Lambda}$ are the eigenvalues of $\mathbf{A}$ and the columns of $\mathbf{P}$ are the corresponding eigenvectors. Now we introduce the coordinate transformation

$$\mathbf{x}(t) = \mathbf{Pu}(t)$$

into the differential equation $\mathbf{x}'(t) = \mathbf{Ax}(t)$. Since $\mathbf{P}$ is independent of t it follows that $\mathbf{x}'(t) = \mathbf{Pu}'(t)$, and hence we find

$$\mathbf{Pu}'(t) = \mathbf{APu}(t).$$

Moreover, since $\mathbf{P}$ is invertible, we have

$$\mathbf{u}'(t) = \mathbf{P}^{-1}\mathbf{APu}(t) = \mathbf{\Lambda u}(t).$$

The initial condition for $\mathbf{u}$ is $\mathbf{u}(t_0) = \mathbf{u}^{(0)}$, where

$$\mathbf{u}^{(0)} = \mathbf{P}^{-1}\mathbf{x}(t_0) = \mathbf{P}^{-1}\mathbf{x}^{(0)}.$$

Thus in terms of the new unknown $\mathbf{u}(t)$ our problem is to solve

$$\begin{aligned} \mathbf{u}'(t) &= \mathbf{\Lambda u}(t) \\ \mathbf{u}(t_0) &= \mathbf{u}^{(0)}, \end{aligned} \qquad (9)$$

where $\mathbf{\Lambda}$ is a diagonal matrix. Writing out the system (9) gives

$$\begin{aligned} u_1'(t) &= \lambda_1 u_1(t) \\ u_2'(t) &= \lambda_2 u_2(t) \\ &\vdots \\ u_n'(t) &= \lambda_n u_n(t). \end{aligned}$$

Note that each of these equations involves only one unknown. That is, the system $\mathbf{x}'(t) = \mathbf{Ax}(t)$ has been *uncoupled* by the use of the coordinate transformation $\mathbf{x} = \mathbf{Pu}$. Each of the problems

$$u_i'(t) = \lambda_i u_i(t), \qquad u_i(t_0) = u_i^{(0)}$$

is a particular case of (6) and hence we can immediately write the solution

$$u_i(t) = u_i^{(0)} e^{\lambda_i(t-t_0)}, \qquad 1 \le i \le n.$$

Thus the solution of (9) is given by

$$\mathbf{u}(t) = \begin{bmatrix} u_1^{(0)} e^{\lambda_1(t-t_0)} \\ u_2^{(0)} e^{\lambda_2(t-t_0)} \\ \vdots \\ u_n^{(0)} e^{\lambda_n(t-t_0)} \end{bmatrix}$$

and the solution of the original problem $\mathbf{x}'(t) = \mathbf{A}\mathbf{x}(t)$ is obtained by $\mathbf{x}(t) = \mathbf{P}\mathbf{u}(t)$.

Example 7.17

We solve the initial-value problem

$$\mathbf{x}'(t) = \begin{bmatrix} 3 & 1 \\ 2 & 4 \end{bmatrix}\mathbf{x}(t)$$

$$\mathbf{x}(0) = \begin{bmatrix} 1 \\ 1 \end{bmatrix}$$

by uncoupling the system as described above. The coefficient matrix has characteristic polynomial

$$\begin{aligned} p_{\mathbf{A}}(\lambda) &= \lambda^2 - 7\lambda + 10 \\ &= (\lambda - 5)(\lambda - 2) \end{aligned}$$

and hence the eigenvalues are $\lambda_1 = 2$, $\lambda_2 = 5$. Corresponding eigenvectors are found to be

$$\mathbf{x}^{(1)} = \begin{bmatrix} 1 \\ -1 \end{bmatrix}, \qquad \mathbf{x}^{(2)} = \begin{bmatrix} 1 \\ 2 \end{bmatrix}.$$

Define $\mathbf{P}$ to be the matrix of eigenvectors

$$\mathbf{P} = \begin{bmatrix} 1 & 1 \\ -1 & 2 \end{bmatrix};$$

then

$$\mathbf{P}^{-1} = \frac{1}{3}\begin{bmatrix} 2 & -1 \\ 1 & 1 \end{bmatrix}.$$

Introducing the coordinate transformation $\mathbf{x}(t) = \mathbf{P}\mathbf{u}(t)$, we obtain the new initial-value problem:

$$\begin{aligned} \mathbf{u}'(t) &= \mathbf{P}^{-1}\mathbf{A}\mathbf{P}\mathbf{u}(t) = \mathbf{\Lambda}\mathbf{u}(t) \\ \mathbf{u}(0) &= \mathbf{P}^{-1}\mathbf{x}(0) = \mathbf{u}^{(0)}, \end{aligned}$$

where

$$\mathbf{\Lambda} = \begin{bmatrix} 2 & 0 \\ 0 & 5 \end{bmatrix} \quad \text{and} \quad \mathbf{u}^{(0)} = \begin{bmatrix} \frac{1}{3} \\ \frac{2}{3} \end{bmatrix}.$$

The unique solution of the new problem is given by

$$\mathbf{u}(t) = \frac{1}{3}\begin{bmatrix} e^{2t} \\ 2e^{5t} \end{bmatrix}$$

and hence the solution of the original problem is $\mathbf{x}(t) = \mathbf{P}\mathbf{u}(t)$:

$$\begin{aligned} \mathbf{x}(t) &= \frac{1}{3}\begin{bmatrix} e^{2t} + 2e^{5t} \\ -e^{2t} + 4e^{5t} \end{bmatrix} \\ &= \tfrac{1}{3}e^{2t}\mathbf{x}^{(1)} + \tfrac{2}{3}e^{5t}\mathbf{x}^{(2)}. \end{aligned}$$

The solution of the initial-value problem is a linear combination of eigenvectors with exponential function coefficients.

■

Let us return to the problem (8) and derive a formula for the solution which gives additional emphasis to the analogy between the n-dimensional system and the scalar equation (6). The solution of the uncoupled problem is given by

$$\mathbf{u}(t) = \begin{bmatrix} u_1^{(0)} e^{\lambda_1(t-t_0)} \\ u_2^{(0)} e^{\lambda_2(t-t_0)} \\ \vdots \\ u_n^{(0)} e^{\lambda_n(t-t_0)} \end{bmatrix},$$

and hence if we define the diagonal matrix $e^{\Lambda t}$ by

$$e^{\Lambda t} = \mathbf{diag}\,(e^{\lambda_1 t}, \ldots, e^{\lambda_n t}),$$

it follows that

$$\mathbf{u}(t) = e^{\Lambda(t-t_0)}\mathbf{u}^{(0)} = e^{\Lambda(t-t_0)}\mathbf{P}^{-1}\mathbf{x}(t_0).$$

Thus the solution of the original problem is given by

$$\mathbf{x}(t) = \mathbf{P}\mathbf{u}(t) = \mathbf{P}e^{\Lambda(t-t_0)}\mathbf{P}^{-1}\mathbf{x}^{(0)}.$$

Define the matrix $e^{\mathbf{A}(t-t_0)}$ by

$$e^{\mathbf{A}(t-t_0)} = \mathbf{P}e^{\Lambda(t-t_0)}\mathbf{P}^{-1},$$

then the unique solution of (8) is given by $\mathbf{x}(t) = e^{\mathbf{A}(t-t_0)}\mathbf{x}^{(0)}$.

Example 7.18

For the coefficient matrix of Example 7.17 we have

$$e^{\Lambda t} = \begin{bmatrix} e^{2t} & 0 \\ 0 & e^{5t} \end{bmatrix}$$

and

$$e^{\mathbf{A}t} = \mathbf{P}e^{\Lambda t}\mathbf{P}^{-1}$$
$$= \frac{1}{3}\begin{bmatrix} 2e^{2t} + e^{5t} & -e^{2t} + e^{5t} \\ -2e^{2t} + 2e^{5t} & e^{2t} + 2e^{5t} \end{bmatrix}.$$

■

The unique solution of the initial-value problem $\mathbf{x}'(t) = \mathbf{A}\mathbf{x}(t)$, $\mathbf{x}(t_0) = \mathbf{x}^{(0)}$ is given by

$$\mathbf{x}(t) = e^{\mathbf{A}(t-t_0)}\mathbf{x}^{(0)}.$$

In Example 7.17 we observed that the solution could be written as a linear combination of eigenvectors with exponential functions as coefficients. To see that this is true in general, we let $\mathbf{u}^{(0)} = \mathbf{P}^{-1}\mathbf{x}^{(0)}$; then

$$\begin{aligned}\mathbf{x}(t) &= \mathbf{P}e^{\Lambda(t-t_0)}\mathbf{u}^{(0)} \\ &= \mathbf{P}\begin{bmatrix} u_1^{(0)}e^{\lambda_1(t-t_0)} \\ \vdots \\ u_n^{(0)}e^{\lambda_n(t-t_0)} \end{bmatrix} \\ &= \sum_{j=1}^{n} u_j^{(0)}e^{\lambda_j(t-t_0)}\mathbf{x}^{(j)}.\end{aligned}$$

From this equation we can draw some inferences regarding the behavior, as $t \to \infty$, of solutions of $d\mathbf{x}/dt = \mathbf{A}\mathbf{x}$. For a diagonalizable matrix $\mathbf{A}$, every solution can be expressed as a sum of exponential solutions and the long-term behavior of solutions is governed by the eigenvalues of $\mathbf{A}$. Specifically, if every eigenvalue λ of $\mathbf{A}$ satisfies $\lambda < 0$, then $e^{\lambda(t-t_0)} \to 0$ as $t \to \infty$, and we conclude that every solution $\mathbf{x}(t)$ of (6) satisfies

$$\lim_{t\to\infty} \mathbf{x}(t) = \boldsymbol{\theta}.$$

In this case $d\mathbf{x}/dt = \mathbf{A}\mathbf{x}$ is said to be a stable differential equation. If, on the other hand, $\mathbf{A}$ has a positive eigenvalue, say λ_i, then $e^{\lambda_i(t-t_0)} \to \infty$ as $t \to \infty$. The differential equation is unstable if $\mathbf{A}$ has a positive eigenvalue.

Thus far we have shown that a first-order system of differential equations whose coefficient matrix is diagonalizable can be solved by using a coordinate transformation whose effect is to uncouple the differential equations. Moreover, by defining the exponential of a matrix we see the direct analogy between the scalar equation and the system. When the coefficient matrix is not diagonalizable other methods are used. We do not pursue this here because it would take us beyond the scope of the book.

In what follows we shall make use of the simultaneous diagonalization technique, which was introduced at the end of Section 7.4, to solve certain systems of equations that arise in circuit analysis and mechanical vibrations. First we need to demonstrate how solutions of a scalar second-order differential equation are obtained.

Consider the mass–spring system of Example 7.16 in which a mass spring is immersed in a fluid which impedes its motion. We found that the position $x(t)$ satisfies the second-order differential equation

$$m\frac{d^2x}{dt^2} + p\frac{dx}{dt} + kx = 0$$

We divide through by m to give the second-order equation

$$\frac{d^2x}{dt^2} + b\frac{dx}{dt} + cx = 0, \tag{10}$$

where $b = p/m$ and $c = k/m$.

In what follows we shall derive solutions of (10) for arbitrary coefficients b and c. We seek solutions of (10) of the form $w(t) = e^{rt}$. Substitution into (10) gives

$$\frac{d^2w}{dt^2} + b\frac{dw}{dt} + cw = e^{rt}(r^2 + bt + c),$$

and hence w is a solution of (10) if r satisfies

$$r^2 + br + c = 0.$$

Solutions of this equation are given by the quadratic formula and three separate cases must be considered: (1) $b^2 - 4c > 0$, (2) $b^2 - 4c = 0$, and (3) $b^2 - 4c < 0$. This corresponds to two distinct real roots, two equal real roots, and complex roots, respectively.

In case (1) let the roots be

$$r_1 = \frac{-b + \sqrt{b^2 - 4c}}{2}, \qquad r_2 = \frac{-b - \sqrt{b^2 - 4c}}{2}.$$

Then the corresponding solutions of (10) are $e^{r_1 t}$ and $e^{r_2 t}$. The general solution takes the form $x(t) = C_1 e^{r_1 t} + C_2 e^{r_2 t}$.

In case (2) there is a repeated root $r = -b/2$. Then e^{rt} is a solution of (10). Moreover, it can be shown (see Exercise 12) that another solution is given by te^{rt}. The general solution takes the form $x(t) = C_1 e^{rt} + C_2 te^{rt}$.

Finally, in case (3) there are complex roots given by

$$r_1 = \frac{-b + i\sqrt{4c - b^2}}{2}, \qquad r_2 = \frac{-b - i\sqrt{4c - b^2}}{2}.$$

For ease of notation let $\lambda = -b/2$ and $\mu = \sqrt{4c - b^2}/2$; then solutions of (10) are given by the complex exponentials $e^{(\lambda + i\mu)t}$ and $e^{(\lambda - i\mu)t}$. To obtain real solutions of (10) we need a result from complex analysis:

$$e^{(\lambda \pm i\mu)t} = e^{\lambda t}(\cos \mu t \pm i \sin \mu t).$$

It is not difficult to show that the real and imaginary parts of $e^{(\lambda + i\mu)t}$ are solutions (see Exercise 13) of (10). Therefore, real solutions of (10) are given by $e^{\lambda t} \cos \mu t$ and $e^{\lambda t} \sin \mu t$. The general solution takes the form

$$x(t) = C_1 e^{\lambda t} \cos \mu t + C_2 e^{\lambda t} \sin \mu t.$$

Example 7.19

Consider the following second-order differential equations

$$\text{(a)}\ x''(t) - x'(t) - 2x(t) = 0$$
$$\text{(b)}\ y''(t) - 6y'(t) + 9y(t) = 0$$
$$\text{(c)}\ w''(t) - 2w'(t) + 5w(t) = 0$$

For equation (a) the quadratic equation is $r^2 - r - 2 = 0$, which has roots $r_1 = -1$, $r_2 = 2$. The general solution of equation (a) is

$$x(t) = C_1 e^{-t} + C_2 e^{2t}.$$

For equation (b) the quadratic equation is $r^2 - 6r + 9 = 0$, which can be written as $(r - 3)^2 = 0$. Thus there is a repeated root, $r = 3$. The general solution of equation (b) is

$$y(t) = C_1 e^{3t} + C_2 te^{3t}.$$

Finally, we have the quadratic equation $r^2 - 2r + 5 = 0$ for equation (c). The quadratic formula gives the complex roots $r_1 = 1 + 2i$ and $r_2 = 1 - 2i$. The general solution of equation (c) is given by

$$w(t) = e^t(C_1 \cos 2t + C_2 \sin 2t).$$

■

Next we consider a more complicated mass–spring system in which the motion is governed by a pair of second-order differential equations. Consider the system depicted in Figure 7.9. There are two forces acting on the first mass m_1. The first spring exerts the force $-k_1 x_1$, whereas the second spring exerts the force $(x_2 - x_1) k_2$ on m_1. Thus by Newton's law

$$m_1 \frac{d^2 x_1}{dt^2} = -(k_1 + k_2) x_1 + k_2 x_2.$$

Similarly, the second spring exerts the force $-k_2 (x_2 - x_1)$ on mass 2. Therefore, we have

$$m_2 \frac{d^2 x_2}{dt^2} = k_2 x_1 - k_2 x_2.$$

We introduce the vector $x = [x_1 \quad x_2]^{\mathrm{T}}$ and write this pair of second-order differential equations in matrix form:

$$\begin{bmatrix} m_1 & 0 \\ 0 & m_2 \end{bmatrix} \begin{bmatrix} x_1'' \\ x_2'' \end{bmatrix} = \begin{bmatrix} -k_1 - k_2 & k_2 \\ k_2 & -k_2 \end{bmatrix} \begin{bmatrix} x_1 \\ x_2 \end{bmatrix}. \tag{11}$$

The matrix on the left-hand side is called the *mass matrix* and the matrix on the right-hand side is called the *stiffness matrix*.

Next we show how to solve this second-order system using the fact that the mass matrix **M** is positive definite and the stiffness matrix **K** is real symmetric. That is, we consider the system

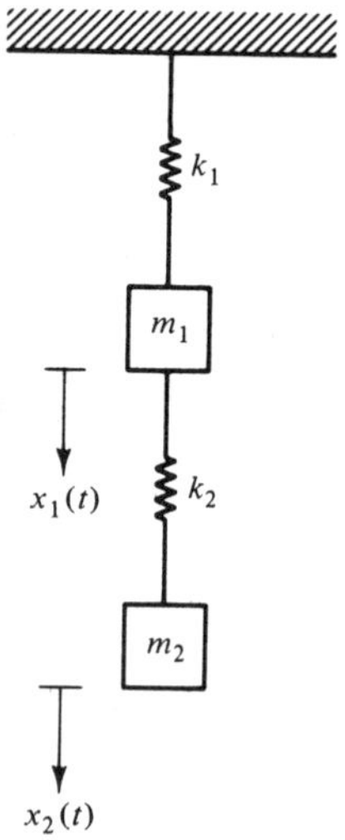

Figure 7.9

$$\mathbf{M}\mathbf{x}''(t) = \mathbf{K}\mathbf{x}(t) \tag{12}$$

with **M** and **K** real symmetric and **M** positive definite. Then, by Theorem 7.6, there is an invertible matrix **P** such that

$$\mathbf{P}^{\mathrm{T}}\mathbf{M}\mathbf{P} = \mathbf{I} \quad \text{and} \quad \mathbf{P}^{\mathrm{T}}\mathbf{K}\mathbf{P} = \boldsymbol{\Lambda}.$$

Recall that the diagonal entries of $\boldsymbol{\Lambda}$ are obtained by solving det $(\mathbf{K} - \lambda\mathbf{M}) = 0$ and the columns of **P** are obtained (up to a constant multiple) by solving $(\mathbf{K} - \lambda_j\mathbf{M})\mathbf{u}^{(j)} = \boldsymbol{\theta}$. By introducing the change of variables

$$\mathbf{x} = \mathbf{P}\mathbf{y},$$

the coupled system $\mathbf{M}\mathbf{x}'' = \mathbf{K}\mathbf{x}$ becomes $\mathbf{M}\mathbf{P}\mathbf{y}'' = \mathbf{K}\mathbf{P}\mathbf{y}$. Multiplication by $\mathbf{P}^{\mathrm{T}}$ then gives the uncoupled system

$$\mathbf{y}''(t) = \boldsymbol{\Lambda}\mathbf{y}(t).$$

The solution of the original system is then

$$\mathbf{x}(t) = \mathbf{P}\mathbf{y}(t) = y_1(t)\mathbf{P}_1 + y_2(t)\mathbf{P}_2,$$

where $\mathbf{P}_1$ and $\mathbf{P}_2$ are the columns of **P**. The columns of **P** are called the *normal modes of vibration* and the solutions of det $(\mathbf{K}-\lambda\mathbf{M}) = 0$ determine the normal frequencies of vibration. The *normal frequencies* are given by $\omega_j = \sqrt{-\lambda_j}$, where λ_j is a solution of det $(\mathbf{K}-\lambda\mathbf{M}) = 0$.

Example 7.20

We return to the two mass–spring system with $k_1 = 2$, $k_2 = 1$, $m_1 = 3$, and $m_2 = \frac{3}{2}$. Then the mass and stiffness matrices are

$$\mathbf{M} = \begin{bmatrix} 3 & 0 \\ 0 & \frac{3}{2} \end{bmatrix}, \qquad \mathbf{K} = \begin{bmatrix} -3 & 1 \\ 1 & -1 \end{bmatrix}.$$

To find the normal frequencies, we solve det $(\mathbf{K}-\lambda\mathbf{M}) = 0$:

$$0 = \begin{vmatrix} -3-3\lambda & 1 \\ 1 & \dfrac{-2-3\lambda}{2} \end{vmatrix} = \frac{(3\lambda + 1)(3\lambda + 4)}{2}$$

and hence $\lambda_1 = -\frac{1}{3}$, $\lambda_2 = -\frac{4}{3}$. The normal modes are found by solving $(\mathbf{K} - \lambda_j\mathbf{M})\mathbf{u}^{(j)} = \boldsymbol{\theta}$ for $j = 1, 2$. This yields

$$\mathbf{u}^{(1)} = \begin{bmatrix} 1 \\ 2 \end{bmatrix}, \qquad \mathbf{u}^{(2)} = \begin{bmatrix} 1 \\ -1 \end{bmatrix}.$$

Thus for some scalars α and β, $\mathbf{P} = [\alpha\mathbf{u}^{(1)} \ \ \beta\mathbf{u}^{(2)}]$. Moreover, the uncoupled equations are

$$\begin{aligned} y_1'' &= -\tfrac{1}{3}y_1 \\ y_2'' &= -\tfrac{4}{3}y_2 \end{aligned}.$$

The solutions of these equations are

$$y_1(t) = C_1 \cos \frac{t}{\sqrt{3}} + C_2 \sin \frac{t}{\sqrt{3}}$$

and

$$y_2(t) = C_3 \cos \frac{2t}{\sqrt{3}} + C_4 \sin \frac{2t}{\sqrt{3}},$$

respectively. Now it follows that the solution of the original system is given by

$$\begin{aligned}\mathbf{x}(t) &= y_1(t)\mathbf{P}_1 + y_2(t)\mathbf{P}_2 \\ &= \left(C_1' \cos \frac{t}{\sqrt{3}} + C_2' \sin \frac{t}{\sqrt{3}}\right)\begin{bmatrix}1\\2\end{bmatrix} \\ &\qquad + \left(C_3' \cos \frac{2t}{\sqrt{3}} + C_4' \sin \frac{2t}{\sqrt{3}}\right)\begin{bmatrix}1\\-1\end{bmatrix},\end{aligned}$$

where we have "absorbed" the scalars α and β in the constants C_j'.

■

In the next example we show how a system of linear differential equations arises in circuit analysis when there is magnetic coupling between two circuits.

Example 7.21

In the circuit of Figure 7.10, energy is transferred from one circuit to the other through magnetic flux linking the two circuits. The mutual inductance between two coils of inductances L_1 and L_2 is given by $M = kL_1 L_2$, where k is the coefficient of coupling. The constant k is a dimensionless quantity and M is given in henrys. The loop equations for the voltages give the two differential equations:

$$-M\frac{di_1}{dt} + (L_2 + L_3)\frac{di_2}{dt} + R_2 i_2 = 0$$

$$-M\frac{di_2}{dt} + L_1\frac{di_1}{dt} + R_1 i_1 = 0.$$

In matrix form this system is

$$\begin{bmatrix}-L_1 & M \\ M & -L_2 - L_3\end{bmatrix}\begin{bmatrix}i_1'(t)\\ i_2'(t)\end{bmatrix} = \begin{bmatrix}R_1 & 0\\ 0 & R_2\end{bmatrix}\begin{bmatrix}i_1(t)\\ i_2(t)\end{bmatrix}.$$

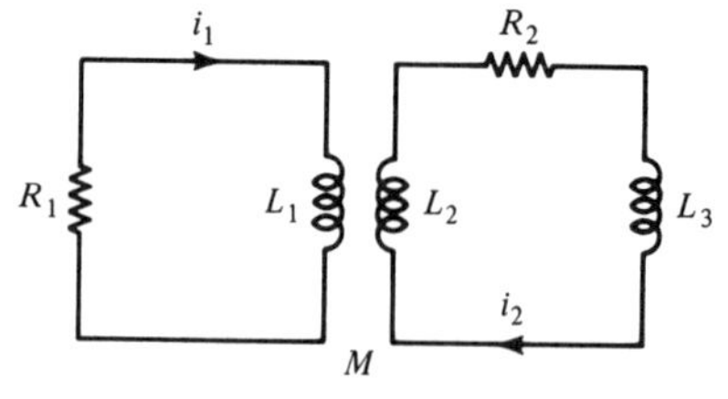

Figure 7.10

This first-order system takes the form $\mathbf{Ax}' = \mathbf{Bx}$, with $\mathbf{A}$ and $\mathbf{B}$ real symmetric. If $\mathbf{A}$ is positive definite, then $\mathbf{A}$ and $\mathbf{B}$ can be simultaneously diagonalized by an appropriate coordinate transformation. Specifically, there is an invertible matrix $\mathbf{P}$ such that $\mathbf{P}^T\mathbf{AP} = \mathbf{I}$ and $\mathbf{P}^T\mathbf{BP} = \mathbf{\Lambda}$. The change of variables $\mathbf{x} = \mathbf{Py}$ gives the uncoupled system

$$\mathbf{y}' = \mathbf{\Lambda y},$$

where $\mathbf{\Lambda}$ is a diagonal matrix. ■

EXERCISES 7.5

1. Solve each of the following initial-value problems.
   (a) $dx/dt + 3x = 0, \quad x(2) = 4$
   (b) $u'(t) = 2u(t), \quad u(0) = 3$
2. For the problems of Exercise 1, determine the limiting value of the solution as $t \to \infty$.
3. Solve each of the following first-order systems, $\mathbf{x}' = \mathbf{Ax}$, $\mathbf{x}(t_0) = \mathbf{x}^{(0)}$.
   (a) $\mathbf{A} = \begin{bmatrix} 5 & 1 \\ 1 & 5 \end{bmatrix}, \quad \mathbf{x}^{(0)} = \begin{bmatrix} -1 \\ 1 \end{bmatrix}, \quad t_0 = 1$
   (b) $\mathbf{A} = \begin{bmatrix} 1 & 3 \\ 3 & 1 \end{bmatrix}, \quad \mathbf{x}^{(0)} = \begin{bmatrix} 1 \\ 0 \end{bmatrix}, \quad t_0 = 0$
   (c) $\mathbf{A} = \begin{bmatrix} 3 & 0 & -2 \\ 1 & 2 & -2 \\ 1 & 3 & -3 \end{bmatrix}, \quad \mathbf{x}^{(0)} = \begin{bmatrix} 2 \\ 1 \\ -1 \end{bmatrix}, \quad t_0 = 0$
4. Consider the circuit of Figure 7.7. If $R_1 = 1/3$, $R_2 = 2$, $L = .25$, and $C = 1$, solve the system of Example 7.15 with initial conditions $i(0) = 0$, $v(0) = 5$.
5. Consider the third-order differential equation

   $$ax''' + bx'' + cx' + dx = 0$$

   with $a \neq 0$. Convert this to a first-order system of differential equations by introduction of the variables $x_1(t) = x(t)$, $x_2(t) = x_1'(t)$, and $x_3(t) = x_2'(t)$. Show that the resultant first-order system has coefficient matrix

   $$\mathbf{A} = \begin{bmatrix} 0 & 1 & 0 \\ 0 & 0 & 1 \\ -\frac{d}{a} & -\frac{c}{a} & -\frac{b}{a} \end{bmatrix}.$$

   Show that the characteristic polynomial of $\mathbf{A}$ is

   $$p_{\mathbf{A}}(\lambda) = \frac{a\lambda^3 + b\lambda^2 + c\lambda + d}{a}.$$

6. Solve the first-order system in Exercise 5 if $a = 1$, $b = -1$, $c = -6$, and $d = 0$.
7. Find the matrix exponential $e^{\mathbf{D}t}$ where $\mathbf{D} = \mathbf{diag}\,(1, 0, -1)$.

8. Find the matrix exponential $e^{\mathbf{A}t}$, where

$$\mathbf{A} = \begin{bmatrix} 2 & -2 & 1 \\ -2 & -1 & 2 \\ 1 & 2 & 2 \end{bmatrix}.$$

9. Solve the following second-order initial-value problems.
(a) $d^2y/dt^2 - 8\,dy/dt - 9y = 0, \quad y(1) = 2, \quad y'(1) = 0$
(b) $x'' - 2x' + x = 0, \quad x(0) = 1, \quad x'(0) = -1$

*10. Given two differentiable functions, u and v, define the Wronskian of u and v at t by

$$W(u, v)(t) = \begin{vmatrix} u(t) & v(t) \\ u'(t) & v'(t) \end{vmatrix}.$$

In each of the following cases, show that the Wronskian of u and v is nonzero for all t.
(a) $u(t) = e^{r_1 t}, \quad v(t) = e^{r_2 t}$ with $r_1 \neq r_2$
(b) $u(t) = e^{rt}, \quad v(t) = te^{rt}$
(c) $u(t) = e^{\lambda t}\cos \mu t, \quad v(t) = e^{\lambda t}\sin \mu t$

11. Solutions of the differential equation

$$x''(t) + bx'(t) + cx(t) = 0$$

take the form $x(t) = C_1 e^{r_1 t} + C_2 e^{r_2 t}$, where r_1 and r_2 are roots of the equation $r^2 + br + c = 0$. Suppose that $b^2 - 4c > 0$ and initial conditions are specified by $x(t_0) = x^{(0)}$ and $x'(t_0) = x^{(1)}$. Show that the constants C_1 and C_2 are uniquely determined by the initial conditions if the Wronskian of $u(t) = e^{r_1 t}$ and $v(t) = e^{r_2 t}$ at t_0 is nonzero.

12. Suppose that $b^2 - 4c = 0$ and let $w(t) = Cte^{-bt/2}$. Show that w is a solution of equation (10).

13. Suppose that $b^2 - 4c < 0$ and let $\lambda = b/2$ and $\mu = \sqrt{4c - b^2}/2$. Show that $w(t) = Ce^{\lambda t}\cos \mu t$ and $v(t) = Ce^{\lambda t}\sin \mu t$ are solutions of equation (10).

14. Find the natural frequencies and normal modes of vibration for the mass–spring system of Figure 7.9 with $k_1 = 4$, $k_2 = 2$, $m_1 = 1$, and $m_2 = \frac{1}{2}$.

15. Solve the initial value problem $\mathbf{x}'' = \mathbf{A}\mathbf{x}$, $\mathbf{x}(0) = \mathbf{x}^{(0)}$, $\mathbf{x}'(0) = \mathbf{x}^{(1)}$, where

$$\mathbf{A} = \begin{bmatrix} 5 & 1 \\ 1 & 5 \end{bmatrix}, \qquad \mathbf{x}^{(0)} = \begin{bmatrix} 0 \\ 2 \end{bmatrix}, \qquad \mathbf{x}^{(1)} = \begin{bmatrix} -2 \\ 2 \end{bmatrix}.$$

16. Solve for the currents in the following circuit:

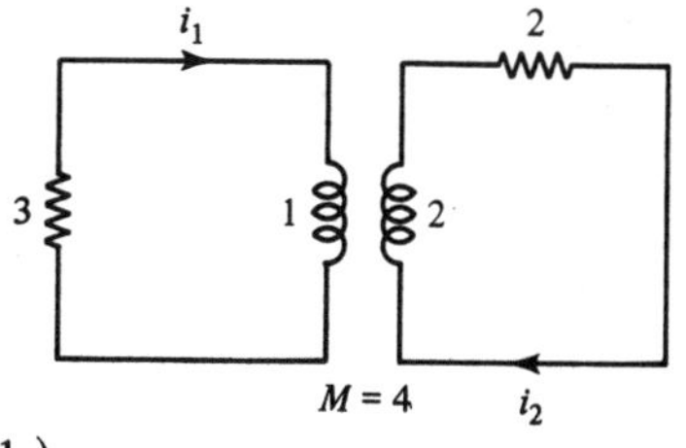

(*Hint:* See Example 7.21.)

17. Solve the system by the use of simultaneous diagonalization.

$$\begin{bmatrix} 2 & -3 \\ -3 & 5 \end{bmatrix}\begin{bmatrix} x_1''(t) \\ x_2''(t) \end{bmatrix} + \begin{bmatrix} 3 & 1 \\ 1 & 21 \end{bmatrix}\begin{bmatrix} x_1(t) \\ x_2(t) \end{bmatrix} = \begin{bmatrix} 0 \\ 0 \end{bmatrix}, \quad \begin{bmatrix} x_1(0) \\ x_2(0) \end{bmatrix} = \begin{bmatrix} 1 \\ -1 \end{bmatrix}.$$

(*Hint:* See Exercise 5 of Section 7.4.)

***18.** Use the method outlined in Example 7.15 to show that the current $i(t)$ and voltage $v(t)$ in the circuit below satisfy the following system of differential equations:

$$\begin{bmatrix} L & 0 \\ 0 & C \end{bmatrix}\begin{bmatrix} i'(t) \\ v'(t) \end{bmatrix} = \frac{1}{R_1 + R_2}\begin{bmatrix} -R_1R_2 & -R_1 \\ R_1 & -1 \end{bmatrix}\begin{bmatrix} i(t) \\ v(t) \end{bmatrix}$$

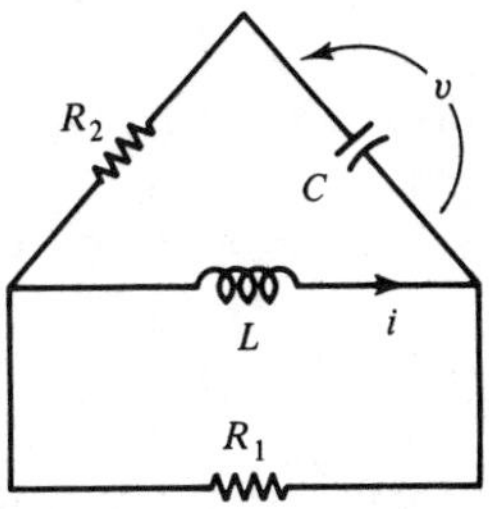

19. For the circuit of Exercise 18, suppose that $L = 2$, $C = \frac{1}{2}$, $R_1 = 2$, and $R_2 = \frac{1}{2}$. If $i(0) = 0$ and $v(0) = 5$, find the solution of the system. Determine the steady-state current and voltage, that is, find the limit as $t \to \infty$ of $i(t)$ and $v(t)$.

20. Find the general solution of the mass–spring system in Figure 7.9 if $m_1 = 1$, $m_2 = \frac{1}{2}$, $k_1 = 4$, and $k_2 = 2$.

21. Consider the mass–spring system shown below, in which there is no friction acting on the masses. This system is governed by the following system of second-order differential equations:

$$m_1\frac{d^2x_1}{dt^2} + (k_1 + k_2)x_1 - k_2x_2 = 0$$

$$m_2\frac{d^2x_2}{dt^2} + (k_2 + k_3)x_2 - k_2x_1 = 0.$$

Solve for the normal modes of vibration if $m_1 = 4$, $m_2 = 2$, $k_1 = k_2 = 2$, and $k_3 = 1$.

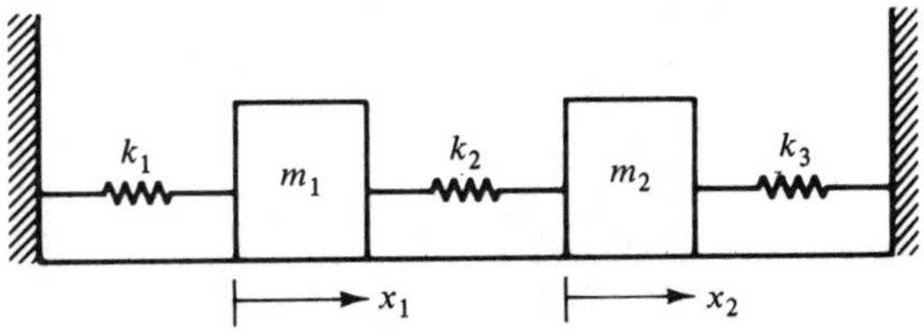

22. For the magnetically coupled circuit of Figure 7.10 let $L_1 = 3$, $L_2 = L_3 = 1$, $M = 4$, $R_1 = 1$, and $R_2 = 2$. Use simultaneous diagonalization to solve the first-order system of differential equations which result in Example 7.21.

GLOSSARY

congruence transformation For a given matrix $\mathbf{A}$, $\mathbf{Q}$ is a congruence transformation matrix if $\mathbf{Q}$ is invertible and $\mathbf{Q}^T\mathbf{AQ}$ takes some desired simple form.

coordinate transformation For a given matrix $\mathbf{A}$, $\mathbf{P}$ is a coordinate transformation matrix if $\mathbf{P}$ is invertible and $\mathbf{P}^{-1}\mathbf{AP}$ takes some desired simple form.

diagonalizable matrix A square matrix that is similar to a diagonal matrix.

plane rotation The rotation of coordinate axes through some angle in the counter-clockwise direction.

positive definite matrix A real symmetric matrix **A** whose quadratic form, $\langle \mathbf{x}, \mathbf{Ax} \rangle$, is positive for all nonzero vectors **x**.

similar matrices Two square matrices, **A** and **B**, are similar if $\mathbf{A} = \mathbf{P}^{-1}\mathbf{BP}$ for some invertible matrix **P**.

simultaneous diagonalization A technique for diagonalization of two symmetric matrices using the same congruence transformation matrix.

translation of axes The xy coordinate axes are translated to the uv coordiate axes by the change of variables $x = u + \alpha$, $y = v + \beta$.

NOTES AND COMMENTS

There is an extensive and well-developed theory of quadratic forms associated with symmetric matrices. For additional information regarding quadratic forms and their connection with intermediate eigenvalues, consult the book by G. Strang, *Linear Algebra and Its Applications,* 2nd edition (New York: Academic Press, 1980) or the book by B. Noble and J. Daniel: *Applied Linear Algebra,* 2nd edition (Englewood Cliffs, N.J.: Prentice-Hall, 1977).

Matrix methods can also be applied to first-order systems of ordinary differential equations when the coefficient matrix is not diagonalizable. One way to treat this general problem is to define and make use of the exponential of a general square matrix. In the text it was sufficient for our purposes to define the matrix exponential of a diagonal matrix. The book by Noble and Daniel develops this theory by making use of the Jordan canonical form of a matrix.

8

Numerical Linear Algebra

In this chapter we consider some numerical aspects of methods for solving the central problems of linear algebra, namely the linear system $\mathbf{Ax} = \mathbf{b}$ and the eigenproblem $\mathbf{Ax} = \lambda\mathbf{x}$. In previous chapters we have studied these problems with the primary aims of explaining the theory and illustrating how these and related problems arise in applications. All of our illustrations have involved low-dimensional problems that could be easily solved with pencil and paper. However, realistic problems are usually much larger and therefore require effective numerical methods for their solution.

Previously, we pointed out certain techniques that are not efficient for the purpose of calculation. For example, the determination of eigenvalues by use of the characteristic equation is not used for matrices of size greater than 4. The difficulty here is the factorization of the characteristic polynomial, that is, there is no algebraic formula for determination of the roots of a polynomial of degree greater than 4. Other examples of impractical techniques include the solution of a linear system by use of the inverse or by Cramer's rule. This is not to say that the inverse or Cramer's rule is useless. Rather, they are theoretical tools. In this chapter we examine some reasonably effective methods for solving problems on a modern digital computer.

The solution of linear systems is a basic component of many more complicated numerical procedures and is perhaps the most frequently encountered task in numerical computing. For most square linear systems the method of choice is the Gaussian elimination procedure discussed in Chapter 2. We examine a slight variation of this method that is more appropriate for use on a digital computer. Moreover, we introduce the concept of the condition number of a square matrix and demonstrate its importance in connection with error estimation. For rectangular systems of equations we examine, in Section 8.4, the method of least squares for obtaining a pseudosolution.

For the eigenproblem we restrict our consideration to diagonalizable matrices in Section 8.3 and examine reasonably effective procedures, the power and inverse power methods, for the determination of eigenvalues.

We begin with a discussion of norms that are appropriate for measuring errors in the context of numerical linear algebra.

8.1 Errors, Norms, and Conditioning

When attempting to find the solution, $\mathbf{x}$, of a given linear system, one computes an approximation, $\hat{\mathbf{x}}$, of $\mathbf{x}$ by means of some numerical method implemented on a digital computer. Consider, for example, the simple linear system

$$\begin{aligned} 3x_1 + \quad x_2 &= 7 \\ 7x_1 + 2.35x_2 &= 16.35. \end{aligned}$$

The exact solution of this system is

$$\mathbf{x} = \begin{bmatrix} 2 \\ 1 \end{bmatrix}.$$

Suppose, however, that we solve this system on a hypothetical computer which keeps only four significant digits and uses rounding. Then, using Guassian elimination, we subtract $\frac{7}{3}$ times the first equation from the second equation, yielding

$$\begin{aligned} 3x_1 + x_2 &= 7 \\ \left(2.35 - \frac{7}{3}\right)x_2 &= 16.35 - \frac{49}{3}. \end{aligned}$$

However, keeping only four digits, the computer finds

$$2.35 - \frac{7}{3} = .017 \quad \text{and} \quad 16.35 - \frac{49}{3} = .02,$$

and hence the computed triangular system which results is

$$\begin{aligned} 3\hat{x}_1 + \hat{x}_2 &= 7 \\ .017\hat{x}_2 &= .02. \end{aligned}$$

Back substitution then gives (keeping only four digits)

$$\hat{x}_2 = \frac{.02}{.017} = 1.177$$

and

$$\hat{x}_1 = \frac{5.823}{3} = 1.941.$$

Thus the computed approximate solution is

$$\hat{\mathbf{x}} = \begin{bmatrix} 1.941 \\ 1.177 \end{bmatrix}.$$

Clearly, $\hat{\mathbf{x}}$ is not the same as $\mathbf{x}$, and it is useful to have a measure of the accuracy of $\hat{\mathbf{x}}$. Generally, the closeness of $\hat{\mathbf{x}}$ to $\mathbf{x}$ is measured in terms on the size of the difference $\mathbf{x} - \hat{\mathbf{x}}$. The smaller the difference $\mathbf{x} - \hat{\mathbf{x}}$, the closer $\hat{\mathbf{x}}$ is to $\mathbf{x}$. In Chapter 5 we introduced a means of measuring the size of a vector in terms of its norm. A **norm** on $\mathbf{R}^n$ is a real-valued function, $\|\cdot\|$, that satisfies the following:

1. $\|\mathbf{u}\| \geq 0$ for all $\mathbf{u} \in \mathbf{R}^n$.
2. $\|\mathbf{u}\| = 0$ if and only if $\mathbf{u} = \boldsymbol{\theta}$.
3. $\|k\mathbf{u}\| = |k|\|\mathbf{u}\|$ for all $\mathbf{u} \in \mathbf{R}^n$ and any scalar k.
4. $\|\mathbf{u} + \mathbf{v}\| \leq \|\mathbf{u}\| + \|\mathbf{v}\|$ for all $\mathbf{u}, \mathbf{v} \in \mathbf{R}^n$.

We have seen in Theorem 5.2 that the usual Euclidean norm

$$\|\mathbf{x}\| = \{x_1^2 + x_2^2 + \cdots + x_n^2\}^{1/2}$$

satisfies these properties. Another useful norm is the so-called "infinity" or max norm given by

$$\|\mathbf{x}\|_\infty = \max_{1\le i\le n} |x_i|.$$

Example 8.1

We measure the error, $\mathbf{x} - \hat{\mathbf{x}}$, for our simple 2×2 system in both the Euclidean and infinity norms:

$$\|\mathbf{x} - \hat{\mathbf{x}}\| = \sqrt{(2 - 1.941)^2 + (1 - 1.177)^2} = 0.1866$$

$$\|\mathbf{x} - \hat{\mathbf{x}}\|_\infty = \max\{|2 - 1.941|, |1 - 1.177|\} = 0.177.$$

■

If $\mathbf{x}$ is some unknown vector and $\hat{\mathbf{x}}$ is an approximation to $\mathbf{x}$, then the quantity $\|\mathbf{x} - \hat{\mathbf{x}}\|$ is called the **absolute error** in the approximation (as measured by the norm $\|\cdot\|$). A small absolute error does not necessarily mean that $\hat{\mathbf{x}}$ is a good approximation to $\mathbf{x}$, since absolute error does not take into account the size of $\mathbf{x}$. It can happen that $\|\mathbf{x} - \hat{\mathbf{x}}\|$ is quite small, yet $\hat{\mathbf{x}}$ is not a very good approximation to $\mathbf{x}$. A more reliable measure of the error is the quantity

$$\frac{\|\mathbf{x} - \hat{\mathbf{x}}\|}{\|\mathbf{x}\|},$$

which is called the **relative error** in $\hat{\mathbf{x}}$.

Example 8.2

Suppose that

$$\mathbf{x} = \begin{bmatrix} .5 \times 10^{-6} \\ 10^{-6} \end{bmatrix} \quad \text{and} \quad \hat{\mathbf{x}} = \begin{bmatrix} 10^{-6} \\ 10^{-6} \end{bmatrix}.$$

Then the absolute error

$$\|\mathbf{x} - \hat{\mathbf{x}}\|_\infty = .5 \times 10^{-6}$$

is quite small, but the relative error is 50%:

$$\frac{\|\mathbf{x} - \hat{\mathbf{x}}\|_\infty}{\|\mathbf{x}\|_\infty} = \frac{.5 \times 10^{-6}}{10^{-6}} = .5.$$

■

Since all the problems we deal with will involve matrices as well as vectors, it will be useful to have a norm that is defined on matrices. By a **matrix norm** we mean a function $\|\cdot\|$ defined on $n \times n$ matrices which satisfies:

1′. $\|\mathbf{A}\| \geq 0$ for all $n \times n$ matrices $\mathbf{A}$.
2′. $\|\mathbf{A}\| = 0$ if and only if $\mathbf{A} = \mathbf{O}$.
3′. $\|k\mathbf{A}\| = |k|\|\mathbf{A}\|$ for all scalars k.
4′. $\|\mathbf{A} + \mathbf{B}\| \leq \|\mathbf{A}\| + \|\mathbf{B}\|$ for all $\mathbf{A}$, $\mathbf{B}$.
5′. $\|\mathbf{AB}\| \leq \|\mathbf{A}\|\|\mathbf{B}\|$ for all $\mathbf{A}$, $\mathbf{B}$.

Note that properties (1′)–(4′) are the same as the properties (1)–(4) of a vector norm. Property (5′) is a special requirement that relates a matrix norm to the matrix product.

We will consider only two special matrix norms. These matrix norms are compatible, in a sense that we make precise below, with the two vector norms $\|\cdot\|_\infty$ and $\|\cdot\|$. The matrix norms will be denoted by the same symbols, but no confusion should arise because the matrix norm can only be applied to a matrix and the vector norm can only be applied to a vector. For $n \times n$ matrices we define the matrix norms

$$\|\mathbf{A}\|_\infty = \max_{1 \leq i \leq n} \sum_{j=1}^{n} |a_{ij}|$$

and

$$\|\mathbf{A}\| = \left(\sum_{i=1}^{n} \sum_{j=1}^{n} a_{ij}^2 \right)^{1/2}.$$

In words, $\|\mathbf{A}\|_\infty$ is equal to the maximum absolute row sum of $\mathbf{A}$.

Example 8.3

Suppose that

$$\mathbf{A} = \begin{bmatrix} -1 & 2 & 3 \\ 2 & 1 & 1 \\ 4 & -3 & 2 \end{bmatrix},$$

then

$$\|\mathbf{A}\|_\infty = \max_{1 \leq i \leq 3} \sum_{j=1}^{3} |a_{ij}| = \max\{6, 4, 9\} = 9$$

$$\|\mathbf{A}\| = \left(\sum_{i=1}^{3} \sum_{j=1}^{3} a_{ij}^2 \right)^{1/2} = \sqrt{49} = 7.$$

■

These matrix norms satisfy an important *compatibility condition* with respect to the corresponding vector norms, namely

$$\|\mathbf{Ax}\| \leq \|\mathbf{A}\|\|\mathbf{x}\|$$

and

$$\|\mathbf{Ax}\|_\infty \leq \|\mathbf{A}\|_\infty\|\mathbf{x}\|_\infty.$$

Note that in each of these inequalities, the norm on the left is a vector norm, the first norm on the right is a matrix norm, and the second norm on the right is a vector norm.

We verify the compatibility condition for the max norm only:

$$\begin{aligned}\|\mathbf{Ax}\|_\infty &= \max_{1\le i\le n} |(\mathbf{Ax})_i| \\ &= \max_{1\le i\le n}\left|\sum_{j=1}^{n} a_{ij}x_j\right| \\ &\le \max_{1\le i\le n}\sum_{j=1}^{n} |a_{ij}||x_j|,\end{aligned}$$

and since $|x_j| \le \|\mathbf{x}\|_\infty$ for each j,

$$\begin{aligned}\|\mathbf{Ax}\|_\infty &\le \max_{1\le i\le n}\sum_{j=1}^{n} |a_{ij}|\,\|\mathbf{x}\|_\infty \\ &\le \|\mathbf{A}\|_\infty\|\mathbf{x}\|_\infty.\end{aligned}$$

Using this result, we can show that the matrix norm $\|\cdot\|_\infty$ satisfies condition (5′). Let $\mathbf{C} = \mathbf{AB}$ and suppose that the maximum of the absolute row sums of $\mathbf{C}$ occurs in the ith row, that is,

$$\|\mathbf{C}\|_\infty = \sum_{j=1}^{n} |c_{ij}|.$$

Choose a vector $\mathbf{x}$ by $x_j = \text{sign}\,(c_{ij})$; that is, $x_j = 1$ if $c_{ij} \ge 0$ and $x_j = -1$ otherwise. Then $\|\mathbf{x}\|_\infty = 1$, and

$$\begin{aligned}\|\mathbf{Cx}\|_\infty &= \max_{1\le k\le n}\left|\sum_{j=1}^{n} c_{kj}\,\text{sign}\,(c_{ij})\right| \\ &= \sum_{j=1}^{n} |c_{ij}| = \|\mathbf{C}\|_\infty.\end{aligned}$$

Therefore, by the compatibility condition,

$$\begin{aligned}\|\mathbf{AB}\|_\infty &= \|\mathbf{C}\|_\infty = \|\mathbf{Cx}\|_\infty = \|\mathbf{ABx}\|_\infty \\ &\le \|\mathbf{A}\|_\infty\|\mathbf{Bx}\|_\infty \le \|\mathbf{A}\|_\infty\|\mathbf{B}\|_\infty\|\mathbf{x}\|_\infty \\ &= \|\mathbf{A}\|_\infty\|\mathbf{B}\|_\infty.\end{aligned}$$

We do not prove the product inequality for the matrix norm $\|\cdot\|$; however, it is illustrated in the following example.

Example 8.4

Suppose that

$$\mathbf{A} = \begin{bmatrix} 1 & 3 & -1 \\ 0 & 1 & 2 \\ -1 & 1 & 1 \end{bmatrix}, \qquad \mathbf{B} = \begin{bmatrix} -2 & 1 & 2 \\ 1 & -1 & 0 \\ 0 & 3 & 1 \end{bmatrix}, \quad \text{and} \quad \mathbf{x} = \begin{bmatrix} 2 \\ 1 \\ -1 \end{bmatrix}.$$

Then

$$\mathbf{Ax} = \begin{bmatrix} 6 \\ -1 \\ -2 \end{bmatrix} \quad \text{and} \quad \mathbf{AB} = \begin{bmatrix} 1 & -5 & 1 \\ 1 & 5 & 2 \\ 3 & 1 & -1 \end{bmatrix}.$$

It follows that

$$\begin{aligned} \|\mathbf{Ax}\|_\infty &= 6 < (5)(2) = \|\mathbf{A}\|_\infty \|\mathbf{x}\|_\infty \\ \|\mathbf{Ax}\| &= \sqrt{41} < (\sqrt{19})(\sqrt{6}) = \|\mathbf{A}\|\|\mathbf{x}\| \\ \|\mathbf{AB}\|_\infty &= 7 < (5)(5) = \|\mathbf{A}\|_\infty \|\mathbf{B}\|_\infty \\ \|\mathbf{AB}\| &= \sqrt{65} < (\sqrt{19})(\sqrt{21}) = \|\mathbf{A}\|\|\mathbf{B}\|. \end{aligned}$$

■

In most of what follows we use the max norm for matrices because it is easier to compute than the norm $\|\cdot\|$. The compatibility condition for a matrix norm gives information on the amount by which a matrix can "stretch" a vector. Specifically, for any nonzero vector $\mathbf{x}$, we have

$$\frac{\|\mathbf{Ax}\|}{\|\mathbf{x}\|} \le \|\mathbf{A}\|.$$

That is, the relative elongation in the vector $\mathbf{x}$ is at most $\|\mathbf{A}\|$. This idea can be used to give an estimate for the change in the exact solution of a linear system which results from a change in the right-hand side. Suppose, for example, that $\mathbf{A}$ is an $n \times n$ invertible matrix and $\mathbf{Ax} = \mathbf{b}$. If the right-hand side is changed to $\hat{\mathbf{b}}$, we have a new solution $\hat{\mathbf{x}}$ of $\mathbf{A}\hat{\mathbf{x}} = \hat{\mathbf{b}}$. The effect of the change from $\mathbf{b}$ to $\hat{\mathbf{b}}$ on the solution can be estimated in the following way:

$$\mathbf{x} = \mathbf{A}^{-1}\mathbf{b}, \qquad \hat{\mathbf{x}} = \mathbf{A}^{-1}\hat{\mathbf{b}}$$

and hence

$$\mathbf{x} - \hat{\mathbf{x}} = \mathbf{A}^{-1}\mathbf{b} - \mathbf{A}^{-1}\hat{\mathbf{b}} = \mathbf{A}^{-1}(\mathbf{b} - \hat{\mathbf{b}}).$$

Therefore,

$$\|\mathbf{x} - \hat{\mathbf{x}}\| = \|\mathbf{A}^{-1}(\mathbf{b} - \hat{\mathbf{b}})\| \le \|\mathbf{A}^{-1}\|\|\mathbf{b} - \hat{\mathbf{b}}\|;$$

that is, the change in the solution is at most $\|\mathbf{A}^{-1}\|$ times the change in the right-hand side. To put it another way, the absolute error in the approximation $\hat{\mathbf{x}}$ is at most $\|\mathbf{A}^{-1}\|$ times the absolute error in the approximate right-hand side $\hat{\mathbf{b}}$.

Example 8.5

Suppose that

$$\mathbf{A} = \begin{bmatrix} 3 & 4 \\ 5 & 7 \end{bmatrix}, \quad \text{then} \quad \mathbf{A}^{-1} = \begin{bmatrix} 7 & -4 \\ -5 & 3 \end{bmatrix},$$

and hence $\|\mathbf{A}^{-1}\|_\infty = 11$. Therefore, in solving $\mathbf{Ax} = \mathbf{b}$, the norm of a change in $\mathbf{x}$ (which results from a change in $\mathbf{b}$) will be at most 11 times the norm of the change in **b**. Suppose, for example, that

$$\mathbf{b} = \begin{bmatrix} 3 \\ 2 \end{bmatrix}, \quad \text{then} \quad \mathbf{x} = \begin{bmatrix} 13 \\ -9 \end{bmatrix},$$

while if

$$\hat{\mathbf{b}} = \begin{bmatrix} 2 \\ 3 \end{bmatrix}, \quad \text{then} \quad \hat{\mathbf{x}} = \begin{bmatrix} 2 \\ -1 \end{bmatrix}.$$

Therefore, a change of $\|\mathbf{b} - \hat{\mathbf{b}}\|_\infty = 1$ in the right-hand side gives rise to a change of

$$\|\mathbf{x} - \hat{\mathbf{x}}\|_\infty = 11 = (11)\|\mathbf{b} - \hat{\mathbf{b}}\|_\infty$$

in the solution. We hasten to add that the factor $\|\mathbf{A}^{-1}\|$ gives only an upper bound on the change in the solution. Indeed, if

$$\mathbf{b} = \begin{bmatrix} 7 \\ 12 \end{bmatrix}, \quad \text{then} \quad \mathbf{x} = \begin{bmatrix} 1 \\ 1 \end{bmatrix},$$

and if

$$\hat{\mathbf{b}} = \begin{bmatrix} 3 \\ 5 \end{bmatrix}, \quad \text{then} \quad \hat{\mathbf{x}} = \begin{bmatrix} 1 \\ 0 \end{bmatrix}.$$

Therefore,

$$\|\mathbf{x} - \hat{\mathbf{x}}\|_\infty = 1 < (11)\|\mathbf{b} - \hat{\mathbf{b}}\|_\infty = 77.$$

■

The inequality

$$\|\mathbf{x} - \hat{\mathbf{x}}\|_\infty \leq \|\mathbf{A}^{-1}\|_\infty \|\mathbf{b} - \hat{\mathbf{b}}\|_\infty,$$

where $\mathbf{Ax} = \mathbf{b}$, $\mathbf{A}\hat{\mathbf{x}} = \hat{\mathbf{b}}$, gives a bound on the absolute error in the solution arising from an error in the right-hand side. We have pointed out that it is generally more suitable to measure errors in a relative sense. A bound similar to that above, but for relative errors, can be given in terms of the condition number of the matrix **A**. If **A** is an invertible matrix, then the **condition number** of **A**, $K_\infty(\mathbf{A})$, is defined by

$$K_\infty(\mathbf{A}) = \|\mathbf{A}\|_\infty \|\mathbf{A}^{-1}\|_\infty.$$

We could also define the condition number with respect to the $\|\cdot\|$ norm, but $K_\infty(\mathbf{A})$ is easier to compute.

Example 8.6

Let

$$\mathbf{A} = \begin{bmatrix} 2 & 4 \\ 6 & 11 \end{bmatrix}; \quad \text{then} \quad \mathbf{A}^{-1} = \frac{1}{2}\begin{bmatrix} -11 & 4 \\ 6 & -2 \end{bmatrix},$$

and therefore

$$K_\infty(\mathbf{A}) = \|\mathbf{A}\|_\infty \|\mathbf{A}^{-1}\|_\infty = 17(\tfrac{15}{2}) = 127.5.$$

■

The following theorem gives the previously mentioned bound for the relative error due to an error in the right-hand side.

Theorem 8.1. Suppose that $\mathbf{Ax} = \mathbf{b}$, and $\mathbf{A\hat{x}} = \hat{\mathbf{b}}$ with $\mathbf{A}$ invertible. Then the following relative error bound holds:

$$\frac{\|\mathbf{x} - \hat{\mathbf{x}}\|_\infty}{\|\mathbf{x}\|_\infty} \leq K_\infty(\mathbf{A})\frac{\|\mathbf{b} - \hat{\mathbf{b}}\|_\infty}{\|\mathbf{b}\|_\infty}.$$

Proof. We recall that

$$\|\mathbf{x} - \hat{\mathbf{x}}\|_\infty \leq \|\mathbf{A}^{-1}\|_\infty \|\mathbf{b} - \hat{\mathbf{b}}\|_\infty.$$

However,

$$\|\mathbf{b}\|_\infty = \|\mathbf{Ax}\|_\infty \leq \|\mathbf{A}\|_\infty \|\mathbf{x}\|_\infty$$

and hence

$$\begin{aligned}\|\mathbf{x} - \hat{\mathbf{x}}\|_\infty &\leq \|\mathbf{A}^{-1}\|_\infty \|\mathbf{b}\|_\infty \frac{\|\mathbf{b} - \hat{\mathbf{b}}\|_\infty}{\|\mathbf{b}\|_\infty} \\ &\leq \|\mathbf{A}^{-1}\|_\infty \|\mathbf{A}\|_\infty \|\mathbf{x}\|_\infty \frac{\|\mathbf{b} - \hat{\mathbf{b}}\|_\infty}{\|\mathbf{b}\|_\infty},\end{aligned}$$

that is,

$$\frac{\|\mathbf{x} - \hat{\mathbf{x}}\|_\infty}{\|\mathbf{x}\|_\infty} \leq K_\infty(\mathbf{A})\frac{\|\mathbf{b} - \hat{\mathbf{b}}\|_\infty}{\|\mathbf{b}\|_\infty}.$$

■

This theorem says that the relative error in the solution is bounded by the condition number of $\mathbf{A}$ times the relative error in the right-hand side. That is, the condition number $K_\infty(\mathbf{A})$ plays the same role for relative errors that the number $\|\mathbf{A}^{-1}\|_\infty$ played for absolute errors.

Example 8.7

Let $\mathbf{A}$ be the matrix of Example 8.6, and suppose that

$$\mathbf{b} = \begin{bmatrix} 2 \\ 5 \end{bmatrix}, \qquad \hat{\mathbf{b}} = \begin{bmatrix} 2.06 \\ 4.92 \end{bmatrix}.$$

Then

$$\mathbf{x} = \begin{bmatrix} -1 \\ 1 \end{bmatrix} \quad \text{and} \quad \hat{\mathbf{x}} = \begin{bmatrix} -1.49 \\ 1.26 \end{bmatrix}.$$

We have $\|\mathbf{x} - \hat{\mathbf{x}}\|_\infty = .49$, $\|\mathbf{b} - \hat{\mathbf{b}}\|_\infty = .08$, and hence

$$.49 = \frac{\|\mathbf{x} - \hat{\mathbf{x}}\|_\infty}{\|\mathbf{x}\|_\infty} \leq K_\infty(\mathbf{A})\frac{\|\mathbf{b} - \hat{\mathbf{b}}\|_\infty}{\|\mathbf{b}\|_\infty} = 127.5\frac{(.08)}{5} = 2.04.$$

■

If a matrix **A** has a relatively large condition number, then the solution of the system $\mathbf{Ax} = \mathbf{b}$ is normally very sensitive to small changes in **b**. A matrix with a large condition number is called an **ill-conditioned matrix.** To illustrate, consider the two systems

$$\mathbf{Ax} = \begin{bmatrix} 1 & 2 \\ 1.001 & 2 \end{bmatrix} \begin{bmatrix} x_1 \\ x_2 \end{bmatrix} = \begin{bmatrix} 3 \\ 3 \end{bmatrix}$$

and

$$\mathbf{A\hat{x}} = \begin{bmatrix} 1 & 2 \\ 1.001 & 2 \end{bmatrix} \begin{bmatrix} \hat{x}_1 \\ \hat{x}_2 \end{bmatrix} = \begin{bmatrix} 3 \\ 3.001 \end{bmatrix}.$$

The solution of the former is $x_1 = 0$, $x_2 = 1.5$, whereas the latter has the solution $\hat{x}_1 = \hat{x}_2 = 1$. For this matrix a change of one unit in the *fourth* digit of b_2 results in a change of one unit in the *first* digit of x_1. The solution $\hat{\mathbf{x}}$ bears no resemblance to $\mathbf{x}$, and moreover, there is no round-off error present here (the solutions are exact). A simple calculation shows that the condition number of **A** is $K_\infty(\mathbf{A}) = 6002$. The important point to realize is this: With an ill-conditioned matrix **A**, no numerical method can be used to compute a "solution" of $\mathbf{Ax} = \mathbf{b}$ with the expectation of good accuracy. The sensitivity of the exact solution (to changes in **b**) is inherited by the computed solution, no matter which numerical method is used. In a package of subroutines for linear algebra, the LINPACK library, the user is provided with an estimate for the condition number so that he or she can judge the quality of a computed solution of a linear system of equations. There is, in general, no cure-all for ill-conditioned linear systems, but the problem solver should be aware of possible difficulties.

Partial Pivoting

We now reexamine the elimination method in the context of computation on a digital computer. Our aim is to demonstrate the need for a pivoting strategy, that is, a means of selecting the pivotal rows. A straightforward use of Gaussian elimination uses row interchanges only to avoid zero pivots. However, in practice, row interchanges may be necessary to avoid the introduction of large errors which result from round-off error. Without such row interchanging the computed solution may be grossly inaccurate.

For the purpose of illustration, consider the following simple example:

$$\mathbf{Ax} = \begin{bmatrix} .001 & 2 \\ 1 & 1 \end{bmatrix} \begin{bmatrix} x_1 \\ x_2 \end{bmatrix} = \begin{bmatrix} 2 \\ 2 \end{bmatrix} = \mathbf{b},$$

whose solution is to be computed on some hypothetical computer. For simplicity suppose that the computer retains only the first two significant digits by rounding. The first step in elimination is to multiply the first equation by $m = 10^3$ and subtract the result from the second equation. This gives

$$.001x_1 + 2x_2 = 2$$

$$(1 - 2 \times 10^3)x_2 = 2(1 - 10^3).$$

However, to two significant digits the computer gives

$$1 - 2 \times 10^3 = -2 \times 10^3 \quad \text{and} \quad 2(1 - 10^3) = -2 \times 10^3,$$

and hence the computed upper triangular system is

$$\begin{aligned} .001\hat{x}_1 + 2\hat{x}_2 &= 2 \\ -2 \times 10^3\hat{x}_2 &= -2 \times 10^3, \end{aligned}$$

and back substitution (with two significant digits) yields $\hat{x}_2 = 1$ and $\hat{x}_1 = 0$. The computed solution is

$$\hat{\mathbf{x}} = \begin{bmatrix} 0 \\ 1 \end{bmatrix},$$

whereas the exact solution is

$$\mathbf{x} = \begin{bmatrix} \dfrac{2}{1.999} \\ \\ \dfrac{1.998}{1.999} \end{bmatrix}.$$

The computed solution is not even close to $\mathbf{x}$ and in fact is in error by 100%:

$$\frac{\|\mathbf{x} - \hat{\mathbf{x}}\|_\infty}{\|\mathbf{x}\|_\infty} = 1.$$

The difficulty in this simple example is the addition of two numbers of vastly different scale, that is,

$$1 + (-2m) = a_{22} + m(-a_{12}) \quad \text{and,} \quad 2 + (-2m) = b_2 + m(-b_1),$$

where $m = 10^3$. Notice that the computed value of $a_{22} + m(-a_{12})$ is unaffected if a_{22} is changed by 100%. That is, if a_{22} is changed from 1 to 2, the computed value of $a_{22} + m(-a_{12})$ is not changed. In fact, the information contained in a_{22} is completely lost. The source of this difficulty is the large value of m, or equivalently, the small value (.001) of the pivot. The remedy should be clear: Choose among all possible pivots in a given column the one with largest magnitude. Then the rows are exchanged accordingly to make this largest value the pivot. This strategy is called **partial pivoting.**

When partial pivoting is applied to our simple example, it is necessary to exchange rows:

$$\begin{aligned} x_1 + x_2 &= 2 \\ .001x_1 + 2x_2 &= 2. \end{aligned}$$

Then the computed upper triangular system is

$$\begin{aligned} \hat{x}_1 + \hat{x}_2 &= 2 \\ 2\hat{x}_2 &= 2, \end{aligned}$$

and hence the computed solution is $\hat{x}_1 = \hat{x}_2 = 1$, whose relative error is only .05%:

$$\frac{\|\mathbf{x} - \hat{\mathbf{x}}\|_\infty}{\|\mathbf{x}\|_\infty} = .0005.$$

EXERCISES 8.1

1. Let $\mathbf{x} = [1 \quad -2 \quad 0 \quad 3]^T$ and calculate: **(a)** $\|\mathbf{x}\|$ and **(b)** $\|\mathbf{x}\|_\infty$.

2. Find vectors $\mathbf{u}$, $\mathbf{v} \in \mathbf{R}^n$ such that

$$\|\mathbf{u}\| = \sqrt{n}\,\|\mathbf{u}\|_\infty$$
$$\|\mathbf{v}\| = \|\mathbf{v}\|_\infty.$$

***3.** Show that for any $\mathbf{x} \in \mathbf{R}^n$,

$$\|\mathbf{x}\|_\infty \le \|\mathbf{x}\| \le \sqrt{n}\,\|\mathbf{x}\|_\infty.$$

4. Determine the condition number of the identity matrix, $K_\infty(\mathbf{I})$.

5. Determine the condition number of $\mathbf{D} = \mathbf{diag}\,(-2, 3, 5, 7)$.

6. Is it true that $K_\infty(\alpha\mathbf{A}) = K_\infty(\mathbf{A})$ for any scalar $\alpha \neq 0$?

7. Show that $K_\infty(\mathbf{A}) \ge 1$ for every invertible matrix $\mathbf{A}$. (*Hint:* Use Exercise 4 and $\mathbf{A}^{-1}\mathbf{A} = \mathbf{I}$.)

8. If

$$\mathbf{x} = \begin{bmatrix} 1 \\ 2 \\ -1 \end{bmatrix} \quad \text{and} \quad \mathbf{A} = \begin{bmatrix} 3 & 1 & 0 \\ 2 & -2 & 4 \\ 1 & 6 & 1 \end{bmatrix},$$

verify the inequality: $\|\mathbf{Ax}\|_\infty \le \|\mathbf{A}\|_\infty \|\mathbf{x}\|_\infty$.

9. For the matrix $\mathbf{A}$ of Exercise 8 find a vector $\mathbf{x}$ such that $\|\mathbf{Ax}\|_\infty = \|\mathbf{A}\|_\infty \|\mathbf{x}\|_\infty$.

10. Find $\|\mathbf{A}\|$ and $\|\mathbf{A}\|_\infty$ for each of the following matrices.

(a) $\mathbf{A} = \begin{bmatrix} 0 & -1 & 2 \\ 1 & 0 & 3 \\ -2 & -3 & 0 \end{bmatrix}$ **(b)** $\mathbf{A} = \begin{bmatrix} 10^{-3} & 0 & 0 \\ 0 & 10^{-3} & 0 \\ 0 & 0 & 10^{-3} \end{bmatrix}$

(c) $\mathbf{A} = \begin{bmatrix} 4 & 1 & 0 & 0 \\ 1 & 4 & 1 & 0 \\ 0 & 1 & 4 & 1 \\ 0 & 0 & 1 & 4 \end{bmatrix}$

11. Suppose that λ is an eigenvalue of an invertible matrix $\mathbf{A}$. Show that

$$|\lambda| \le \|\mathbf{A}\|_\infty \quad \text{and} \quad |\lambda^{-1}| \le \|\mathbf{A}^{-1}\|_\infty.$$

As a consequence, show that

$$\left|\frac{\lambda_1}{\lambda_n}\right| \le K_\infty(\mathbf{A}),$$

where λ_n and λ_1 denote the eigenvalues of $\mathbf{A}$ having least and greatest magnitudes, respectively.

12. Suppose that $\mathbf{A}$ is an invertible matrix. Show that for all $\mathbf{x} \neq \boldsymbol{\theta}$,

$$\frac{\|\mathbf{x}\|_\infty}{\|\mathbf{Ax}\|_\infty} \leq \|\mathbf{A}^{-1}\|_\infty,$$

and hence

$$\frac{\|\mathbf{A}\|_\infty \|\mathbf{x}\|_\infty}{\|\mathbf{Ax}\|_\infty} \leq K_\infty(\mathbf{A}).$$

13. Solve the following system of linear equations by using Gaussian elimination with partial pivoting. Perform all arithmetic operations using rounding and retain five significant digits.

$$\begin{aligned} 16x_1 - 9x_2 + x_3 &= 38 \\ -2x_1 + 1.124x_2 + 8x_3 &= -12.871 \\ 4x_1 + 3x_2 + x_3 &= 14 \end{aligned}$$

Repeat your computations without partial pivoting and compare the computed solutions.

14. Repeat the calculations of Exercise 13 for the system

$$\begin{aligned} 4x_1 - 3x_2 + x_3 &= 1 \\ -x_1 + 2x_2 - 2x_3 &= 1 \\ 2x_1 + x_2 - x_3 &= 3 \end{aligned}$$

whose exact solution is $\mathbf{x} = [1 \quad 1 \quad 0]^T$.

8.2 *Iterative Solution of Linear Systems*

Even the most powerful of today's modern supercomputers is hard pressed to solve a linear system of equations involving several thousand equations in several thousand unknowns by Gaussian elimination if the coefficient matrix is full (a matrix is full if it contains very few zeros). However, the numerical solution of differential equations, especially partial differential equations, frequently requires the solution of very large square matrix problems. Fortunately, these matrices consist mostly of zero entries. Such matrices are called *sparse*. Because of the difficulty of solving large sparse matrix problems by Gaussian elimination, several iterative methods for obtaining an approximate solution have evolved. In this section we describe two of the simplest iterative methods, the Jacobi and Gauss–Seidel methods. By way of contrast with Gaussian elimination, we mention that these iterative methods will never (except for the unlikely choice $\mathbf{x}^{(0)} = \mathbf{A}^{-1}\mathbf{b}$) give the exact solution of $\mathbf{Ax} = \mathbf{b}$, even when no round-off errors occur.

General Iterative Methods

We describe the basic idea of iterative methods for the approximate solution of

$$\mathbf{Ax} = \mathbf{b}, \tag{1}$$

where $\mathbf{A}$ is an invertible matrix. One begins by introducing a splitting of $\mathbf{A}$. That is,

$\mathbf{A}$ is written as the difference of two matrices, say $\mathbf{A} = \mathbf{M} - \mathbf{K}$. Then (1) can be rewritten as

$$\mathbf{Mx} = \mathbf{b} + \mathbf{Kx}.$$

Now for any initial guess $\mathbf{x}^{(0)}$ one generates the sequence of vectors $\{\mathbf{x}^{(k)}\}$ by

$$\mathbf{Mx}^{(k)} = \mathbf{b} + \mathbf{Kx}^{(k-1)}, \qquad k \geq 1. \tag{2}$$

If $\mathbf{M}$ and $\mathbf{K}$ are suitably chosen, the kth vector, $\mathbf{x}^{(k)}$, should provide an improved approximation from the previous approximation, $\mathbf{x}^{(k-1)}$. Thus for k sufficiently large, the vector $\mathbf{x}^{(k)}$ should be a good approximation to $\mathbf{x}$.

The choice of $\mathbf{M}$ and $\mathbf{K}$ is certainly not arbitrary. For the method to be of any value, we need:

1. $\mathbf{M}$ must be invertible so that $\mathbf{x}^{(k)}$ is well defined. Moreover, $\mathbf{x}^{(k)}$ should be very easy to compute. The computation of $\mathbf{x}^{(k)}$ will be fast if $\mathbf{M}$ is a triangular matrix or, better yet, a diagonal matrix.
2. The sequence $\{\mathbf{x}^{(k)}\}$ should converge to the solution $\mathbf{x} = \mathbf{A}^{-1}\mathbf{b}$ of problem (1).

Assuming the invertibility of $\mathbf{M}$, let us examine the question of convergence. We say that the sequence $\{\mathbf{x}^{(k)}\} \subset \mathbf{R}^n$ converges to $\mathbf{x}$ if

$$\lim_{k\to\infty} \|\mathbf{x} - \mathbf{x}^{(k)}\|_\infty = 0.$$

This means that each component of $\{\mathbf{x}^{(k)}\}$ converges to the corresponding component of $\mathbf{x}$, that is, for i $= 1, \ldots, n$,

$$\lim_{k\to\infty} x_i^{(k)} = x_i.$$

We have defined convergence in terms of the max norm; however, this is equivalent to convergence with respect to the Euclidean norm $\|\cdot\|$ (see Exercise 15).

To analyze the convergence of the iteration method, we subtract equation (2) from equation (1) to find the following formula for the error $\boldsymbol{\epsilon}^{(k)} = \mathbf{x} - \mathbf{x}^{(k)}$:

$$\mathbf{M}\boldsymbol{\epsilon}^{(k)} = \mathbf{K}\boldsymbol{\epsilon}^{(k-1)}.$$

Hence

$$\boldsymbol{\epsilon}^{(k)} = \mathbf{M}^{-1}\mathbf{K}\boldsymbol{\epsilon}^{(k-1)}.$$

This is a vector difference equation (see Chapter 7) whose solution is given by

$$\boldsymbol{\epsilon}^{(k)} = (\mathbf{M}^{-1}\mathbf{K})^k\boldsymbol{\epsilon}^{(0)}.$$

It is desirable for the method to be convergent for any initial guess, $\mathbf{x}^{(0)}$. We find the norm of the error in the kth iteration satisfies

$$\begin{aligned}\|\boldsymbol{\epsilon}^{(k)}\|_\infty &= \|(\mathbf{M}^{-1}\mathbf{K})^k\boldsymbol{\epsilon}^{(0)}\|_\infty \\ &\leq \|(\mathbf{M}^{-1}\mathbf{K})^k\|_\infty\|\boldsymbol{\epsilon}^{(0)}\|_\infty.\end{aligned}$$

But by the product property of matrix norms (5′) we have

$$\|(\mathbf{M}^{-1}\mathbf{K})^k\|_\infty \leq \|\mathbf{M}^{-1}\mathbf{K}\|_\infty^k.$$

Hence we find

$$\|\boldsymbol{\epsilon}^{(k)}\|_\infty \leq \|\mathbf{M}^{-1}\mathbf{K}\|_\infty^k \|\boldsymbol{\epsilon}^{(0)}\|_\infty.$$

From this it follows that $\lim_{k\to\infty} \|\boldsymbol{\epsilon}^{(k)}\|_\infty = 0$ if $\|\mathbf{M}^{-1}\mathbf{K}\|_\infty < 1$. Exactly the same analysis shows that $\lim_{k\to\infty} \|\boldsymbol{\epsilon}^{(k)}\| = 0$ if $\|\mathbf{M}^{-1}\mathbf{K}\| < 1$. Thus a sufficient condition for convergence of the iteration (2) for any $\mathbf{x}^{(0)}$ is: The norm of the iteration matrix $\mathbf{M}^{-1}\mathbf{K}$ is less than 1.

If $\|\mathbf{M}^{-1}\mathbf{K}\|_\infty = \alpha < 1$, it is possible to find an upper bound for the size of the error, $\|\mathbf{x} - \mathbf{x}^{(k)}\|_\infty$, without knowing $\mathbf{x}$. By the triangle inequality [property (4) of vector norms] we have

$$\|\mathbf{x} - \mathbf{x}^{(k)}\|_\infty \leq \|\mathbf{x}^{(k+1)} - \mathbf{x}^{(k)}\|_\infty + \|\mathbf{x} - \mathbf{x}^{(k+1)}\|_\infty.$$

But $\mathbf{x} - \mathbf{x}^{(k+1)} = \mathbf{M}^{-1}\mathbf{K}(\mathbf{x} - \mathbf{x}^{(k)})$ and hence

$$\|\mathbf{x} - \mathbf{x}^{(k+1)}\|_\infty \leq \|\mathbf{M}^{-1}\mathbf{K}\|_\infty \|\mathbf{x} - \mathbf{x}^{(k)}\|_\infty = \alpha\|\mathbf{x} - \mathbf{x}^{(k)}\|_\infty.$$

Combining these two inequalities gives

$$\|\mathbf{x} - \mathbf{x}^{(k)}\|_\infty \leq \|\mathbf{x}^{(k+1)} - \mathbf{x}^{(k)}\|_\infty + \alpha\|\mathbf{x} - \mathbf{x}^{(k)}\|_\infty,$$

or equivalently,

$$\|\mathbf{x} - \mathbf{x}^{(k)}\|_\infty \leq \frac{\|\mathbf{x}^{(k+1)} - \mathbf{x}^{(k)}\|_\infty}{1 - \alpha}. \tag{3}$$

If α is known, then (3) gives a computable upper bound for the error in the kth iterate. We illustrate the use of (3) in Example 8.9.

It is possible to establish necessary and sufficient conditions for the convergence of (2) with an arbitrary starting vector. This is the content of the following theorem.

Theorem 8.2. Suppose that $\mathbf{A} = \mathbf{M} - \mathbf{K}$ with $\mathbf{A}$ and $\mathbf{M}$ invertible. The iteration defined by (2) converges to $\mathbf{x} = \mathbf{A}^{-1}\mathbf{b}$ for any $\mathbf{x}^{(0)}$ if and only if the eigenvalue of $\mathbf{M}^{-1}\mathbf{K}$ having largest magnitude, call it λ_1, satisfies

$$|\lambda_1| < 1.$$

■

The size of the largest magnitude eigenvalue of $\mathbf{A}$, $|\lambda_1|$, is called the **spectral radius** of $\mathbf{A}$ and is denoted by $\rho(\mathbf{A})$. That is,

$$\rho(\mathbf{A}) = \max\{|\lambda| : \mathbf{A}\mathbf{x} = \lambda\mathbf{x} \text{ for some } \mathbf{x} \neq \boldsymbol{\theta}\}.$$

Example 8.8

For

$$\mathbf{A} = \begin{bmatrix} -1 & -2 & 4 \\ -2 & 2 & 2 \\ 4 & 2 & -1 \end{bmatrix},$$

the eigenvalues are $\lambda_1 = -6$, $\lambda_2 = 3$, $\lambda_3 = 3$. Thus $\rho(\mathbf{A}) = \max\{|-6|, |3|, |3|\} = 6$.

■

The proof of Theorem 8.2 is beyond the scope of this book. At the end of the section we prove the theorem for the case when $\mathbf{M}^{-1}\mathbf{K}$ is diagonalizable.

Jacobi Iteration

The Jacobi iteration results from the following choice of splitting. Choose

$$\mathbf{M} = \mathbf{diag}\,(a_{11}, a_{22}, \ldots, a_{nn})$$

and

$$\mathbf{K} = \mathbf{M} - \mathbf{A}.$$

With this splitting, the equations for the kth vector $\mathbf{x}^{(k)}$ are as follows:

$$\begin{aligned} a_{11}x_1^{(k)} &= b_1 - (a_{12}x_2^{(k-1)} + \cdots + a_{1n}x_n^{(k-1)}) \\ a_{22}x_2^{(k)} &= b_2 - (a_{21}x_1^{(k-1)} + \cdots + a_{2n}x_n^{(k-1)}) \\ &\vdots \\ a_{nn}x_n^{(k)} &= b_n - (a_{n1}x_1^{(k-1)} + \cdots + a_{n,n-1}x_{n-1}^{(k-1)}). \end{aligned}$$

If all the diagonal entries of $\mathbf{A}$ are nonzero, the vector $\mathbf{x}^{(k)}$ is easily determined. Moreover, if $\mathbf{A}$ is sparse, most of the terms on the right-hand side are zero and the calculations are fast.

Example 8.9

We illustrate Jacobi's method for the problem

$$\mathbf{A}\mathbf{x} = \begin{bmatrix} 4 & 1 & 0 \\ 1 & 4 & 1 \\ 0 & 1 & 4 \end{bmatrix} \begin{bmatrix} x_1 \\ x_2 \\ x_3 \end{bmatrix} = \begin{bmatrix} 2 \\ -6 \\ 2 \end{bmatrix} = \mathbf{b}.$$

The Jacobi iteration is given by

$$\begin{bmatrix} 4 & 0 & 0 \\ 0 & 4 & 0 \\ 0 & 0 & 4 \end{bmatrix} \mathbf{x}^{(k)} = \begin{bmatrix} 2 \\ -6 \\ 2 \end{bmatrix} - \begin{bmatrix} 0 & 1 & 0 \\ 1 & 0 & 1 \\ 0 & 1 & 0 \end{bmatrix} \mathbf{x}^{(k-1)},$$

or equivalently,

$$\mathbf{x}^{(k)} = \frac{1}{2}\begin{bmatrix} 1 \\ -3 \\ 1 \end{bmatrix} - \frac{1}{4}\begin{bmatrix} 0 & 1 & 0 \\ 1 & 0 & 1 \\ 0 & 1 & 0 \end{bmatrix} \mathbf{x}^{(k-1)}.$$

If we choose $\mathbf{x}^{(0)} = \begin{bmatrix} 1 \\ 1 \\ 1 \end{bmatrix}$, then the first four Jacobi steps give

$$\mathbf{x}^{(1)} = \begin{bmatrix} \frac{1}{4} \\ -2 \\ \frac{1}{4} \end{bmatrix}, \quad \mathbf{x}^{(2)} = \begin{bmatrix} 1 \\ -\frac{13}{8} \\ 1 \end{bmatrix}, \quad \mathbf{x}^{(3)} = \begin{bmatrix} \frac{29}{32} \\ -2 \\ \frac{29}{32} \end{bmatrix}, \quad \mathbf{x}^{(4)} = \begin{bmatrix} 1 \\ -\frac{125}{64} \\ 1 \end{bmatrix}.$$

The exact solution is $\mathbf{x} = [1 \quad -2 \quad 1]^T$. For this example it is easy to see that

$$\mathbf{M}^{-1}\mathbf{K} = -\frac{1}{4}\begin{bmatrix} 0 & 1 & 0 \\ 1 & 0 & 1 \\ 0 & 1 & 0 \end{bmatrix},$$

and hence $\|\mathbf{M}^{-1}\mathbf{K}\|_\infty = .5 < 1$. Using this we can verify the upper bound (3) for the error in the third iterate:

$$\frac{3}{32} = \|\mathbf{x} - \mathbf{x}^{(3)}\|_\infty \leq \left(\frac{1}{.5}\right)\|\mathbf{x}^{(4)} - \mathbf{x}^{(3)}\|_\infty = \frac{3}{16}.$$

■

For Example 8.9, convergence of the Jacobi iteration is guaranteed since the norm of the iteration matrix is less than 1. In the following theorem we state sufficient conditions for the convergence of Jacobi's method. First we need to introduce the notion of diagonal dominance. An $n \times n$ matrix $\mathbf{A}$ is **row diagonally dominant** if

$$|a_{ii}| > \sum_{\substack{j=1 \\ j\neq i}}^{n} |a_{ij}| \qquad \text{for } i = 1, \ldots, n.$$

In words, $\mathbf{A}$ is row diagonally dominant if, for each row, the magnitude of the diagonal entry is greater than the sum of the magnitudes of the off-diagonal entries. Note that the diagonal entries must be nonzero in a row diagonally dominant matrix.

Theorem 8.3. Suppose that $\mathbf{A}$ is row diagonally dominant; then the splitting $\mathbf{A} = \mathbf{M} - \mathbf{K}$ for Jacobi iteration satisfies $\|\mathbf{M}^{-1}\mathbf{K}\|_\infty < 1$; hence the iteration converges for any initial approximation $\mathbf{x}^{(0)}$.

Proof. The absolute row sum for the ith row of $\mathbf{M}^{-1}\mathbf{K}$ is given by

$$\sum_{\substack{j=1 \\ j\neq i}}^{n} \frac{|a_{ij}|}{|a_{ii}|},$$

which is less than 1 since $\mathbf{A}$ is diagonally dominant. Hence the maximum absolute row sum of $\mathbf{M}^{-1}\mathbf{K}$ is less than 1. That is,

$$\|\mathbf{M}^{-1}\mathbf{K}\|_\infty < 1.$$

The convergence of the method for any initial approximation is now guaranteed by Theorem 8.2.

■

We emphasize that this theorem only provides a sufficient condition for convergence. The method might also converge in some cases when $\|\mathbf{M}^{-1}\mathbf{K}\|_\infty \geq 1$.

Gauss–Seidel Iteration

The Gauss–Seidel method is a slight modification of the Jacobi method. The idea is very simple and natural: Start using each component of the new vector $\mathbf{x}^{(k)}$ as soon as it is computed. To elaborate, we write out the component equations. The first component equation remains the same:

$$a_{11}x_1^{(k)} = b_1 - (a_{12}x_2^{(k-1)} + a_{13}x_3^{(k-1)} + \cdots + a_{1n}x_n^{(k-1)}).$$

But in the next and succeeding equations we use $x_1^{(k)}$ in place of $x_1^{(k-1)}$ on the right-hand side. That is, the second component equation is

$$a_{22}x_2^{(k)} = b_2 - a_{21}x_1^{(k)} - (a_{23}x_3^{(k-1)} + \cdots + a_{2n}x_n^{(k-1)}).$$

Now that $x_1^{(k)}$ and $x_2^{(k)}$ are computed, we use them in the right-hand side of the remaining equations. The third equation is

$$a_{33}x_3^{(k)} = b_3 - a_{31}x_1^{(k)} - a_{32}x_2^{(k)} - (a_{34}x_4^{(k-1)} + \cdots + a_{3n}x_n^{(k-1)}),$$

and so on. The splitting of **A** for this method consists in choosing **M** to be the lower triangular part of **A**, and **K** = **M** − **A**. Then **K** is the negative of the strictly upper triangular part of **A**.

Example 8.10

We repeat the solution of the problem from Example 8.9 using the same initial guess and Gauss–Seidel iteration. The vector $\mathbf{x}^{(k)}$ is given by

$$\begin{bmatrix} 4 & 0 & 0 \\ 1 & 4 & 0 \\ 0 & 1 & 4 \end{bmatrix} \mathbf{x}^{(k)} = \begin{bmatrix} 2 \\ -6 \\ 2 \end{bmatrix} - \begin{bmatrix} 0 & 1 & 0 \\ 0 & 0 & 1 \\ 0 & 0 & 0 \end{bmatrix} \mathbf{x}^{(k-1)},$$

and hence the component equations are

$$\begin{aligned} 4x_1^{(k)} &= 2 - x_2^{(k-1)} \\ 4x_2^{(k)} &= -6 - x_1^{(k)} - x_3^{(k-1)} \\ 4x_3^{(k)} &= 2 - x_2^{(k)}. \end{aligned}$$

The first two Gauss–Seidel iterations give

$$\mathbf{x}^{(1)} = \begin{bmatrix} \frac{1}{4} \\ -\frac{29}{16} \\ \frac{61}{64} \end{bmatrix}, \qquad \mathbf{x}^{(2)} = \begin{bmatrix} \frac{61}{64} \\ -\frac{253}{128} \\ \frac{509}{512} \end{bmatrix}.$$

For the Jacobi method we found $\|\mathbf{x} - \mathbf{x}^{(4)}\|_\infty = \frac{3}{64}$, whereas $\|\mathbf{x} - \mathbf{x}^{(2)}\|_\infty = \frac{3}{64}$ for the Gauss–Seidel method. For this example, Gauss–Seidel produces the same accuracy with half the number of iterations as Jacobi.

■

For many applications Gauss–Seidel converges much faster than Jacobi and is generally preferred. However, there are matrices for which Jacobi converges but Gauss–Seidel does not, and vice versa. In some specific applications, most notably to the numerical solution of partial differential equations, there is a considerably faster method. This method is called successive overrelaxation (SOR) and depends on a parameter, ω. We give a very brief description of the method.

Let **D**, **L**, and **U** denote the diagonal, the strictly lower triangular part, and strictly upper triangular part of **A**, respectively. The SOR iteration is

$$[\mathbf{D} + \omega\mathbf{L}]\mathbf{x}^{(k)} = [(1 - \omega)\mathbf{D} - \omega\mathbf{U}]\mathbf{x}^{(k-1)} + \omega\mathbf{b},$$

where ω is a parameter satisfying $0 < \omega < 2$. Notice that the choice $\omega = 1$ corresponds to Gauss–Seidel. When this method is used to solve the system of equations that arises in the numerical treatment of differential equations, frequently the best choice of parameter is near 2. The optimal choice of the parameter ω can be determined in terms of the spectral radius of the iteration matrix for the Jacobi iteration. See the book of Young, which is cited in the notes and comments.

Example 8.11

For the same system of equations as in Examples 8.9 and 8.10, we illustrate SOR with $\omega = 1.03$. The sequence of vectors $\{\mathbf{x}^{(k)}\}$ satisfies

$$\begin{bmatrix} 4 & 0 & 0 \\ 1.03 & 4 & 0 \\ 0 & 1.03 & 4 \end{bmatrix}\mathbf{x}^{(k)} = -\begin{bmatrix} .12 & 1.03 & 0 \\ 0 & .12 & 1.03 \\ 0 & 0 & .12 \end{bmatrix}\mathbf{x}^{(k-1)} + \begin{bmatrix} 2.06 \\ -6.18 \\ 2.06 \end{bmatrix}.$$

For $\mathbf{x}^{(0)} = [1 \quad 1 \quad 1]^{\mathrm{T}}$, we find the following terms of the sequence to five significant digits:

$$\mathbf{x}^{(1)} = \begin{bmatrix} .22750 \\ -1.8911 \\ .97195 \end{bmatrix}, \qquad \mathbf{x}^{(2)} = \begin{bmatrix} .99513 \\ -1.9948 \\ .99950 \end{bmatrix}, \qquad \mathbf{x}^{(3)} = \begin{bmatrix} .99881 \\ -1.9997 \\ .99994 \end{bmatrix}.$$

Note that the error in $\mathbf{x}^{(2)}$ is $\|\mathbf{x} - \mathbf{x}^{(2)}\|_\infty = .0052$, whereas the error in the fourth iterate for Gauss–Seidel is $\frac{3}{64} = .0469$. Thus the sequence from SOR converges much more rapidly.

■

Optional

In what follows we demonstrate the proof of Theorem 8.2 in the case where $\mathbf{M}^{-1}\mathbf{K}$ is diagonalizable. Thus we suppose that

$$\mathbf{M}^{-1}\mathbf{K} = \mathbf{P}\mathbf{D}\mathbf{P}^{-1}$$

with $\mathbf{D} = \mathbf{diag}(\lambda_1, \ldots, \lambda_n)$. Then we find

$$(\mathbf{M}^{-1}\mathbf{K})^k = \mathbf{P}\mathbf{D}^k\mathbf{P}^{-1},$$

and hence the error in the kth iteration of (2) is given by

$$\boldsymbol{\epsilon}^{(k)} = \mathbf{P}\mathbf{D}^k\mathbf{P}^{-1}\boldsymbol{\epsilon}^{(0)}.$$

It is not difficult to show that $\lim_{k\to\infty}\|\boldsymbol{\epsilon}^{(k)}\|_\infty = 0$ if and only if $\lim_{k\to\infty}\|\mathbf{w}^{(k)}\|_\infty = 0$, where $\mathbf{w}^{(k)} = \mathbf{P}^{-1}\boldsymbol{\epsilon}^{(k)}$ (see Exercise 19). But $\mathbf{w}^{(k)}$ satisfies

$$\begin{aligned} \mathbf{w}^{(k)} &= \mathbf{D}^k\mathbf{w}^{(0)} \\ &= \sum_{j=1}^{n} \lambda_j^k w_j^{(0)}\mathbf{e}^{(j)}. \end{aligned}$$

If $|\lambda_j| < |\lambda_1|$ for $j = 2, \ldots, n$, then from

$$\mathbf{w}^{(k)} = (\lambda_1)^k w_1^{(0)}\mathbf{e}^{(1)} + (\lambda_2)^k w_2^{(0)}\mathbf{e}^{(2)} + \cdots + (\lambda_n)^k w_n^{(0)}\mathbf{e}^{(n)}$$

it is clear that for $|\lambda_1| < 1$ and any $\mathbf{w}^{(0)}$,

$$\lim_{k\to\infty}(\lambda_j)^k w_j^{(0)}\mathbf{e}^{(j)} = \boldsymbol{\theta}, \qquad 1 \le j \le n.$$

Conversely, if $\lim_{k\to\infty} \mathbf{w}^{(k)} = \boldsymbol{\theta}$ for any $\mathbf{w}^{(0)}$, then the choice

$$\mathbf{w}^{(0)} = \mathbf{e}^{(1)} = \text{first standard basis vector}$$

gives

$$\mathbf{w}^{(k)} = \lambda_1^{(k)}\mathbf{e}^{(1)}.$$

Thus $\lim_{k\to\infty} \mathbf{w}^{(k)} = \boldsymbol{\theta}$ for any $\mathbf{w}^{(0)}$ implies that $|\lambda_1| < 1$.

EXERCISES 8.2

For Exercises 1–3, solve the system of equations approximately by Jacobi iteration. Compute four iterations and round each computation to three significant digits.

1. $\begin{aligned} 3x_1 - x_2 &= -2 \\ x_1 + 4x_2 &= -5 \end{aligned}$ $\quad \mathbf{x}^{(0)} = \begin{bmatrix} 1 \\ 0 \end{bmatrix}$, solution is $\mathbf{x} = \begin{bmatrix} -1 \\ -1 \end{bmatrix}$.

2. $\begin{aligned} 4x_1 - 2x_2 + x_3 &= 3 \\ x_1 + 3x_2 - x_3 &= -4 \\ -2x_1 + 6x_3 &= 6 \end{aligned}$ $\quad \mathbf{x}^{(0)} = \begin{bmatrix} 1 \\ 0 \\ 0 \end{bmatrix}$, solution is $\mathbf{x} = \begin{bmatrix} 0 \\ -1 \\ 1 \end{bmatrix}$

3. $\begin{aligned} 8x_1 - 2x_2 + 4x_3 &= 10 \\ -2x_1 + 10x_2 + 5x_3 &= 1 \\ 2x_1 - 3x_2 + 6x_3 &= -5 \end{aligned}$ $\quad \mathbf{x}^{(0)} = \begin{bmatrix} 1 \\ 0 \\ 0 \end{bmatrix}$, solution is $\mathbf{x} = \begin{bmatrix} 2 \\ 1 \\ -1 \end{bmatrix}$

4. For the systems of Exercises 1–3, compute $\|\mathbf{M}^{-1}\mathbf{K}\|_\infty = \alpha$ using the Jacobi splitting of $\mathbf{A}$. If $\alpha < 1$, estimate the absolute error in $\mathbf{x}^{(3)}$ by the bound given in equation (3). Compare with the actual error $\|\mathbf{x} - \mathbf{x}^{(3)}\|_\infty$.

5. Repeat Exercise 1 using Gauss–Seidel iteration.

6. Repeat Exercise 2 using Gauss–Seidel iteration.

7. Repeat Exercise 3 using Gauss–Seidel iteration.

8. Repeat Exercise 4 using Gauss–Seidel iteration.

9. Determine which of the following matrices is row diagonally dominant.

(a) $\mathbf{A} = \begin{bmatrix} 2 & -1.1 & .8 \\ -3 & 4 & .5 \\ 4 & -1 & 6 \end{bmatrix}$ **(b)** $\mathbf{A} = \begin{bmatrix} -4 & 5 \\ 2 & -3 \end{bmatrix}$

(c) $\mathbf{A} = \begin{bmatrix} 7.6 & -5.1 & 2.5 \\ -1 & 4 & 3.6 \\ 2.7 & -1.3 & 4.1 \end{bmatrix}$

10. The coefficient matrix for the system

$$\begin{aligned} 2x_1 + 10x_2 - 5x_3 &= 7 \\ -2x_1 - 3x_2 + 6x_3 &= 1 \\ 8x_1 - 2x_2 + 4x_3 &= 10 \end{aligned}$$

is not row diagonally dominant. Can you rearrange the order of the equations to make the coefficient matrix diagonally dominant? How would you apply Jacobi's iteration to the system? Explain.

11. Determine the spectral radius, $\rho(\mathbf{M}^{-1}\mathbf{K})$, for the Jacobi splitting applied to the coefficient matrix of Exercise 1.

12. Repeat Exercise 11 for the Gauss–Seidel splitting.

13. Write a computer program to solve the 20×20 system below using Jacobi iteration. Terminate the iteration when $\|\mathbf{x}^{(k+1)} - \mathbf{x}^{(k)}\|_\infty < 10^{-5}\|\mathbf{x}^{(k+1)}\|_\infty$.

$$\begin{bmatrix} 2 & -1 & & & & \\ -1 & 2 & -1 & & \bigcirc & \\ & -1 & 2 & -1 & & \\ & & \cdot & \cdot & \cdot & \\ & & & \cdot & \cdot & \cdot \\ & \bigcirc & & \cdot & \cdot & \cdot \\ & & & -1 & 2 & -1 \\ & & & & -1 & 2 \end{bmatrix} \begin{bmatrix} x_1 \\ x_2 \\ x_3 \\ \cdot \\ \cdot \\ \cdot \\ x_{n-1} \\ x_n \end{bmatrix} = \begin{bmatrix} 3 \\ -4 \\ 4 \\ \cdot \\ \cdot \\ -4 \\ 4 \\ -3 \end{bmatrix}$$

The solution is $\mathbf{x} = [1 \quad -1 \quad 1 \quad -1 \quad \cdots \quad 1 \quad -1]^{\mathrm{T}}$.

14. Repeat Exercise 13 for Gauss–Seidel iteration. Compare the number of iterations required with Exercise 13.

15. Use Exercise 3 of Section 8.1 to show that

$$\lim_{k\to\infty} \|\mathbf{x} - \mathbf{x}^{(k)}\|_\infty = 0$$

if and only if

$$\lim_{k\to\infty} \|\mathbf{x} - \mathbf{x}^{(k)}\| = 0.$$

Thus convergence in the max norm is equivalent to convergence with respect to the Euclidean norm $\|\cdot\|$.

***16.** Suppose that $\mathbf{A}$ is row diagonally dominant. Use Gershgorin's theorem (see Section 6.2) to show that 0 is not an eigenvalue of $\mathbf{A}$ and hence $\mathbf{A}$ is invertible.

17. Find the spectral radius of $\mathbf{M}^{-1}\mathbf{K}$ for the Jacobi splitting applied to

$$\mathbf{A} = \begin{bmatrix} 2 & -1 & 0 \\ -1 & 2 & -1 \\ 0 & -1 & 2 \end{bmatrix}.$$

18. Repeat Exercise 17 for the Gauss–Seidel splitting.

***19.** Let $\boldsymbol{\epsilon}^{(k)}$ denote the error in the kth step of iteration (2) and set $\mathbf{w}^{(k)} = \mathbf{P}^{-1}\boldsymbol{\epsilon}^{(k)}$.

(a) Show that

$$\frac{\|\mathbf{w}^{(k)}\|_\infty}{\|\mathbf{P}^{-1}\|_\infty} \le \|\boldsymbol{\epsilon}^{(k)}\|_\infty \le \|\mathbf{P}\|_\infty \|\mathbf{w}^{(k)}\|_\infty.$$

(b) Use part (a) to conclude that $\lim_{k\to\infty} \boldsymbol{\epsilon}^{(k)} = \boldsymbol{\theta}$ if and only if $\lim_{k\to\infty} \mathbf{w}^{(k)} = \boldsymbol{\theta}$.

20. Modify your program for Exercise 14 to perform SOR. Try several values of the parameter ω on the system of Exercise 13. Were you able to find value of ω to accelerate the convergence?

8.3 Eigenvalue Calculation

In this section we consider a class of methods for the calculation of eigenvalues. These procedures, called *power methods,* are conceptually simple and reasonably effective for many problems. Throughout this section we assume that the matrix whose eigenvalues we seek is diagonalizable.

Power Method

The simple power method is iterative in nature. Starting with an initial vector $\mathbf{w}^{(0)}$, one generates the sequence

$$\mathbf{w}^{(1)} = \mathbf{A}\mathbf{w}^{(0)}$$

$$\mathbf{w}^{(2)} = \mathbf{A}\mathbf{w}^{(1)}$$

and in general,

$$\mathbf{w}^{(k)} = \mathbf{A}\mathbf{w}^{(k-1)} .$$

After the kth step we obtain $\mathbf{w}^{(k)} = \mathbf{A}^k\mathbf{w}^{(0)}$. Since $\mathbf{A}$ is diagonalizable, $\mathbf{w}^{(0)}$ can be expressed as a linear combination of the eigenvectors of $\mathbf{A}$. Denoting the eigenvalues of $\mathbf{A}$ by λ_j and corresponding eigenvectors by $\mathbf{u}^{(j)}$, we have

$$\mathbf{w}^{(0)} = c_1\mathbf{u}^{(1)} + c_2\mathbf{u}^{(2)} + \cdots + c_n\mathbf{u}^{(n)},$$

and hence $\mathbf{w}^{(k)}$ is given by

$$\mathbf{w}^{(k)} = \mathbf{A}^k\mathbf{w}^{(0)} = c_1\lambda_1^k\mathbf{u}^{(1)} + c_2\lambda_2^k\mathbf{u}^{(2)} + \cdots + c_n\lambda_n^k\mathbf{u}^{(n)}. \tag{4}$$

Now suppose that the eigenvalues of $\mathbf{A}$ are arranged in decreasing order of magnitude and that there is a *unique* eigenvalue of largest magnitude. That is,

$$|\lambda_1| > |\lambda_2| \geq \cdots \geq |\lambda_n|.$$

An eigenvalue λ_1 of $\mathbf{A}$ is called the **dominant eigenvalue** of $\mathbf{A}$ if its magnitude is strictly greater than that of any other eigenvalue of $\mathbf{A}$. Thus, if $\mathbf{A}$ is a 3×3 matrix with eigenvalues -3, 1, 3 then $\mathbf{A}$ has no dominant eigenvalue since $|-3| = |3|$.

If $c_1 \neq 0$, we see that the term $c_1\lambda_1^k\mathbf{u}^{(1)}$ will be the dominant term in the sum (4):

$$\mathbf{w}^{(k)} = \lambda_1^k\left\{c_1\mathbf{u}^{(1)} + c_2\left(\frac{\lambda_2}{\lambda_1}\right)^k\mathbf{u}^{(2)} + \cdots + c_n\left(\frac{\lambda_n}{\lambda_1}\right)^k\mathbf{u}^{(n)}\right\}.$$

The directions of the vectors $\mathbf{w}^{(k)}$ approach the direction of $\mathbf{u}^{(1)}$ as $k \to \infty$ since, for $2 \leq j \leq n$,

$$\lim_{k\to\infty} c_j\left(\frac{\lambda_j}{\lambda_1}\right)^k\mathbf{u}^{(j)} = \boldsymbol{\theta}.$$

Hence $\mathbf{w}^{(k)}/\lambda_1^k \to c_1\mathbf{u}^{(1)}$ as $k \to \infty$. Now, if $\mathbf{v}$ is any vector such that $\langle\mathbf{w}^{(k-1)}, \mathbf{v}\rangle \neq 0$, the ratio

$$\beta_k = \frac{\langle\mathbf{w}^{(k)}, \mathbf{v}\rangle}{\langle\mathbf{w}^{(k-1)}, \mathbf{v}\rangle} \tag{5}$$

approaches λ_1 as $k \to \infty$. A common choice of $\mathbf{v}$ in practice is as follows: Let i_k be the index of the component of $\mathbf{w}^{(k-1)}$ with largest magnitude, that is,

$$|w_{i_k}^{(k-1)}| = \|\mathbf{w}^{(k-1)}\|_\infty.$$

Choose $\mathbf{v}$ to be the i_kth standard basis vector: $\mathbf{v} = \mathbf{e}^{(i_k)}$. Then we get $\langle \mathbf{w}^{(k)}, \mathbf{v} \rangle = \langle \mathbf{w}^{(k)}, \mathbf{e}^{(i_k)} \rangle = w_{i_k}^{(k)}$ and hence

$$\beta_k = \frac{w_{i_k}^{(k)}}{w_{i_k}^{(k-1)}}.$$

That is, β_k is a ratio of corresponding components of successive iterates.

Another common choice of $\mathbf{v}$ is $\mathbf{v} = \mathbf{w}^{(k-1)}$, in which case the ratio

$$\gamma_k = \frac{\langle \mathbf{w}^{(k)}, \mathbf{w}^{(k-1)} \rangle}{\|\mathbf{w}^{(k-1)}\|^2} = \frac{\langle \mathbf{A}\mathbf{w}^{(k-1)}, \mathbf{w}^{(k-1)} \rangle}{\|\mathbf{w}^{(k-1)}\|^2},$$

called a **Rayleigh quotient,** provides an approximation to the dominant eigenvalue.

For sufficiently large k the ratio β_k should be a good approximation to the dominant eigenvalue of $\mathbf{A}$, and $\mathbf{w}^{(k)}$ should be a good approximation to an eigenvector corresponding to the dominant eigenvalue. The rate of convergence is governed by the factor $|\lambda_2/\lambda_1|$. The smaller this quantity, the faster the convergence of the power method.

The previous discussion gives the basic idea of the power method; however, to use the method in practice it is necessary to modify the procedure. First, we point out that the kth vector, $\mathbf{w}^{(k)}$, should not be computed by $\mathbf{w}^{(k)} = \mathbf{A}^k\mathbf{w}^{(0)}$. It is far more efficient to multiply successively by $\mathbf{A}$; that is, we compute

$$\mathbf{w}^{(1)} = \mathbf{A}\mathbf{w}^{(0)},$$

$$\mathbf{w}^{(2)} = \mathbf{A}\mathbf{w}^{(1)}, \text{ and so on.}$$

When implemented on a computer, it is necessary to introduce a scaling factor to prevent $\mathbf{w}^{(k)}$ from growing too large ($|\lambda_1| > 1$), or from becoming too small ($|\lambda_1| < 1$). This scaling prevents machine overflow or underflow, respectively. Given $\mathbf{w}^{(0)}$, the procedure is as follows:

Compute $\mathbf{z}^{(1)} = \mathbf{A}\mathbf{w}^{(0)}$ and set $\mathbf{v}^{(1)} = \mathbf{z}^{(1)}/\|\mathbf{z}^{(1)}\|_\infty$;
compute $\mathbf{z}^{(2)} = \mathbf{A}\mathbf{v}^{(1)}$ and set $\mathbf{v}^{(2)} = \mathbf{z}^{(2)}/\|\mathbf{z}^{(2)}\|_\infty$;
and so on.

In this way, the vector $\mathbf{v}^{(k)}$ is always a unit vector with respect to the max norm. In the remainder of the section, we always denote vectors in the power methods without scaling by $\mathbf{w}$, the intermediate vectors by $\mathbf{z}$, and the scaled vectors by $\mathbf{v}$.

The approximate eigenvalue is then given as follows. Let i_k be such that $|z_{i_k}^{(k-1)}| = \|\mathbf{z}^{(k-1)}\|_\infty$, and compute

$$\gamma_k = \frac{\langle \mathbf{z}^{(k)}, \mathbf{e}^{(i_k)} \rangle}{\langle \mathbf{v}^{(k-1)}, \mathbf{e}^{(i_k)} \rangle} = \frac{z_{i_k}^{(k)}}{v_{i_k}^{(k-1)}}.$$

The index i_k simply refers to the component of $\mathbf{z}^{(k-1)}$ with largest magnitude. Before giving an example of the procedure, we point out that scaling does not affect the approximate eigenvalue. That is,

$$\gamma_k = \beta_k,$$

when $\mathbf{v}$ is chosen as $\mathbf{e}^{(i_k)}$ in (5). The proof of this fact is given at the end of the section.

In the following example, we use the scaled power method to produce the first few approximations to the dominant eigenvalue.

Example 8.12

Let

$$\mathbf{A} = \begin{bmatrix} 4 & -1 & 0 \\ -1 & 5 & -1 \\ 0 & -1 & 4 \end{bmatrix}, \quad \text{and choose} \quad \mathbf{w}^{(0)} = \begin{bmatrix} 0 \\ 1 \\ 0 \end{bmatrix}.$$

Then

$$\mathbf{z}^{(1)} = \mathbf{A}\mathbf{w}^{(0)} = \begin{bmatrix} -1 \\ 5 \\ -1 \end{bmatrix}, \quad \mathbf{v}^{(1)} = \frac{1}{5}\begin{bmatrix} -1 \\ 5 \\ -1 \end{bmatrix}, \quad \text{and} \quad \gamma_1 = \frac{z_2^{(1)}}{w_2^{(0)}} = 5$$

$$\mathbf{z}^{(2)} = \mathbf{A}\mathbf{v}^{(1)} = \frac{1}{5}\begin{bmatrix} -9 \\ 27 \\ -9 \end{bmatrix}, \quad \mathbf{v}^{(2)} = \frac{1}{27}\begin{bmatrix} -9 \\ 27 \\ -9 \end{bmatrix}, \quad \text{and} \quad \gamma_2 = \frac{z_2^{(2)}}{v_2^{(1)}} = \frac{27}{5}$$

$$\mathbf{z}^{(3)} = \mathbf{A}\mathbf{v}^{(2)} = \frac{1}{3}\begin{bmatrix} -7 \\ 17 \\ -7 \end{bmatrix}, \quad \mathbf{v}^{(3)} = \frac{1}{17}\begin{bmatrix} -7 \\ 17 \\ -7 \end{bmatrix}, \quad \text{and} \quad \gamma_3 = \frac{z_2^{(3)}}{v_2^{(2)}} = \frac{17}{3}$$

$$\mathbf{z}^{(4)} = \mathbf{A}\mathbf{v}^{(3)} = \frac{1}{17}\begin{bmatrix} -45 \\ 99 \\ -45 \end{bmatrix}, \quad \mathbf{v}^{(4)} = \frac{1}{99}\begin{bmatrix} -45 \\ 99 \\ -45 \end{bmatrix}, \quad \text{and} \quad \gamma_4 = \frac{z_2^{(4)}}{v_2^{(3)}} = \frac{99}{17}. \quad .$$

Thus the sequence of approximations, $\{\gamma_k\}$, to the dominant eigenvalue is 5., 5.4, 5.666666, 5.823529, ...

■

In Example 8.12 observe that the second entry of each $\mathbf{v}^{(k)}$, for $k \geq 1$, is the largest in magnitude. In fact, this always happens for sufficiently large k. Thus, for large k, the same entry of $\mathbf{v}^{(k)}$ will have the largest magnitude. Consequently, $i_j = i_k$ for $j \geq k$, and we see that

$$|v_{i_k}^{(k)}| = \frac{|z_{i_k}^{(k)}|}{\|\mathbf{z}^{(k)}\|_\infty} = \frac{\|\mathbf{z}^{(k)}\|_\infty}{\|\mathbf{z}^{(k)}\|_\infty} = 1,$$

and hence

$$\gamma_{k+1} = \frac{z_{i_k}^{(k+1)}}{v_{i_k}^{(k)}} = z_{i_k}^{(k+1)}.$$

Thus $\lim_{k\to\infty} \|\mathbf{z}^{(k)}\|_\infty = |\lambda_1|$. Note that $1/\|\mathbf{z}^{(k)}\|_\infty$ is the scale factor which is used in the kth step of the procedure.

The matrix of Example 8.12 was examined previously in Example 6.7. The eigenvalues of $\mathbf{A}$ are $\lambda_1 = 6 > \lambda_2 = 4 > \lambda_3 = 3$ with corresponding eigenvectors

$$\mathbf{u}^{(1)} = \begin{bmatrix} -1 \\ 2 \\ -1 \end{bmatrix}, \qquad \mathbf{u}^{(2)} = \begin{bmatrix} -1 \\ 0 \\ 1 \end{bmatrix}, \qquad \mathbf{u}^{(3)} = \begin{bmatrix} 1 \\ 1 \\ 1 \end{bmatrix}.$$

The choice of $\mathbf{w}^{(0)}$ from Example 8.12 can be written as

$$\mathbf{w}^{(0)} = \tfrac{1}{3}\mathbf{u}^{(1)} + \tfrac{1}{3}\mathbf{u}^{(3)},$$

and hence the kth vector in the power method (without scaling) is given by

$$\mathbf{w}^{(k)} = \mathbf{A}^k\mathbf{w}^{(0)} = \frac{6^k\mathbf{u}^{(1)} + 3^k\mathbf{u}^{(3)}}{3}.$$

Therefore, the choice $\mathbf{v} = \mathbf{w}^{(k-1)}$ in (5) gives

$$\frac{\langle \mathbf{w}^{(k)}, \mathbf{w}^{(k-1)}\rangle}{\|\mathbf{w}^{(k-1)}\|^2} = \frac{6^{2k} + 3^{2k}}{6^{2k-1} + 3^{2k-1}}$$

$$= 6\frac{\{1 + (\frac{1}{2})^{2k}\}}{1 + (\frac{1}{2})^{2k-1}}.$$

From this it is clear that the sequence γ_k converges to the dominant eigenvalue, $\lambda_1 = 6$, of $\mathbf{A}$.

Inverse Power Method

The power method, as described previously, determines a sequence that converges to the dominant eigenvalue. The unique eigenvalue (if it exists) of smallest magnitude of an invertible matrix $\mathbf{A}$ is called the **minimal eigenvalue** of $\mathbf{A}$. Thus the reciprocal of the dominant eigenvalue of $\mathbf{A}^{-1}$ (if it exists) is the minimal eigenvalue of $\mathbf{A}$. If the minimal eigenvalue is desired, one applies the power method to $\mathbf{A}^{-1}$. Let $\mathbf{w}^{(0)}$ be an initial vector and generate a sequence of vectors by application of the power method to $\mathbf{A}^{-1}$:

$$\mathbf{z}^{(1)} = \mathbf{A}^{-1}\mathbf{w}^{(0)}, \qquad \mathbf{v}^{(1)} = \frac{\mathbf{z}^{(1)}}{\|\mathbf{z}^{(1)}\|_\infty}$$

$$\mathbf{z}^{(2)} = \mathbf{A}^{-1}\mathbf{v}^{(1)}, \qquad \mathbf{v}^{(2)} = \frac{\mathbf{z}^{(2)}}{\|\mathbf{z}^{(2)}{}_\infty\|}, \qquad \text{and so on.}$$

Then, if $\mathbf{w}^{(0)} = c_1\,\mathbf{u}^{(1)} + c_2\mathbf{u}^{(2)} + \cdots + c_n\mathbf{u}^{(n)}$ with $c_n \neq 0$, and $|\lambda_n| < |\lambda_{n-1}|$ it follows that the ratio

$$\gamma_k = \frac{z_{i_k}^{(k)}}{v_{i_k}^{(k-1)}}$$

converges to the dominant eigenvalue of $\mathbf{A}^{-1}$; that is, the ratio γ_k converges to $1/\lambda_n$

as $k \to \infty$. It is important to realize that the vectors $\mathbf{z}^{(k)}$ should not be determined using multiplication by $\mathbf{A}^{-1}$. Rather one should solve the system

$$\mathbf{A}\mathbf{z}^{(k)} = \mathbf{v}^{(k-1)}$$

by using the factorization $\mathbf{PA} = \mathbf{LU}$, which is determined once and for all prior to the first iteration.

Example 8.13

We apply the inverse power method to

$$\mathbf{A} = \begin{bmatrix} 4 & -1 & 0 \\ -1 & 5 & -1 \\ 0 & -1 & 4 \end{bmatrix}.$$

Using Gaussian elimination we find the LU factorization of $\mathbf{A}$:

$$\mathbf{A} = \mathbf{LU} = \begin{bmatrix} 1 & 0 & 0 \\ -\frac{1}{4} & 1 & 0 \\ 0 & -\frac{4}{19} & 1 \end{bmatrix} \begin{bmatrix} 4 & -1 & 0 \\ 0 & \frac{19}{4} & -1 \\ 0 & 0 & \frac{72}{19} \end{bmatrix}.$$

Let

$$\mathbf{w}^{(0)} = \begin{bmatrix} 0 \\ 1 \\ 0 \end{bmatrix};$$

then we solve $\mathbf{LUz}^{(1)} = \mathbf{w}^{(0)}$ by

$$\mathbf{Ly} = \mathbf{w}^{(0)} \quad \text{and} \quad \mathbf{Uz}^{(1)} = \mathbf{y}.$$

This gives

$$\mathbf{y} = \begin{bmatrix} 0 \\ 1 \\ \frac{4}{19} \end{bmatrix} \quad \text{and} \quad \mathbf{z}^{(1)} = \begin{bmatrix} \frac{1}{18} \\ \frac{4}{18} \\ \frac{1}{18} \end{bmatrix}.$$

Scaling gives $\|\mathbf{z}^{(1)}\|_\infty = \frac{2}{9}$, and hence

$$\mathbf{v}^{(1)} = \frac{\mathbf{z}^{(1)}}{\|\mathbf{z}^{(1)}\|_\infty} = \frac{1}{4}\begin{bmatrix} 1 \\ 4 \\ 1 \end{bmatrix}.$$

Continuing in this way, we find

$$\mathbf{z}^{(2)} = \begin{bmatrix} \frac{1}{8} \\ \frac{1}{4} \\ \frac{1}{8} \end{bmatrix}, \quad \|\mathbf{z}^{(2)}\|_\infty = \frac{1}{4}, \quad \text{and} \quad \mathbf{v}^{(2)} = \frac{1}{2}\begin{bmatrix} 1 \\ 2 \\ 1 \end{bmatrix}$$

$$\mathbf{z}^{(3)} = \begin{bmatrix} \frac{7}{36} \\ \frac{5}{18} \\ \frac{7}{36} \end{bmatrix}, \quad \|\mathbf{z}^{(3)}\|_\infty = \frac{5}{18}, \quad \text{and} \quad \mathbf{v}^{(3)} = \frac{1}{10}\begin{bmatrix} 7 \\ 10 \\ 7 \end{bmatrix}$$

$$\mathbf{z}^{(4)} = \begin{bmatrix} \frac{1}{4} \\ \frac{3}{10} \\ \frac{1}{4} \end{bmatrix}, \quad \|\mathbf{z}^{(4)}\|_\infty = \frac{3}{10}, \quad \text{and} \quad \mathbf{v}^{(4)} = \frac{1}{12}\begin{bmatrix} 10 \\ 12 \\ 10 \end{bmatrix}.$$

The sequence $\|\mathbf{z}^{(1)}\|_\infty = \frac{2}{9}$, $\|\mathbf{z}^{(2)}\|_\infty = \frac{1}{4}$, $\|\mathbf{z}^{(3)}\|_\infty = \frac{10}{36}$, $\|\mathbf{z}^{(4)}\|_\infty = \frac{3}{10}$ is converging to the reciprocal of the minimal eigenvalue of $\mathbf{A}$. That is, the sequence is converging to $\frac{1}{3}$, or equivalently the sequence $\frac{9}{2} = 4.5$, 4, $\frac{36}{10} = 3.6$, $\frac{10}{3} = 3.333$ converges to 3, the minimal eigenvalue of $\mathbf{A}$.

■

Shifted Inverse Power Method

In the power method, the rate of convergence is determined by the ratio $|\lambda_2|/|\lambda_1|$, and in the inverse power method, by the ratio $|\lambda_n|/|\lambda_{n-1}|$. In either case, we are assuming that the appropriate ratio is less than 1, that is, $\mathbf{A}$ has a dominant and minimal eigenvalue, respectively. However, if the ratio is close to 1, say .95, the convergence will be quite slow. This is a serious criticism of the power and inverse power methods. Fortunately, there is a procedure for accelerating convergence. Suppose that α is a good approximation to some eigenvalue λ_j of $\mathbf{A}$. Then the eigenvalues of $\mathbf{A} - \alpha\mathbf{I}$ are of the form $\lambda - \alpha$, where λ is an eigenvalue of $\mathbf{A}$. That is, each eigenvalue of $\mathbf{A}$ is shifted by α. It follows that the eigenvalues of $(\mathbf{A} - \alpha\mathbf{I})^{-1}$ take the form $1/(\lambda - \alpha)$. Now, if α is very close to some eigenvalue of $\mathbf{A}$, then the convergence of the inverse power method will be tremendously accelerated. For simplicity, suppose that α is very close to λ_1. Then the kth step of the shifted inverse power method solves the system

$$(\mathbf{A} - \alpha\mathbf{I})\mathbf{z}^{(k)} = \mathbf{v}^{(k-1)},$$

and sets $\mathbf{v}^{(k)} = \mathbf{z}^{(k)}/\|\mathbf{z}^{(k)}\|_\infty$. A few, say three, iterations should suffice to give an accurate approximation γ_3 to $1/(\lambda_1 - \alpha)$. Then we have

$$\lambda_1 \doteq \alpha + \frac{1}{\gamma_3}.$$

Example 8.14

We illustrate how the shifted inverse power method is applied to

$$\mathbf{A} = \begin{bmatrix} 4 & -1 & 0 \\ -1 & 5 & -1 \\ 0 & -1 & 4 \end{bmatrix}.$$

Recall that the dominant eigenvalue of $\mathbf{A}$ is $\lambda_1 = 6$. In Example 8.11 we obtained the approximation of $\frac{17}{3}$ to λ_1. We choose $\alpha = \frac{17}{3}$ and apply the inverse power method $\mathbf{A} - \frac{17}{3}\mathbf{I}$. First, we calculate the LU factorization of $\mathbf{A} - \frac{17}{3}\mathbf{I}$. We find

$$\mathbf{A} - \tfrac{17}{3}\mathbf{I} = \mathbf{LU} = \begin{bmatrix} 1 & 0 & 0 \\ \frac{3}{5} & 1 & 0 \\ 0 & 15 & 1 \end{bmatrix}\begin{bmatrix} -\frac{5}{3} & -1 & 0 \\ 0 & -\frac{1}{15} & -1 \\ 0 & 0 & \frac{40}{3} \end{bmatrix}.$$

For

$$\mathbf{w}^{(0)} = \begin{bmatrix} 1 \\ 0 \\ 0 \end{bmatrix},$$

we compute $\mathbf{LUz}^{(1)} = \mathbf{w}^{(0)}$, $\|\mathbf{z}^{(1)}\|_\infty$, and $\mathbf{v}^{(1)} = \mathbf{z}^{(1)}/\|\mathbf{z}^{(1)}\|_\infty$. Then, for $k > 1$, we compute $\mathbf{LUz}^{(k)} = \mathbf{v}^{(k-1)}$, $\|\mathbf{z}^{(k)}\|_\infty$, and $\mathbf{v}^{(k)} = \mathbf{z}^{(k)}/\|\mathbf{z}^{(k)}\|_\infty$. This yields, to four significant digits, the following:

$$\mathbf{z}^{(1)} = \begin{bmatrix} .075 \\ -1.125 \\ .675 \end{bmatrix}, \qquad \|\mathbf{z}^{(1)}\|_\infty = 1.125, \qquad \mathbf{v}^{(1)} = \begin{bmatrix} .06667 \\ -1. \\ .6000 \end{bmatrix}$$

$$\mathbf{z}^{(2)} = \begin{bmatrix} 1.535 \\ -2.625 \\ 1.215 \end{bmatrix}, \qquad \|\mathbf{z}^{(2)}\|_\infty = 2.625, \qquad \mathbf{v}^{(2)} = \begin{bmatrix} .5848 \\ -1. \\ .4629 \end{bmatrix}$$

$$\mathbf{z}^{(3)} = \begin{bmatrix} 1.476 \\ -3.045 \\ 1.554 \end{bmatrix}, \qquad \|\mathbf{z}^{(3)}\|_\infty = 3.045, \qquad \mathbf{v}^{(3)} = \begin{bmatrix} .4847 \\ -1. \\ .5088 \end{bmatrix}$$

$$\mathbf{z}^{(4)} = \begin{bmatrix} 1.509 \\ -3.000 \\ 1.491 \end{bmatrix}, \qquad \|\mathbf{z}^{(4)}\|_\infty = 3.000, \qquad \mathbf{v}^{(4)} = \begin{bmatrix} .503 \\ -1. \\ .4970 \end{bmatrix}.$$

The sequence $\|\mathbf{z}^{(1)}\|_\infty = 1.125$, $\|\mathbf{z}^{(2)}\|_\infty = 2.625$, $\|\mathbf{z}^{(3)}\|_\infty = 3.045$, $\|\mathbf{z}^{(4)}\|_\infty = 3.000, \ldots$ is converging to the dominant eigenvalue of $[\mathbf{A} - (\frac{17}{3})\mathbf{I}]^{-1}$. Thus $1/(\lambda_1 - \frac{17}{3}) \doteq \|\mathbf{z}^{(k)}\|_\infty$; in particular,

$$\lambda_1 \doteq \frac{17}{3} + \frac{1}{\|\mathbf{z}^{(4)}\|_\infty} = 6.000$$

to four significant digits. Thus the shifted inverse power method provides a much improved approximation to $\lambda_1 = 6$.

■

Optional

We demonstrate that scaling does not affect the approximate eigenvalue in the power method. That is, we show that

$$\beta_k = \gamma_k$$

when $\mathbf{v}$ is chosen as $\mathbf{e}^{(i_k)}$ in (5). Since $\mathbf{w}^{(0)}$ is used as the starting vector in the power method with or without scaling, we have

$$\mathbf{z}^{(1)} = \mathbf{w}^{(1)},$$

and hence $\mathbf{v}^{(1)} = \mathbf{w}^{(1)}/\|\mathbf{w}^{(1)}\|_\infty$. Then $\mathbf{z}^{(2)} = \mathbf{Av}^{(1)} = \mathbf{A}(\mathbf{w}^{(1)}/\|\mathbf{w}^{(1)}\|_\infty)$, and thus

$$\mathbf{z}^{(2)} = \frac{\mathbf{w}^{(2)}}{\|\mathbf{w}^{(1)}\|_\infty}.$$

Therefore,

$$\mathbf{v}^{(2)} = \frac{\mathbf{z}^{(2)}}{\|\mathbf{z}^{(2)}\|_\infty} = \frac{\mathbf{w}^{(2)}}{\|\mathbf{w}^{(2)}\|_\infty}.$$

In general, we find that $\mathbf{z}^{(k)} = \mathbf{w}^{(k)}/\|\mathbf{w}^{(k-1)}\|_\infty$ and $\mathbf{v}^{(k)} = \mathbf{w}^{(k)}/\|\mathbf{w}^{(k)}\|_\infty$. Now it is clear that the approximate eigenvalue from the scaled power method satisfies

$$\begin{aligned}\gamma_k &= \frac{z_{i_k}^k}{v_{i_k}^{(k-1)}} = \frac{(\mathbf{A}\mathbf{v}^{(k-1)})_{i_k}}{v_{i_k}^{(k-1)}} \\ &= \frac{(\mathbf{A}\mathbf{w}^{(k-1)})_{i_k}/\|\mathbf{w}^{(k-1)}\|_\infty}{w_{i_k}^{(k-1)}/\|\mathbf{w}^{k-1)}\|_\infty} \\ &= \frac{(\mathbf{A}\mathbf{w}^{(k-1)})_{i_k}}{w_{i_k}^{(k-1)}} = \beta_k.\end{aligned}$$

It is also true that scaling in the inverse and shifted inverse power methods does not affect the approximate eigenvalue calculation.

EXERCISES 8.3

1. For each of the following matrices, find the spectral radius and the dominant eigenvalue (if it exists).

(a) $\mathbf{A} = \begin{bmatrix} 10 & -2 & -2 \\ -2 & 7 & 1 \\ -2 & 1 & 7 \end{bmatrix}$ **(b)** $\mathbf{A} = \begin{bmatrix} 5 & 3 \\ -3 & -1 \end{bmatrix}$ **(c)** $\mathbf{A} = \begin{bmatrix} 2 & 4 & 7 \\ -1 & -3 & -7 \\ 4 & 4 & 5 \end{bmatrix}$

(d) $\mathbf{A} = \begin{bmatrix} 1 & 1 \\ 1 & 1 \end{bmatrix}$ **(e)** $\mathbf{A} = \begin{bmatrix} 2 & 1 & 0 \\ 0 & 2 & 0 \\ 0 & 0 & 1 \end{bmatrix}$ **(f)** $\mathbf{A} = \begin{bmatrix} -5 & 3 \\ 3 & 3 \end{bmatrix}$

2. Let $\mathbf{A} = \begin{bmatrix} 2 & 1 \\ -5 & -4 \end{bmatrix}$. Use the power method with scaling to approximate the dominant eigenvalue of $\mathbf{A}$. Calculate only three iterations and round your calculations to three significant digits. Use $\mathbf{w}^{(0)} = [1 \quad 0]^T$. Find the exact dominant eigenvalue, λ_1, and calculate the relative error:

$$\frac{|\lambda_1 - \gamma_3|}{|\lambda_1|}.$$

3. For the matrix of Exercise 2, apply the inverse power method with scaling to approximate the minimal eigenvalue of $\mathbf{A}$. Calculate three iterations and round your calculations to three significant digits. Calculate the relative error in λ_3:

$$\frac{|\lambda_3 - 1/\gamma_3|}{|\lambda_3|}.$$

4. Repeat Exercise 2 for the matrix

$$\mathbf{A} = \begin{bmatrix} 1 & -4 & 2 \\ -4 & 1 & -2 \\ 2 & -2 & -2 \end{bmatrix}.$$

5. Repeat Exercise 3 for matrix **A** of Exercise 6.

6. Let

$$\mathbf{A} = \begin{bmatrix} 4 & -1 & -1 \\ -1 & 4 & -1 \\ -1 & -1 & 4 \end{bmatrix}$$

and let $\mathbf{A} = \mathbf{M} - \mathbf{K}$ be the Jacobi splitting of **A**. Use the power method to approximate the spectral radius of $\mathbf{M}^{-1}\mathbf{K}$, $\rho(\mathbf{M}^{-1}\mathbf{K})$. Compute four iterations and round your calculations to three significant digits.

7. Repeat Exercise 6 for the Gauss–Seidel splitting of **A**.

8. The matrix

$$\mathbf{A} = \begin{bmatrix} 2 & -1 & 0 \\ -1 & 2 & -1 \\ 0 & -1 & 2 \end{bmatrix}$$

has dominant eigenvalue $\lambda_1 \doteq 3.4$. Apply the shifted inverse power method with $\alpha = 3.4$ to obtain an improved approximation to λ_1. Use $\mathbf{w}^{(0)} = [1 \quad 0 \quad 0]^T$ and compute two iterations. Round your calculations to three significant digits. Determine the exact value of λ_1 and calculate the relative error in your approximation.

9. Use the power method to compute an approximation to the dominant eigenvalue of the 3×3 Hilbert matrix:

$$\mathbf{H} = \begin{bmatrix} 1 & \frac{1}{2} & \frac{1}{3} \\ \frac{1}{2} & \frac{1}{3} & \frac{1}{4} \\ \frac{1}{3} & \frac{1}{4} & \frac{1}{5} \end{bmatrix}.$$

Do four iterations and round your computations to three digits.

10. Perform the same calculations as in Exercise 9 using the inverse power method to approximate the minimal eigenvalue of **H**.

11. Use the results of Exercises 9 and 10 to give a (approximate) lower bound for $K_\infty(\mathbf{H})$. (*Hint:* See Exercise 12 of Section 8.1.)

8.4 Overdetermined Systems and Least-Squares Solutions

Imagine a laboratory experiment in which a quantity, y, that is known to depend linearly on several variables, is measured for several values of the independent variables. To be more specific, suppose it is known that

$$y = c_1x_1 + c_2x_2 + \cdots + c_nx_n,$$

but the coefficients $c_1, c_2, \ldots, c_n$ are unknown. By selecting the values of $x_1, x_2, \ldots, x_n$ and measuring y the experimenter would like to determine the values of the coefficients $c_1, c_2, \ldots, c_n$. If n experiments are performed, there results n

equations of the type above in the n unknowns $c_1, c_2, \ldots, c_n$ which presumably has a unique solution. Unfortunately, the measurements are not exact; that is, there is some measurement error in the recorded values of y. Based on the belief that more information will enable a better determination of the unknown coefficients, the experimenter performs m such experiments with $m > n$. He or she must then solve a system of equations with more equations than unknowns. Such a system is called **overdetermined.** In this section we snow how to obtain a least-squares solution of an overdetermined system.

Consider the overdetermined linear system of equations

$$\mathbf{Au} = \mathbf{b} \tag{6}$$

with $\mathbf{A}$ an $m \times n$ matrix. By the rank-nullity theorem we have

$$n = r(\mathbf{A}) + n(\mathbf{A}),$$

and hence $r(\mathbf{A}) \leq n$ since $n(\mathbf{A}) \geq 0$. Thus, since $r(\mathbf{A}) < m$, there are vectors $\mathbf{b}$ in $\mathbf{R}^m$ such that $\mathbf{Au} = \mathbf{b}$ has no solution. In other words, the existence of a solution of an overdetermined system is not guaranteed. Nonetheless, in applications it is useful to introduce the notion of a "pseudosolution" of $\mathbf{Au} = \mathbf{b}$. We say that $\hat{\mathbf{u}}$ is a **least-squares solution** of $\mathbf{Au} = \mathbf{b}$ if

$$\|\mathbf{A}\hat{\mathbf{u}} - \mathbf{b}\| = \min_{\mathbf{u} \in \mathbf{R}^n} \|\mathbf{Au} - \mathbf{b}\|.$$

Thus among all possible $\mathbf{u}$, $\hat{\mathbf{u}}$ comes closest to satisfying the equation (6) in the sense that the **residual** $\hat{\mathbf{r}} = \mathbf{b} - \mathbf{A}\hat{\mathbf{u}}$ has the smallest Euclidean norm among all possible residuals $\mathbf{r} = \mathbf{b} - \mathbf{Au}$. The next theorem characterizes least-squares solutions.

Theorem 8.4. The vector $\hat{\mathbf{u}}$ is a least-squares solution of (6) if and only if

$$\mathbf{A}^{\mathrm{T}}\mathbf{A}\hat{\mathbf{u}} = \mathbf{A}^{\mathrm{T}}\mathbf{b}.$$

Proof. First, we suppose that $\hat{\mathbf{u}}$ is a least-squares solution and define, for an arbitrary but fixed vector $\mathbf{u}$ and scalar t, the function

$$F(t) = \|\mathbf{b} - \mathbf{A}(\hat{\mathbf{u}} - t\mathbf{u})\|^2 = \langle \mathbf{b} - \mathbf{A}(\hat{\mathbf{u}} - t\mathbf{u}), \mathbf{b} - \mathbf{A}(\hat{\mathbf{u}} - t\mathbf{u})\rangle.$$

Expanding the inner product gives

$$F(t) = \|\mathbf{b} - \mathbf{A}\hat{\mathbf{u}}\|^2 - 2t\langle \mathbf{b} - \mathbf{A}\hat{\mathbf{u}}, \mathbf{Au}\rangle + t^2\|\mathbf{Au}\|^2.$$

Notice that $F(0) = \|\mathbf{b} - \mathbf{A}\hat{\mathbf{u}}\|^2$ and hence the minimum value of F occurs at $t = 0$. From calculus it follows that $F'(0) = 0$. We find

$$F'(t) = -2\langle \mathbf{b} - \mathbf{A}\hat{\mathbf{u}}, \mathbf{Au}\rangle + 2t\|\mathbf{Au}\|^2$$

and hence

$$F'(0) = 0 = -2\langle \mathbf{b} - \mathbf{A}\hat{\mathbf{u}}, \mathbf{Au}\rangle.$$

It follows that

$$\langle \mathbf{A}^{\mathrm{T}}(\mathbf{b} - \mathbf{A}\hat{\mathbf{u}}), \mathbf{u}\rangle = 0.$$

Since $\mathbf{u}$ is arbitrary we conclude that $\mathbf{A}^{\mathrm{T}}(\mathbf{b} - \mathbf{A}\hat{\mathbf{u}}) = \boldsymbol{\theta}$, or equivalently $\mathbf{A}^{\mathrm{T}}\mathbf{A}\hat{\mathbf{u}} = \mathbf{A}^{\mathrm{T}}\mathbf{b}$.

Conversely, suppose that $\mathbf{A}^T\mathbf{A}\hat{\mathbf{u}} = \mathbf{A}^T\mathbf{b}$. Then we must show that $\hat{\mathbf{u}}$ is a least-squares solution of (6). For any $\mathbf{u}$ we have

$$\begin{aligned}\|\mathbf{b} - \mathbf{A}\mathbf{u}\|^2 &= \|\mathbf{b} - \mathbf{A}\hat{\mathbf{u}} + \mathbf{A}\hat{\mathbf{u}} - \mathbf{A}\mathbf{u}\|^2 \\ &= \|\mathbf{b} - \mathbf{A}\hat{\mathbf{u}}\|^2 + 2\langle \mathbf{b} - \mathbf{A}\hat{\mathbf{u}}, \mathbf{A}(\hat{\mathbf{u}} - \mathbf{u})\rangle + \|\mathbf{A}(\hat{\mathbf{u}} - \mathbf{u})\|^2 \\ &= \|\mathbf{b} - \mathbf{A}\hat{\mathbf{u}}\|^2 + 2\langle \mathbf{A}^T(\mathbf{b} - \mathbf{A}\hat{\mathbf{u}}), \hat{\mathbf{u}} - \mathbf{u}\rangle + \|\mathbf{A}(\hat{\mathbf{u}} - \mathbf{u})\|^2.\end{aligned}$$

But $\langle \mathbf{A}^T(\mathbf{b} - \mathbf{A}\hat{\mathbf{u}}), \hat{\mathbf{u}} - \mathbf{u}\rangle = 0$ since $\mathbf{A}^T\mathbf{A}\hat{\mathbf{u}} = \mathbf{A}^T\mathbf{b}$. Therefore, we have

$$\begin{aligned}\|\mathbf{b} - \mathbf{A}\mathbf{u}\|^2 &= \|\mathbf{b} - \mathbf{A}\hat{\mathbf{u}}\|^2 + \|\mathbf{A}(\hat{\mathbf{u}} - \mathbf{u})\|^2 \\ &\geq \|\mathbf{b} - \mathbf{A}\hat{\mathbf{u}}\|^2,\end{aligned}$$

which completes the proof.

■

The equations

$$\mathbf{A}^T\mathbf{A}\mathbf{u} = \mathbf{A}^T\mathbf{b}$$

are called the **normal equations.** Note that $\mathbf{A}^T\mathbf{A}$ is a square symmetric matrix, and solutions of the normal equations coincide with least-squares solutions of (6).

Example 8.15

Consider the overdetermined system

$$\mathbf{A}\mathbf{u} = \begin{bmatrix} 1 & -2 & 1 \\ 3 & 0 & 4 \\ 2 & 1 & -2 \\ -3 & 1 & 0 \\ 4 & -2 & -1 \end{bmatrix} \begin{bmatrix} u_1 \\ u_2 \\ u_3 \end{bmatrix} = \mathbf{b} = \begin{bmatrix} 2 \\ 1 \\ -1 \\ 0 \\ 6 \end{bmatrix}.$$

Least-squares solutions of $\mathbf{A}\mathbf{u} = \mathbf{b}$ are determined by solving the normal equations $\mathbf{A}^T\mathbf{A}\hat{\mathbf{u}} = \mathbf{A}^T\mathbf{b}$. A simple calculation gives

$$\mathbf{A}^T\mathbf{A} = \begin{bmatrix} 39 & -11 & 5 \\ -11 & 10 & -2 \\ 5 & -2 & 22 \end{bmatrix} \quad \text{and} \quad \mathbf{A}^T\mathbf{b} = \begin{bmatrix} 27 \\ -17 \\ 2 \end{bmatrix},$$

so the normal equations are

$$\begin{bmatrix} 39 & -11 & 5 \\ -11 & 10 & -2 \\ 5 & -2 & 22 \end{bmatrix} \begin{bmatrix} \hat{u}_1 \\ \hat{u}_2 \\ \hat{u}_3 \end{bmatrix} = \begin{bmatrix} 27 \\ -17 \\ 2 \end{bmatrix}.$$

■

Next we show that $\mathbf{A}^T\mathbf{A}$ is invertible if the $m \times n$ matrix $\mathbf{A}$, with $m \geq n$, has maximal rank, that is, if $r(\mathbf{A}) = n$. First we need to establish that $\mathbf{A}$ and $\mathbf{A}^T\mathbf{A}$ have the same nullspace. Clearly, $N(\mathbf{A}) \subseteq N(\mathbf{A}^T\mathbf{A})$ since

$$\mathbf{A}\mathbf{x} = \boldsymbol{\theta} \quad \text{implies that} \quad \mathbf{A}^T(\mathbf{A}\mathbf{x}) = \mathbf{A}^T\boldsymbol{\theta} = \boldsymbol{\theta}.$$

Conversely, suppose that $\mathbf{A}^T\mathbf{A}\mathbf{x} = \boldsymbol{\theta}$; then we have

$$0 = \langle \mathbf{A}^T\mathbf{A}\mathbf{x}, \mathbf{x}\rangle = \langle \mathbf{A}\mathbf{x}, \mathbf{A}\mathbf{x}\rangle = \|\mathbf{A}\mathbf{x}\|^2$$

and it follows that $\mathbf{A}\mathbf{x} = \boldsymbol{\theta}$. Thus we have shown that

$$N(\mathbf{A}) = N(\mathbf{A}^T\mathbf{A}).$$

Using this we are able to show that $\mathbf{A}^T\mathbf{A}$ is invertible if $r(\mathbf{A}) = n$. By the rank-nullity theorem, we have

$$\begin{aligned} r(\mathbf{A}^T\mathbf{A}) &= n - n(\mathbf{A}^T\mathbf{A}) \\ &= n - n(\mathbf{A}) = r(\mathbf{A}) = n. \end{aligned}$$

But an $n \times n$ matrix with rank n is invertible (see Theorem 4.10). Therefore, we have established the following.

Theorem 8.5. Suppose that $\mathbf{A}$ is an $m \times n$ matrix with $m \geq n = r(\mathbf{A})$; then $\mathbf{A}^T\mathbf{A}$ is invertible.

■

As a consequence of this theorem, it follows that there is a unique least-squares solution of (6) given by

$$\hat{\mathbf{u}} = (\mathbf{A}^T\mathbf{A})^{-1}\mathbf{A}^T\mathbf{b}$$

provided that $\mathbf{A}$ has maximal rank.

Example 8.16

Consider the system of three equations in two unknowns:

$$\begin{aligned} -u_1 + 5u_2 &= 1 \\ 2u_1 - u_2 &= 0 \\ 4u_1 + 3u_2 &= -1. \end{aligned}$$

It is not difficult to see that the coefficient matrix for this system,

$$\mathbf{A} = \begin{bmatrix} -1 & 5 \\ 2 & -1 \\ 4 & 3 \end{bmatrix},$$

has maximal rank, that is, $r(A) = 2$. Thus there is a unique least-squares solution given by

$$\hat{\mathbf{u}} = (\mathbf{A}^T\mathbf{A})^{-1}\mathbf{A}^T\begin{bmatrix} 1 \\ 0 \\ -1 \end{bmatrix}.$$

Note that $\mathbf{A}^T\mathbf{A} = \begin{bmatrix} 21 & 5 \\ 5 & 35 \end{bmatrix}$ and $\mathbf{A}^T\mathbf{b} = \begin{bmatrix} -5 \\ 2 \end{bmatrix}$. To find $\hat{\mathbf{u}}$, we solve the normal equations by Gaussian elimination:

$$\left[\begin{array}{cc|c} 21 & 5 & -5 \\ 5 & 35 & 2 \end{array}\right] \longrightarrow \left[\begin{array}{cc|c} 21 & 5 & -5 \\ 0 & \frac{710}{21} & \frac{67}{21} \end{array}\right],$$

and hence by back substitution we obtain

$$\hat{u}_2 = .094366$$

$$\hat{u}_1 = -.26056$$

to five significant digits. ■

If $\mathbf{A}$ is an $m \times n$ matrix with rank $r(\mathbf{A}) = n$, then the matrix

$$\mathbf{A}^{\dagger} = (\mathbf{A}^{\mathrm{T}}\mathbf{A})^{-1}\mathbf{A}^{\mathrm{T}}$$

is called a **pseudoinverse** of $\mathbf{A}$. This matrix shares several of the properties of the inverse. Note, however, that $\mathbf{A}^{\dagger}$ is generally not a square matrix. To illustrate, consider the property of inverses

$$\mathbf{B}^{-1}\mathbf{B} = \mathbf{I}.$$

For $\mathbf{A}^{\dagger}$, we see that

$$\begin{aligned} \mathbf{A}^{\dagger}\mathbf{A} &= [(\mathbf{A}^{\mathrm{T}}\mathbf{A})^{-1}\mathbf{A}^{\mathrm{T}}]\mathbf{A} \\ &= (\mathbf{A}^{\mathrm{T}}\mathbf{A})^{-1}(\mathbf{A}^{\mathrm{T}}\mathbf{A}) = \mathbf{I}, \end{aligned}$$

and hence $\mathbf{A}^{\dagger}$ is a left inverse for $\mathbf{A}$. However, unless $m = n$ (in which case $\mathbf{A}^{\dagger} = \mathbf{A}^{-1}$), $\mathbf{A}^{\dagger}$ is not a right inverse for $\mathbf{A}$. That is, $\mathbf{A}\mathbf{A}^{\dagger} \neq \mathbf{I}$.

Example 8.17

We find the pseudoinverse of the matrix $\mathbf{A}$ in Example 8.16. We have $\mathbf{A}^{\mathrm{T}}\mathbf{A} = \begin{bmatrix} 21 & 5 \\ 5 & 35 \end{bmatrix}$, and hence

$$(\mathbf{A}^{\mathrm{T}}\mathbf{A})^{-1} = \frac{1}{710}\begin{bmatrix} 35 & -5 \\ -5 & 21 \end{bmatrix}.$$

Finally, we get

$$\mathbf{A}^{\dagger} = (\mathbf{A}^{\mathrm{T}}\mathbf{A})^{-1}\mathbf{A}^{\mathrm{T}} = \frac{1}{710}\begin{bmatrix} -60 & 75 & 125 \\ 110 & -31 & 43 \end{bmatrix}.$$

From this, the least-squares solution found in Example 8.16 to five significant digits can be determined exactly by

$$\hat{\mathbf{u}} = \mathbf{A}^{\dagger}\mathbf{b} = \mathbf{A}^{\dagger}\begin{bmatrix} 1 \\ 0 \\ -1 \end{bmatrix} = \frac{1}{710}\begin{bmatrix} -185 \\ 67 \end{bmatrix}.$$

Note also that $\mathbf{A}^{\dagger}\mathbf{A} = \begin{bmatrix} 1 & 0 \\ 0 & 1 \end{bmatrix} = \mathbf{I}$, but

$$\mathbf{AA}^{\dagger} = \frac{1}{710}\begin{bmatrix} 610 & -230 & 90 \\ -230 & 181 & 207 \\ 90 & 207 & 629 \end{bmatrix} \neq \mathbf{I}.$$

■

For many problems, the normal equations are ill-conditioned; that is, the matrix $\mathbf{A}^{\mathrm{T}}\mathbf{A}$ has a large condition number. The following simple but illuminating example of Noble and Daniel illustrates this.*

Example 8.18

Suppose that the least-squares solutions of

$$\mathbf{Au} = \begin{bmatrix} 1 & 1 \\ 1 & 1 \\ 1 & 1.001 \end{bmatrix}\begin{bmatrix} u_1 \\ u_2 \end{bmatrix} = \mathbf{b} = \begin{bmatrix} 2 \\ 3 \\ 2 \end{bmatrix}$$

is to be calculated on a computer which retains four significant digits. The exact normal equations are

$$\begin{bmatrix} 3 & 3.001 \\ 3.001 & 3.002001 \end{bmatrix}\begin{bmatrix} \hat{u}_1 \\ \hat{u}_2 \end{bmatrix} = \begin{bmatrix} 7 \\ 7.002 \end{bmatrix}$$

whose exact solution is $\hat{u}_1 = 502.5$, $\hat{u}_2 = -500$. The computed normal equations are

$$\begin{bmatrix} 3 & 3.001 \\ 3.001 & 3.002 \end{bmatrix}\begin{bmatrix} u_1 \\ u_2 \end{bmatrix} = \begin{bmatrix} 7 \\ 7.002 \end{bmatrix},$$

whose solution obtained on our hypothetical computer is given by

$$u_1 = .3327, \qquad u_2 = 2.$$

Clearly, the computed least-squares solution bears no resemblance to the exact least-squares solution. Note that the residual of the computed solution is

$$\begin{bmatrix} 2 \\ 3 \\ 2 \end{bmatrix} - \begin{bmatrix} 1 & 1 \\ 1 & 1 \\ 1 & 1.001 \end{bmatrix}\begin{bmatrix} .3327 \\ 2 \end{bmatrix} = \begin{bmatrix} -.3327 \\ .6673 \\ -.3347 \end{bmatrix},$$

whereas the exact residual is

$$\hat{\mathbf{r}} = \mathbf{b} - \mathbf{A}\hat{\mathbf{u}} = \begin{bmatrix} -.5 \\ .5 \\ 0 \end{bmatrix}.$$

■

Because the ill-conditioned nature of $\mathbf{A}^{\mathrm{T}}\mathbf{A}$, it is usually preferable to find a least-squares solution by some means other than Gaussian elimination applied to the

*Ben Noble and James W. Daniel, *Applied Linear Algebra,* 2nd ed., © 1977, p. 333. Reprinted by permission of Prentice-Hall, Inc., Englewood Cliffs, New Jersey.

normal equations. One such method is based on the singular value decomposition of a matrix. This procedure applies to a general matrix, but it is beyond the scope of this book to present it here. See the notes and comments at the end of the chapter. Another method, which applies to maximal rank matrices, is based on the *QR* factorization, which was discussed in connection with the Gram–Schmidt process in Chapter 5.

For an $m \times n$ matrix $\mathbf{A}$ with $m > n = r(\mathbf{A})$, we recall Theorem 5.6, which says that there exists an $m \times n$ matrix $\mathbf{Q}$ having orthonormal columns and an $n \times n$ invertible upper triangular matrix $\mathbf{R}$ such that

$$\mathbf{A} = \mathbf{QR}.$$

Then the normal equation for $\mathbf{Au} = \mathbf{b}$ can be written as

$$(\mathbf{QR})^{\mathrm{T}}(\mathbf{QR})\hat{\mathbf{u}} = (\mathbf{QR})^{\mathrm{T}}\mathbf{b}.$$

Using the reverse order rule for transposes gives

$$(\mathbf{QR})^{\mathrm{T}}(\mathbf{QR}) = \mathbf{R}^{\mathrm{T}}\mathbf{Q}^{\mathrm{T}}\mathbf{QR},$$

but $\mathbf{Q}^{\mathrm{T}}\mathbf{Q} = \mathbf{I}$ since $\mathbf{Q}$ has orthonormal columns, and hence the normal equation reduces to

$$\mathbf{R}^{\mathrm{T}}\mathbf{R}\hat{\mathbf{u}} = \mathbf{R}^{\mathrm{T}}\mathbf{Q}^{\mathrm{T}}\mathbf{b}.$$

Finally, we note that $\mathbf{R}^{\mathrm{T}}$ is invertible since $(\mathbf{R}^{\mathrm{T}})^{-1} = (\mathbf{R}^{-1})^{\mathrm{T}}$, and hence

$$\mathbf{R}\hat{\mathbf{u}} = \mathbf{Q}^{\mathrm{T}}\mathbf{b}. \tag{7}$$

Thus, solving the normal equations is equivalent to solving equation (7). But (7) is easy to solve by back substitution as $\mathbf{R}$ is an upper triangular matrix. We emphasize that $\mathbf{A}$ must have maximal rank for the determination of $\mathbf{Q}$ and $\mathbf{R}$.

Most of the work involved in this procedure is in the determination of $\mathbf{Q}$ and $\mathbf{R}$, as will be evident in our next example.

Example 8.19

We find the least-squares solution of

$$\begin{bmatrix} 1 & -2 \\ 1 & 1 \\ -2 & -2 \end{bmatrix}\begin{bmatrix} u_1 \\ u_2 \end{bmatrix} = \begin{bmatrix} 1 \\ \frac{1}{2} \\ -\frac{4}{3} \end{bmatrix}$$

by using the *QR* factorization of the coefficient matrix, $\mathbf{A}$. We denote the columns of $\mathbf{A}$ by $\mathbf{w}^{(1)}$ and $\mathbf{w}^{(2)}$, respectively, and apply the Gram–Schmidt process to find an orthonormal basis for $S[\mathbf{w}^{(1)}, \mathbf{w}^{(2)}]$. First we normalize $\mathbf{w}^{(1)}$:

$$\mathbf{q}^{(1)} = \frac{\mathbf{w}^{(1)}}{\|\mathbf{w}^{(1)}\|} = \frac{1}{\sqrt{6}}\mathbf{w}^{(1)}.$$

Then we calculate

$$\mathbf{y} = \mathbf{w}^{(2)} - \langle \mathbf{w}^{(2)}, \mathbf{q}^{(1)} \rangle \mathbf{q}^{(1)}$$

$$= \begin{bmatrix} -2 \\ 1 \\ -2 \end{bmatrix} - \frac{3}{\sqrt{6}} \frac{1}{\sqrt{6}} \begin{bmatrix} 1 \\ 1 \\ -2 \end{bmatrix}$$

$$= \frac{1}{2} \begin{bmatrix} -5 \\ 1 \\ -2 \end{bmatrix}.$$

Normalization of $\mathbf{y}$ gives

$$\mathbf{q}^{(2)} = \frac{\mathbf{y}}{\|\mathbf{y}\|} = \frac{1}{\sqrt{30}} \begin{bmatrix} -5 \\ 1 \\ -2 \end{bmatrix}.$$

By solving for $\mathbf{w}^{(1)}$ and $\mathbf{w}^{(2)}$ in terms of $\mathbf{q}^{(1)}$ and $\mathbf{q}^{(2)}$, we find

$$\mathbf{w}^{(1)} = \sqrt{6}\,\mathbf{q}^{(1)}$$

$$\mathbf{w}^{(2)} = \frac{\sqrt{6}}{2}\mathbf{q}^{(1)} + \frac{\sqrt{30}}{2}\mathbf{q}^{(2)},$$

and hence

$$\mathbf{A} = \begin{bmatrix} 1 & -2 \\ 1 & 1 \\ -2 & -2 \end{bmatrix} = \mathbf{QR} = \begin{bmatrix} 1/\sqrt{6} & -5/\sqrt{30} \\ 1/\sqrt{6} & 1/\sqrt{30} \\ -2/\sqrt{6} & -2/\sqrt{30} \end{bmatrix} \begin{bmatrix} \sqrt{6} & \sqrt{6}/2 \\ 0 & \sqrt{30}/2 \end{bmatrix}.$$

From this, we must solve $\mathbf{R}\hat{\mathbf{u}} = \mathbf{Q}^{\mathrm{T}}\mathbf{b}$:

$$\begin{bmatrix} \sqrt{6} & \dfrac{\sqrt{6}}{2} \\ 0 & \dfrac{\sqrt{30}}{2} \end{bmatrix} \begin{bmatrix} \hat{u}_1 \\ \hat{u}_2 \end{bmatrix} = \begin{bmatrix} \dfrac{37}{6\sqrt{6}} \\ \dfrac{-71}{6\sqrt{30}} \end{bmatrix}.$$

Using back substitution, we find the unique least-squares solution

$$\hat{u}_2 = -\frac{71}{90} = .7888$$

$$\hat{u}_1 = \frac{37}{36} + \frac{71}{180} = 1.4222.$$

■

Data Fitting

Now we apply our methods to the problem of data fitting, which was discussed briefly in the opening paragraph of this section. The simplest and one of the more common problems in data fitting is described as follows. Suppose that we are given several experimentally determined values of two variables, say x and y, which are tabulated as follows:

x	y
x_1	y_1
x_2	y_2
$\cdot$	$\cdot$
$\cdot$	$\cdot$
$\cdot$	$\cdot$
x_n	y_n

Then we try to obtain a functional relationship $y = f(x)$ between the variables. Based on a plot of the data or on theoretical considerations, we choose a form for f which involves some unknown coefficients. For example, one may choose a straight line $f(x) = mx + b$, or a quadratic polynomial $f(x) = a + bx + cx^2$, or an exponential function $f(x) = ae^{bx}$. See Figure 8.1 for a typical sketch of these choices and the corresponding data.

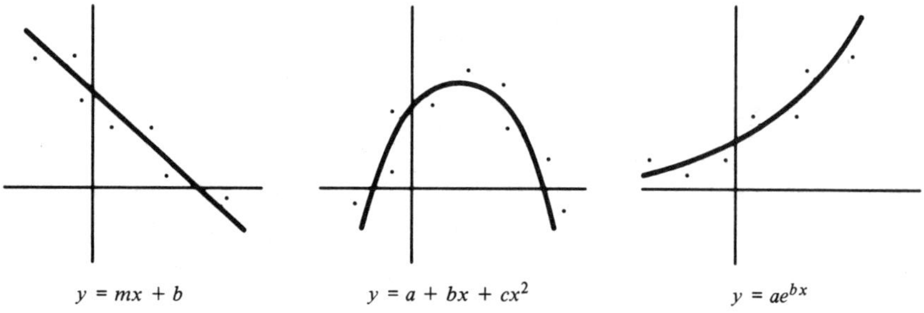

Figure 8.1

Due to inexact measurement, x or y may be in error and it is usually impossible to find a function, f, of the desired form, whose graph will pass through all the data points. Therefore, we choose that f, by determination of its coefficients, which "best fits" the data.

To be more precise, consider a straight-line fit of data, that is, $f(x) = mx + b$. In the absence of measurement error, each of the n data points would lie on the line. Hence the unknown coefficients, m and b, would satisfy the equations

$$\begin{aligned} y_1 &= mx_1 + b \\ y_2 &= mx_2 + b \\ &\vdots \\ y_n &= mx_n + b. \end{aligned}$$

In matrix form, these equations are written as

$$\mathbf{Au} = \begin{bmatrix} x_1 & 1 \\ x_2 & 1 \\ \vdots & \vdots \\ x_n & 1 \end{bmatrix} \begin{bmatrix} m \\ b \end{bmatrix} = \begin{bmatrix} y_1 \\ y_2 \\ \vdots \\ y_n \end{bmatrix} = \mathbf{y}.$$

Because of errors in measurement, the data points (x_i, y_i) will generally be scattered and will not lie on a line. Hence the equation $\mathbf{Au} = \mathbf{y}$ will have no solution. To determine the "best fit" by a striaght line, we choose the coefficient vector $\mathbf{u} = [m \quad b]^T$ to be the least-squares solution of $\mathbf{Au} = \mathbf{y}$. That is, we choose $\hat{\mathbf{u}}$ to satisfy the normal equations

$$\mathbf{A}^T\mathbf{A}\hat{\mathbf{u}} = \mathbf{A}^T\mathbf{y}.$$

This choice, $\hat{\mathbf{u}} = (\mathbf{A}^T\mathbf{A})^{-1}\mathbf{A}^T\mathbf{y}$ minimizes

$$\|\mathbf{y} - \mathbf{Au}\| = \left\{\sum_{i=1}^{n} [y_i - (mx_i + b)]^2\right\}^{1/2}$$

over all possible choices of m and b. The quantity

$$r_i = |y_i - (mx_i + b)|$$

is called the ith residual and is the vertical distance (see Figure 8.2) between the point (x_i, y_i) and the line $y = mx + b$.

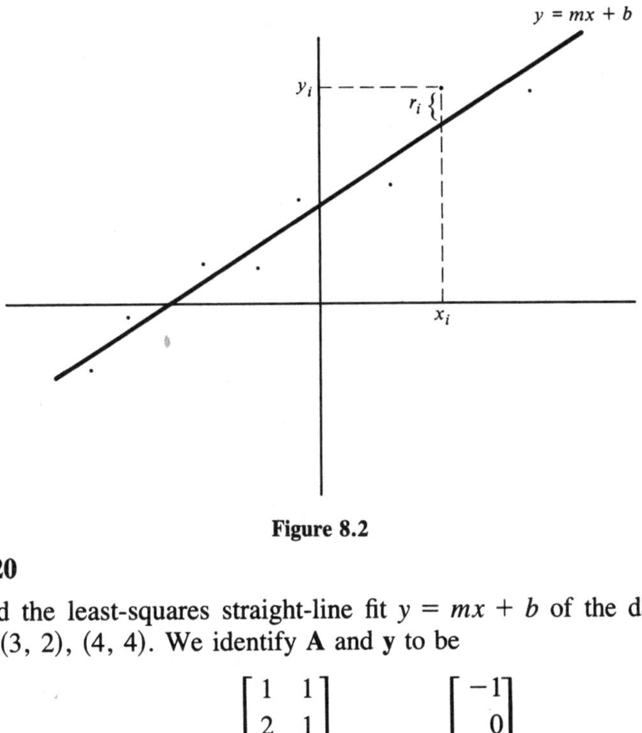

Figure 8.2

Example 8.20

We find the least-squares straight-line fit $y = mx + b$ of the data: $(1, -1)$, $(2, 0)$, $(3, 2)$, $(4, 4)$. We identify $\mathbf{A}$ and $\mathbf{y}$ to be

$$\mathbf{A} = \begin{bmatrix} 1 & 1 \\ 2 & 1 \\ 3 & 1 \\ 4 & 1 \end{bmatrix}, \qquad \mathbf{y} = \begin{bmatrix} -1 \\ 0 \\ 2 \\ 4 \end{bmatrix}.$$

Then we calculate

$$\mathbf{A}^T\mathbf{A} = \begin{bmatrix} 30 & 10 \\ 10 & 4 \end{bmatrix}, \qquad \mathbf{A}^T\mathbf{y} = \begin{bmatrix} 21 \\ 5 \end{bmatrix},$$

and hence $\hat{\mathbf{u}} = [\hat{m} \quad \hat{b}]^{\mathrm{T}}$ satisfies

$$\begin{bmatrix} 30 & 10 \\ 10 & 5 \end{bmatrix} \begin{bmatrix} \hat{m} \\ \hat{b} \end{bmatrix} = \begin{bmatrix} 21 \\ 5 \end{bmatrix}.$$

A simple calculation gives $\hat{m} = 1.1$ and $\hat{b} = -1.2$. Thus the unique least-squares straight-line fit of the data is

$$y = 1.1x - 1.2.$$

■

In the next example, we find the least-squares quadratic polynomial fit of given data.

Example 8.21

It is known that the following data are related by a quadratic polynomial $y = a + bx + cx^2$.

x	y
−3	46
−1	11
0	3
2	4
4	26

Due to measurement error, no quadratic will exactly fit the data, and hence we find a least-squares fit. That is, we find a least-squares solution of the over-determined system

$$\begin{aligned} a - 3b + 9c &= 46 \\ a - b + c &= 11 \\ a &= 3 \\ a + 2b + 4c &= 4 \\ a + 4b + 16c &= 26. \end{aligned}$$

The matrix

$$\mathbf{A} = \begin{bmatrix} 1 & -3 & 9 \\ 1 & -1 & 1 \\ 1 & 0 & 0 \\ 1 & 2 & 4 \\ 1 & 4 & 16 \end{bmatrix}$$

has maximal rank, and hence there is a unique least-squares solution given by the solution of the normal equation

$$\mathbf{A}^{\mathrm{T}}\mathbf{A}\hat{\mathbf{u}} = \mathbf{A}^{\mathrm{T}}\mathbf{y},$$

where $\hat{\mathbf{u}} = [\hat{a} \quad \hat{b} \quad \hat{c}]^{\mathrm{T}}$. The normal equation is

$$\begin{bmatrix} 5 & 2 & 30 \\ 2 & 30 & 44 \\ 30 & 44 & 354 \end{bmatrix} \begin{bmatrix} \hat{a} \\ \hat{b} \\ \hat{c} \end{bmatrix} = \begin{bmatrix} 90 \\ -37 \\ 857 \end{bmatrix},$$

and it follows that the least-squares coefficients are (to five significant digits) $\hat{a} = 3.1075$, $\hat{b} = -5.6315$, and $\hat{c} = 2.8575$. Therefore, the least-squares quadratic fit of the data is given by

$$y = 3.1075 - 5.6315x + 2.8575x^2.$$

■

In our final example, we show how to find an exponential fit of data by introducing a change of variables that reduces the problem to a straight-line least-squares problem.

Example 8.22

Suppose it is known that the data

x	y
$-.5$	1.7
.5	9.4
2.0	119.9
3.0	656.1

are related (in the absence of error) by $y = ae^{bx}$, for some unknown coefficients a and b. By taking natural logarithms, we find

$$\ln(y) = \ln(a) + bx.$$

Thus if we let $w = \ln(y)$ and $c = \ln(a)$, then

$$w = bx + c,$$

which suggests that we find a straight-line fit of data:

x	$w = \ln(y)$
$-.5$	.5306
.5	2.241
2.0	4.787
3.0	6.486

where, for simplicity, we have rounded the logarithms to four significant digits. For $\hat{\mathbf{u}} = [\hat{b} \quad \hat{c}]^{\mathrm{T}}$ the normal equation is

$$\begin{bmatrix} \frac{27}{2} & 5 \\ 5 & 4 \end{bmatrix} \begin{bmatrix} \hat{b} \\ \hat{c} \end{bmatrix} = \begin{bmatrix} 29.89 \\ 14.04 \end{bmatrix}$$

and we find $\hat{c} = 1.385$ and $\hat{b} = 1.701$ to four significant digits. Finally, we recall that $c = \ln(a)$ and hence $a = 3.995$. The resultant exponential fit of data is

$$y = 3.995e^{1.701x}$$

■

EXERCISES 8.4

In Exercises 1–4, find the normal equations for $\mathbf{Au} = \mathbf{b}$.

1. $\mathbf{A} = \begin{bmatrix} 5 & -2 & 1 \\ 2 & 3 & 3 \\ -2 & 4 & 1 \\ 4 & -3 & 6 \end{bmatrix}$, $\mathbf{b} = \begin{bmatrix} 7 \\ -2 \\ 3 \\ 5 \end{bmatrix}$

2. $\mathbf{A} = \begin{bmatrix} -1 & 0 & 2 \\ 3 & 4 & -2 \end{bmatrix}$, $\mathbf{b} = \begin{bmatrix} 3 \\ 6 \end{bmatrix}$

3. $\mathbf{A} = \begin{bmatrix} 1 & 3 \\ 1 & 4 \\ 1 & 6 \\ 1 & 8 \end{bmatrix}$, $\mathbf{b} = \begin{bmatrix} -2 \\ 0 \\ 5 \\ 10 \end{bmatrix}$

4. $\mathbf{A} = \begin{bmatrix} -4 & 3 & 5 & 1 \\ 2 & 5 & 1 & 0 \\ 6 & 2 & 2 & 4 \\ 2 & -1 & 3 & 5 \end{bmatrix}$, $\mathbf{b} = \begin{bmatrix} -1 \\ 1 \\ -1 \\ 1 \end{bmatrix}$

5. In Exercises 1–4, which of the systems $\mathbf{Au} = \mathbf{b}$ is overdetermined?
6. Determine if the matrix $\mathbf{A}$ of Exercise 1 has maximal rank.
7. Determine if the matrix $\mathbf{A}$ of Exercise 2 has maximal rank.
8. Determine if the matrix $\mathbf{A}$ of Exercise 3 has maximal rank.
9. Determine if the matrix $\mathbf{A}$ of Exercise 4 has maximal rank.
10. For each of the systems in Exercises 1–4, find all least-squares solutions by solving the normal equations. Does each have a unique least-squares solution?
11. Find the pseudoinverse of

$$\mathbf{A} = \begin{bmatrix} 3 & -2 \\ 1 & 4 \\ 2 & 5 \end{bmatrix}.$$

12. Suppose that $\mathbf{A}$ is an invertible matrix. Show that

$$\mathbf{A}^{\dagger} = \mathbf{A}^{-1}.$$

13. Show that $\mathbf{AA}^{\dagger}\mathbf{y} = \mathbf{y}$ for all $\mathbf{y} \in R(\mathbf{A})$.
14. Use the QR factorization to solve the normal equations for $\mathbf{Au} = \mathbf{b}$.

$$\mathbf{A} = \begin{bmatrix} -2 & -1 \\ 1 & 2 \\ 3 & 1 \end{bmatrix}, \qquad \mathbf{b} = \begin{bmatrix} 2 \\ 5 \\ 6 \end{bmatrix}$$

15. Compute the least-squares straight-line fit of the following data.

x	y
−1	.8
−.5	2.1
0	3.2
.5	3.9
1	5.1

16. Compute an exponential fit, $y = ae^{bx}$, of the following data.

x	y
−.5	.554
0	1.509
.5	4.086
1.0	11.3
1.5	31.07

Carefully plot the data and your exponential fit of data.

17. Find the quadratic polynomial that best fits the following data.

x	y
1.1	−.17
1.2	.39
1.3	.98
1.4	2.44
1.5	3.81

GLOSSARY

absolute error An approximation $\hat{\mathbf{x}}$ to a vector $\mathbf{x}$ has an absolute error given by $\|\mathbf{x} - \hat{\mathbf{x}}\|$.

condition number An invertible matrix $\mathbf{A}$ has condition number $K_\infty(\mathbf{A}) = \|\mathbf{A}^{-1}\|_\infty \|\mathbf{A}\|_\infty$.

dominant eigenvalue $\mathbf{A}$ has dominant eigenvalue λ_1 if $|\lambda_1| > |\lambda|$ for any other eigenvalue λ of $\mathbf{A}$.

ill-conditioned matrix A square matrix that has a large condition number.

matrix norm A real-valued function that measures the "size" of square matrices.

max norm A vector $\mathbf{x} \in \mathbf{R}^n$ has max norm $\|\mathbf{x}\|_\infty = \max_{1 \le i \le n} |x_i|$.

normal equations For $\mathbf{Au} = \mathbf{b}$, the normal equations are given in matrix form by $\mathbf{A}^T\mathbf{A}\hat{\mathbf{u}} = \mathbf{A}^T\mathbf{b}$.

overdetermined system A linear system of equations having more equations than unknowns.

pivoting strategy A means of selecting the pivotal rows in Gaussian elimination.

pseudoinverse A maximal rank matrix $\mathbf{A}$ has pseudoinverse given by $\mathbf{A}^{\dagger} = (\mathbf{A}^{T}\mathbf{A})^{-1}\mathbf{A}^{T}$.

relative error The relative error in $\hat{\mathbf{x}}$ as an approximation to the vector $\mathbf{x}$ is given by $\|\mathbf{x} - \hat{\mathbf{x}}\|/\|\mathbf{x}\|$.

row diagonal dominance A square matrix is row diagonally dominant if, for each row, the magnitude of the diagonal entry is larger than the sum of the magnitudes of the off-diagonal entries.

spectral radius $\rho(\mathbf{A}) = \max\{|\lambda| : \mathbf{A}\mathbf{x} = \lambda\mathbf{x}\}$.

NOTES AND COMMENTS

There are several vector and matrix norms which we have not developed in this book. Many of these are primarily of theoretical interest and others are useful for numerical linear algebra. We refer the reader to the book, *Numerical Analysis,* 2nd edition (Reading, Mass.: Addison-Wesley, 1982) by L. W. Johnson and R. D. Riess. This book also presents a readable discussion of pivoting strategies. The standard reference for round-off errors in the elimination method is by J. H. Wilkinson, *Rounding Errors in Algebraic Processes* (Englewood Cliffs, N.J.: Prentice-Hall, 1963).

The method of successive overrelaxation (SOR) was analyzed by David Young in his thesis (1950). He developed, for a class of matrices, a formula connecting the eigenvalues of the iteration matrix for SOR to the eigenvalues of the iteration matrix for Jacobi iteration. For a complete discussion, see the book by Young; *Iterative Solution for Large Linear Systems* (New York: Academic Press, 1971). Another book devoted to iterative methods is by R. Varga; *Matrix Iterative Analysis* (Englewood Cliffs, N.J.: Prentice-Hall, 1962).

Our discussion of numerical methods for the eigenproblem has been limited to an introductory presentation of the family of power methods. The powerful *QR* algorithm is presented in the book by A. Ralston and P. Rabinowitz: *A First Course in Numerical Analysis,* 2nd edition (New York: McGraw-Hill, 1978). This area of numerical linear algebra is well developed and we refer the reader to the classic work by J. H. Wilkinson, *The Algebraic Eigenvalue Problem* (Oxford: Oxford University Press, 1965), and the more recent work of J. H. Wilkinson and C. Reinsch, *Linear Algebra: Handbook for Automatic Computation,* Vol. II (New York: Springer-Verlag, 1971).

For a general survey of many of the important practical considerations in numerical linear algebra, we suggest the book by J. R. Rice, *Matrix Computations and Mathematical Software* (New York: McGraw-Hill, 1981).

The method of least squares has important applications in the field of statistics, where it is usually referred to as linear regression. The idea of least squares can be viewed as a special case of Fourier analysis. For an elementary introduction to this point of view, see the book by C. Rorres and H. Anton: *Applications of Linear Algebra,* 2nd edition (New York: Wiley, 1979). For an elementary discussion of least-squares methods in numerical analysis, see the book by J. T. King, *Introduction to Numerical Computation* (New York: McGraw-Hill, 1984).

Answers to Selected Odd-Numbered Exercises

Chapter 1

Exercises 1.1, page 5

1. $\begin{bmatrix} -1 & 0 & 7 \\ 2 & -3 & 1 \\ 0 & 1 & -2 \end{bmatrix} \begin{bmatrix} x_1 \\ x_2 \\ x_3 \end{bmatrix} = \begin{bmatrix} 1 \\ 7 \\ -4 \end{bmatrix}$

5. $x_1 = 2, x_2 = -\frac{5}{2}$

7. $x_4 = 4, x_3 = 3, x_2 = 2, x_1 = 1$

9. $x_2 = 1$, by the third equation. The second equation then gives $x_1 = 1$. But then the first equation gives $4 - 2 = 3$. Impossible!

Exercises 1.2, page 16

1. $x = -2, y = 4, z = 5, w = 2$

3. $\begin{bmatrix} 2 & 1 & 3 \\ 4 & 0 & 7 \\ 16 & 2 & 0 \end{bmatrix}$

5. $\mathbf{X} = \begin{bmatrix} 2 & 14 \\ 4 & -2 \end{bmatrix}$

7. (a) $\begin{bmatrix} -3 & -16 \\ 5 & -3 \end{bmatrix}$ (b) $\begin{bmatrix} 12 & -5 \\ 8 & -2 \end{bmatrix}$ (c) same as (b)

9. (a) $\begin{bmatrix} -7 & -3 \\ 4 & 0 \end{bmatrix}$ (b) $\begin{bmatrix} 2 & -3 \\ 10 & -9 \end{bmatrix}$ (c) and (d) $\begin{bmatrix} -31 & -14 \\ 16 & 8 \end{bmatrix}$

(e) and (f) $\begin{bmatrix} -12 & -7 \\ -12 & -8 \end{bmatrix}$ (g) and (h) $\begin{bmatrix} -2 & 1 \\ 20 & 8 \end{bmatrix}$

11. $\begin{bmatrix} 1 & 2 & 3 \\ 4 & 5 & 6 \\ 7 & 8 & 9 \end{bmatrix}$

13. $\mathbf{AB} = \mathbf{BA} = \begin{bmatrix} \alpha\gamma - \delta\beta & \alpha\delta + \beta\gamma \\ -\beta\gamma - \alpha\delta & -\beta\delta + \alpha\gamma \end{bmatrix}$

17. $\mathbf{A} = \begin{bmatrix} 1 & 0 \\ 1 & 2 \end{bmatrix}$

19. (a) $\begin{bmatrix} x_1 + 6x_2 + 3x_3 \\ -x_1 \qquad + 5x_3 \\ 2x_2 + 3x_3 \end{bmatrix}$ (b) $\begin{bmatrix} 2a_{11} + a_{12} + 3a_{13} \\ 2a_{21} + a_{22} + 3a_{23} \end{bmatrix}$

23. (a) $mk(n + l), nl(m + k)$
(b) 20,000 vs. 200

25. $\begin{bmatrix} \pm 1 & 0 \\ 0 & \pm 1 \end{bmatrix}$, $\begin{bmatrix} \pm 1 & c \\ 0 & \mp 1 \end{bmatrix}$ any c, $\begin{bmatrix} -d & \dfrac{1 - d^2}{c} \\ c & d \end{bmatrix}$ for $c \neq 0$, any d.

Exercises 1.3, page 26

5. (d) $(\mathbf{A}+\mathbf{B})^2 = \mathbf{A}^2 + \mathbf{AB} + \mathbf{BA} + \mathbf{B}^2$
$(\mathbf{A}-\mathbf{B})^2 = \mathbf{A}^2 - \mathbf{AB} - \mathbf{BA} + \mathbf{B}^2$

15. No **17.** $\mathbf{AB} = \begin{bmatrix} 2 & -3 & 13 & 10 \\ 0 & -4 & 14 & 5 \\ 0 & 0 & 4 & 3 \\ 0 & 0 & 0 & 3 \end{bmatrix}$ $\mathbf{BA} = \begin{bmatrix} 2 & 0 & 7 & 15 \\ 0 & -4 & 5 & 6 \\ 0 & 0 & 4 & 5 \\ 0 & 0 & 0 & 3 \end{bmatrix}$

19. $\mathbf{D}^2 = \begin{bmatrix} 1 & 0 & 0 \\ 0 & 4 & 0 \\ 0 & 0 & 9 \end{bmatrix}$ $\mathbf{D}^3 = \begin{bmatrix} 1 & 0 & 0 \\ 0 & 8 & 0 \\ 0 & 0 & 27 \end{bmatrix}$ **23.** (b) $\begin{bmatrix} 1 & -1 & 0 \\ 0 & 1 & -1 \\ 0 & 0 & 1 \end{bmatrix}$

27. $\mathbf{B}^T = (\mathbf{A}^T\mathbf{A})^T = \mathbf{A}^T(\mathbf{A}^T)^T = \mathbf{A}^T\mathbf{A} = \mathbf{B}$, similarly for $\mathbf{C}$

29. $\mathbf{B} = \mathbf{B}^T$, $\mathbf{C} = (\mathbf{A} - \mathbf{A}^T)/2$, $\mathbf{C}^T = -\mathbf{C}$

Exercises 1.4, page 38

1. (a) $\frac{1}{.47}\begin{bmatrix} 10 \\ 21 \end{bmatrix}$ (b) $\frac{1}{.47}\begin{bmatrix} .975 \\ 1.05 \end{bmatrix}$

3. (a) $\begin{bmatrix} 1 & -6 \\ 0 & 1 \end{bmatrix}\begin{bmatrix} 1 & 0 \\ \frac{-1}{5} & 1 \end{bmatrix}\begin{bmatrix} 1 & -4 \\ 0 & 1 \end{bmatrix}\begin{bmatrix} 1 & 0 \\ \frac{-1}{3} & 1 \end{bmatrix}\begin{bmatrix} 1 & -2 \\ 0 & 1 \end{bmatrix} = \frac{1}{15}\begin{bmatrix} 107 & -436 \\ -12 & 51 \end{bmatrix}$

(b) $\frac{1}{15}\begin{bmatrix} 26.8 \\ 1.2 \end{bmatrix}$

5. (a) $\begin{bmatrix} 0 & 1 & 0 & 0 & 0 \\ 1 & 0 & 1 & 0 & 1 \\ 0 & 1 & 0 & 1 & 0 \\ 0 & 0 & 0 & 0 & 1 \\ 0 & 1 & 0 & 0 & 0 \end{bmatrix}$ (b) $\begin{bmatrix} 0 & 1 & 0 & 1 \\ 1 & 0 & 1 & 0 \\ 0 & 1 & 0 & 1 \\ 1 & 1 & 1 & 0 \end{bmatrix}$

Chapter 2

Exercises 2.1, page 45

1. $\begin{bmatrix} 2 \\ 3 \end{bmatrix}$ **3.** $.5\begin{bmatrix} 1 \\ 1 \\ 1 \end{bmatrix}$ **5.** $\begin{bmatrix} -1 \\ 0 \\ -1 \\ 1 \end{bmatrix}$

Exercises 2.2, page 50

1. (a) $\begin{bmatrix} 2 & 1 & 1 & 1 \\ 4 & 3 & 6 & -3 \\ 1 & -1 & 2 & 0 \end{bmatrix}$ (b) $\begin{bmatrix} 2 & 1 & 1 & 1 \\ 1 & -1 & 2 & 0 \\ 6 & 1 & 10 & -3 \end{bmatrix}$ (c) $\begin{bmatrix} 2 & 1 & 1 & 1 \\ 4 & 3 & 6 & -3 \\ 9 & 5 & 14 & -6 \end{bmatrix}$

(d) $\begin{bmatrix} 2 & 1 & 1 & 1 \\ 6 & 1 & 10 & -3 \\ 1 & -1 & 2 & 0 \end{bmatrix}$

3. (a) $\mathbf{P}_{24} = \begin{bmatrix} 1 & 0 & 0 & 0 \\ 0 & 0 & 0 & 1 \\ 0 & 0 & 1 & 0 \\ 0 & 1 & 0 & 0 \end{bmatrix}$ (b) $\mathbf{P}_{13} = \begin{bmatrix} 0 & 0 & 1 & 0 \\ 0 & 1 & 0 & 0 \\ 1 & 0 & 0 & 0 \\ 0 & 0 & 0 & 1 \end{bmatrix}$

(c) $\mathbf{P}_{42}^{-1} = \begin{bmatrix} 1 & 0 & 0 & 0 \\ 0 & 0 & 0 & 1 \\ 0 & 0 & 1 & 0 \\ 0 & 1 & 0 & 0 \end{bmatrix}$ (d) $\mathbf{P} = \begin{bmatrix} 0 & 0 & 1 & 0 \\ 0 & 0 & 0 & 1 \\ 1 & 0 & 0 & 0 \\ 0 & 1 & 0 & 0 \end{bmatrix}$

5. (a) $\mathbf{P} = \mathbf{P}_{13}$ (b) $\mathbf{Q} = \mathbf{P}_{14}$ **7.** Columns, rather than rows, are interchanged.

Exercises 2.3, page 61

1. $\begin{bmatrix} 3 & 2 \\ 7 & 5 \end{bmatrix} = \begin{bmatrix} 1 & 0 \\ \frac{7}{3} & 1 \end{bmatrix}\begin{bmatrix} 3 & 2 \\ 0 & \frac{1}{3} \end{bmatrix}$ **3.** $\mathbf{L} = \begin{bmatrix} 1 & 0 & 0 \\ 2 & 1 & 0 \\ 3 & 4 & 1 \end{bmatrix}$ $\mathbf{U} = \begin{bmatrix} 2 & 1 & 0 \\ 0 & 1 & 2 \\ 0 & 0 & 0 \end{bmatrix}$

7. (a) For any 2×2 matrix $\mathbf{X}$, the 1, 1-entry of $\mathbf{XA}$ is 0. Therefore, $\mathbf{XA} = \mathbf{I}$ for no matrix $\mathbf{X}$.

9. (a) $\mathbf{L} = \begin{bmatrix} 1 & 0 & 0 \\ 2 & 1 & 0 \\ -1 & 3 & 1 \end{bmatrix}$ $\mathbf{U} = \begin{bmatrix} 2 & 1 & -1 \\ 0 & 1 & 3 \\ 0 & 0 & 2 \end{bmatrix}$ (b) $\mathbf{x} = \begin{bmatrix} -1 \\ 0 \\ 1 \end{bmatrix}$ (c) $\mathbf{x} = \begin{bmatrix} 1 \\ 2 \\ 0 \end{bmatrix}$

11. $\mathbf{P} = \begin{bmatrix} 1 & 0 & 0 \\ 0 & 0 & 1 \\ 0 & 1 & 0 \end{bmatrix}$ $\mathbf{L} = \begin{bmatrix} 1 & 0 & 0 \\ -1 & 1 & 0 \\ 3 & 0 & 1 \end{bmatrix}$ $\mathbf{U} = \begin{bmatrix} 1 & 2 & 3 \\ 0 & 4 & 4 \\ 0 & 0 & -8 \end{bmatrix}$

Exercises 2.4, page 66

1. $\mathbf{A}^{-1} = \begin{bmatrix} 1 & 2 \\ 3 & 4 \end{bmatrix}$ **3.** $\begin{bmatrix} 1 & 1 & 0 \\ -1 & -1 & 1 \\ 2 & 0 & 1 \end{bmatrix}$ **5.** $\begin{bmatrix} 1 & 0 & 0 & 0 \\ \frac{1}{2} & \frac{1}{2} & 0 & 0 \\ \frac{7}{2} & \frac{1}{2} & -1 & 0 \\ -\frac{16}{3} & -\frac{2}{3} & \frac{4}{3} & \frac{1}{3} \end{bmatrix}$

7. $\begin{bmatrix} 1 & 0 & -1 & 0 \\ 2 & -1 & -2 & 1 \\ -1 & 0 & 3 & -1 \\ -1 & 1 & 0 & 0 \end{bmatrix}$

Exercises 2.5, page 69

3. In general, Gauss-Jordan elimination on $\mathbf{Ax} = \mathbf{b}$ requires $(n - 1)(n + 1)(n + 2)/2 + 1$ multiplicative operations.

7.

$$\mathbf{L} = \begin{bmatrix} 1 & 0 & 0 & 0 & 0 \\ \frac{1}{2} & 1 & 0 & 0 & 0 \\ 0 & \frac{2}{3} & 1 & 0 & 0 \\ 0 & 0 & \frac{3}{4} & 1 & 0 \\ 0 & 0 & 0 & \frac{4}{5} & 1 \end{bmatrix} \qquad \mathbf{U} = \begin{bmatrix} 2 & 1 & 0 & 0 & 0 \\ 0 & \frac{3}{2} & 1 & 0 & 0 \\ 0 & 0 & \frac{4}{3} & 1 & 0 \\ 0 & 0 & 0 & \frac{5}{4} & 1 \\ 0 & 0 & 0 & 0 & \frac{6}{5} \end{bmatrix}$$

Exercises 2.6, page 74

1. $\frac{1}{7}\begin{bmatrix} 19 \\ -4 \\ 0 \end{bmatrix}, \begin{bmatrix} 2 \\ 0 \\ 1 \end{bmatrix}, \frac{1}{5}\begin{bmatrix} 0 \\ 8 \\ 19 \end{bmatrix}$

3. Let x_1 = number of bags of Mocha Delight
x_2 = number of bags of Hava Java

$$\begin{aligned} \text{Maximize:} \quad & 3x_1 + 2.5x_2 \\ \text{subject to the constraints:} \quad & x_1 + 2x_2 \le 10{,}000 \\ & 1.5x_1 + 1.5x_2 \le 9{,}000 \\ & .5x_1 + .2x_2 \le 2{,}250 \\ & x_1 \ge 0, \quad x_2 \ge 0 \end{aligned}$$

5. The refinery should produce 1000 barrels each of gasoline and heating oil.

Chapter 3

Exercises 3.1, page 86

1. -13 **3.** 14,400 **5.** $A_{23} = 1$, $A_{22} = 13$

11. (a) The number of multiplicative operations is

$$\sum_{k=1}^{n-1} \frac{n!}{(n-k)!} \ge n!$$

(b) Over \$300 billion (c) $n^3/3 + 2n/3 - 1$ (d) Less than 1 cent

15. $-\lambda^3 + 3\lambda^2 + 12\lambda - 25$ **17.** 0

Exercises 3.2, page 92

1. The fourth row is twice the first row. **3.** -2 **5.** $\begin{bmatrix} -4 & 0 & 1 \\ -5 & 1 & 1 \\ 3 & 0 & -1 \end{bmatrix}$

7.

$$\mathbf{A}^{-1} = \begin{bmatrix} 68 & 41 & -13 \\ -31 & -19 & 6 \\ -5 & -3 & 1 \end{bmatrix}$$

Exercises 3.3, page 95

1. The population faces extinction. **3.** $d_1 = \frac{1}{9}$

Chapter 4

Exercises 4.1, page 103

5. The general solution can be written as

$$\begin{bmatrix} 0 \\ 1 \\ 1 \\ 0 \end{bmatrix} + t \begin{bmatrix} \frac{8}{3} \\ \frac{2}{3} \\ -\frac{14}{3} \\ 1 \end{bmatrix}$$

9. (a) $\mathbf{x} = \boldsymbol{\theta}$ (b) $x_1 \begin{bmatrix} 1 \\ 0 \\ 0 \\ 0 \end{bmatrix} + x_4 \begin{bmatrix} 0 \\ -\frac{5}{2} \\ \frac{1}{2} \\ 1 \end{bmatrix}$ for any x_1 and x_4

(c) $\mathbf{x} = \boldsymbol{\theta}$ (d) $x_3 \begin{bmatrix} -1 \\ 1 \\ 1 \end{bmatrix}$ for any x_3

13. $N(\mathbf{A})$ is a line through the origin which is perpendicular to the vector $[1 \quad 2]^T$.

Exercises 4.2, page 112

1. (a) No (b) yes (c) yes (d) no **3.** All vectors in $\mathbf{R}^3$
5. (a) No (b) yes (c) no
7. The three vectors are linearly independent; therefore, none of the vectors is a linear combination of the other two.
9. (a) $\begin{bmatrix} 6 & -1 & 2 & 1 \\ 0 & \frac{4}{3} & \frac{10}{3} & \frac{2}{3} \\ 0 & 0 & \frac{13}{4} & -\frac{1}{4} \end{bmatrix}$ (b) $\begin{bmatrix} 1 & 1 & 1 \\ 0 & 1 & -1 \\ 0 & 0 & 8 \\ 0 & 0 & 0 \end{bmatrix}$ (c) $\begin{bmatrix} 1 & 0 & 0 \\ 0 & 1 & 0 \\ 0 & 0 & 1 \end{bmatrix}$

(d) $\begin{bmatrix} 1 & -3 & 4 \\ 0 & 0 & -7 \end{bmatrix}$

11. The four columns are in $\mathbf{R}^3$; therefore, they are linearly dependent.

Exercises 4.3, page 119

5. $r(\mathbf{A}) = 3$; a basis for $R(\mathbf{A})$ is e.g.,

$$\left\{ \begin{bmatrix} 2 \\ 0 \\ 0 \end{bmatrix}, \begin{bmatrix} -1 \\ 7 \\ 0 \end{bmatrix}, \begin{bmatrix} 6 \\ -10 \\ \frac{55}{7} \end{bmatrix} \right\}$$

7. $r(\mathbf{A}) = 2$, $n(\mathbf{A}) = 2$, a basis for $N(\mathbf{A})$ is e.g.,

$$\left\{\begin{bmatrix}1\\0\\0\\0\end{bmatrix}, \begin{bmatrix}0\\\frac{2}{3}\\1\\0\end{bmatrix}\right\}$$

11. $n(\mathbf{A}) = 6 - 4 = 2$; a basis for $N(\mathbf{A})$ is, e.g.,

$$\left\{\begin{bmatrix}-2\\1\\0\\0\\0\\0\end{bmatrix}, \begin{bmatrix}-1\\0\\0\\1\\1\\0\end{bmatrix}\right\}$$

Exercises 4.4, page 127

1. (a) $\begin{bmatrix}1\\0\\0\\1\end{bmatrix}$ (b) $\begin{bmatrix}1\\1\\0\end{bmatrix}$ (c) $\begin{bmatrix}1\\0\end{bmatrix}$

3. (a) Even (b) even (c) odd **5.** No

9.

$$\mathbf{M} = \begin{bmatrix}1&0&0&0\\0&1&0&0\\0&0&1&0\\0&0&0&1\\1&1&1&0\\1&0&1&1\\1&1&0&1\end{bmatrix} \qquad \mathbf{H} = \begin{bmatrix}1&1&1&0&1&0&0\\1&0&1&1&0&1&0\\1&1&0&1&0&0&1\end{bmatrix}$$

11. The sixth

Chapter 5

Exercises 5.1, page 135

1. $\sqrt{5}$, $\sqrt{20}$, $\sqrt{10}$, $\sqrt{17}$, $\sqrt{13}$, 1 **3.** -5 **5.** Take $\theta = \pi/2$ radians (i.e., 90°)
11. Yes **13.** $t = 3$

Exercises 5.2, page 142

1. 12, 47, 53, 124
5. If $\mathbf{x} = t\mathbf{y}$, then

$$\begin{aligned}|\langle \mathbf{x}, \mathbf{y}\rangle| &= |t||\langle \mathbf{y}, \mathbf{y}\rangle|\\ &= |t|\|\mathbf{y}\|\|\mathbf{y}\|\\ &= \|t\mathbf{y}\|\|\mathbf{y}\|\\ &= \|\mathbf{x}\|\|\mathbf{y}\|\end{aligned}$$

7. Both equal 9. **11.** $t = -3$ or -2

17. $\left\{ \begin{bmatrix} \frac{1}{\sqrt{2}} \\ 0 \\ \frac{1}{\sqrt{2}} \\ 0 \end{bmatrix}, \begin{bmatrix} \frac{1}{\sqrt{3}} \\ \frac{1}{\sqrt{3}} \\ \frac{-1}{\sqrt{3}} \\ 0 \end{bmatrix}, \begin{bmatrix} 0 \\ 0 \\ 0 \\ 1 \end{bmatrix} \right\}$

19. (a) $\mathbf{y} = \frac{7}{\sqrt{2}}\left(\frac{1}{\sqrt{2}}\right)\begin{bmatrix} 1 \\ 1 \end{bmatrix} + \frac{1}{\sqrt{2}}\left(\frac{1}{\sqrt{2}}\right)\begin{bmatrix} -1 \\ 1 \end{bmatrix}$

(b) $\mathbf{y} = \frac{1}{\sqrt{5}}\left(\frac{1}{\sqrt{5}}\right)\begin{bmatrix} 2 \\ 1 \\ 0 \end{bmatrix} - \frac{1}{3\sqrt{5}}\left(\frac{1}{3\sqrt{5}}\right)\begin{bmatrix} -2 \\ 4 \\ 5 \end{bmatrix} + \frac{5}{3}\left(\frac{1}{6}\right)\begin{bmatrix} 2 \\ -4 \\ 4 \end{bmatrix}$

(c) $\mathbf{y} = 5\left(\frac{1}{5}\right)\begin{bmatrix} 1 \\ 2 \\ -2 \\ 4 \end{bmatrix} + \sqrt{19}\left(\frac{1}{\sqrt{19}}\right)\begin{bmatrix} 4 \\ 1 \\ 1 \\ -1 \end{bmatrix}$

Exercises 5.3, page 150

1. $\mathbf{R}_{\pi/4}\mathbf{y} = \begin{bmatrix} \frac{-1}{\sqrt{2}} \\ \frac{3}{\sqrt{2}} \end{bmatrix}, \quad \mathbf{R}_{-\pi/4}\mathbf{y} = \begin{bmatrix} \frac{3}{\sqrt{2}} \\ \frac{-1}{\sqrt{2}} \end{bmatrix}$

7. $\lambda^2\|\mathbf{x}\|^2 = \langle\lambda\mathbf{x}, \lambda\mathbf{x}\rangle = \langle\mathbf{Qx}, \mathbf{Qx}\rangle = \|\mathbf{x}\|^2$; hence $|\lambda| = 1$ **9.** Yes

11. $\mathbf{A} = \begin{bmatrix} \frac{1}{\sqrt{3}} & \frac{-1}{\sqrt{2}} \\ \frac{1}{\sqrt{3}} & 0 \\ \frac{1}{\sqrt{3}} & \frac{1}{\sqrt{2}} \end{bmatrix} \begin{bmatrix} \sqrt{3} & \sqrt{3} \\ 0 & \sqrt{2} \end{bmatrix}$

Exercises 5.4, page 155

7. (b) $\mathbf{AB} = \begin{bmatrix} -1 - 3i & 5 + i \\ 5 - 4i & 4 + 3i \end{bmatrix}$

$(\mathbf{AB})^* = \begin{bmatrix} -1 + 3i & 5 + 4i \\ 5 - i & 4 - 3i \end{bmatrix}$

13. $\langle\mathbf{x}, \lambda\mathbf{y}\rangle = \sum_{k=1}^{n} x_k\overline{(\lambda y)_k} = \bar{\lambda}\sum_{k=1}^{n} x_k\bar{y}_k = \bar{\lambda}\langle\mathbf{x}, \mathbf{y}\rangle$ **17.** $a = \pm 1/\sqrt{2}$

Exercises 5.5, page 159

3. (b) $\|\mathbf{Ax}\|_\infty = \max_{1\le i\le 2} \left| \sum_{j=1}^{2} a_{ij}x_j \right| \le \max_{1\le i\le 2} \sum_{j=1}^{2} |a_{ij}||x_j| \le \|\mathbf{x}\|_\infty \max_{1\le i\le 2} \sum_{j=1}^{2} |a_{ij}|$

5.

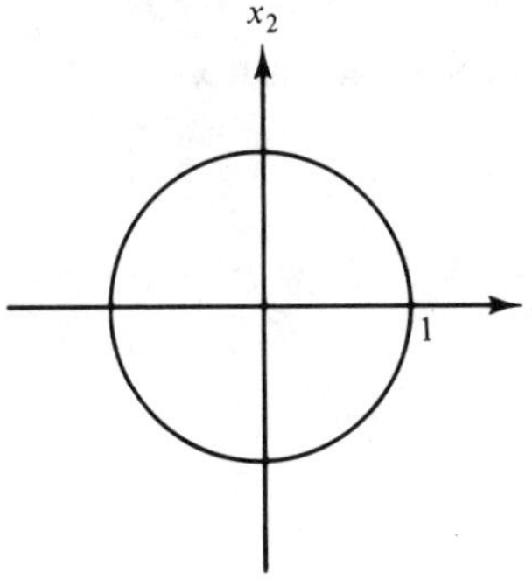

7.

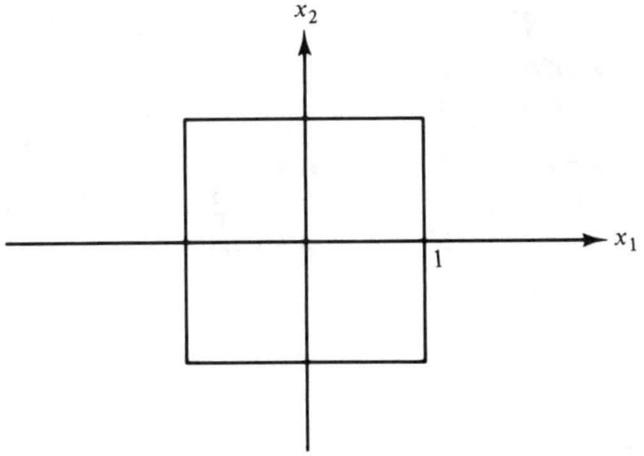

9. $[1\ \ 0\ \ 0\ \ 0]^T$, $[0\ \ 1\ \ 1\ \ 0]^T$

Exercises 5.6, page 165

3. No **5.** The coordinate matrix (to two decimal places) is

$$\begin{bmatrix} 0 & 2 & 2 & -2 & -2 \\ 1 & .73 & -2.73 & -2.73 & .73 \\ 1.73 & -.73 & -2.73 & -2.73 & -.73 \end{bmatrix}$$

7. The coordinate matrix (to two decimal places) is

$$\begin{bmatrix} .78 & -.73 & -.12 & -.83 & -1.44 \\ .27 & 1.64 & -1.11 & -2.17 & .58 \\ .43 & .68 & .18 & -1.55 & -1.05 \end{bmatrix}$$

9. (a) Reflection through the *yz* plane gives the coordinate matrix

$$\begin{bmatrix} 0 & -2 & -2 & 2 & 2 \\ 1 & .73 & -2.73 & -2.73 & .73 \\ 1.73 & -.73 & -2.73 & -2.73 & -.73 \end{bmatrix}$$

(b) Reflection through the *xy* plane gives the coordinate matrix

$$\begin{bmatrix} 0 & 2 & 2 & -2 & -2 \\ 1 & .73 & -2.73 & -2.73 & .73 \\ -1.73 & .73 & 2.73 & 2.73 & .73 \end{bmatrix}$$

(c) Reflection throug the *xz* plane gives the coordinate matrix

$$\begin{bmatrix} 0 & 2 & 2 & -2 & -2 \\ -1 & -.73 & 2.73 & 2.73 & -.73 \\ 1.73 & -.73 & -2.73 & -2.73 & -.73 \end{bmatrix}$$

CHAPTER 6

Exercises 6.1, page 179

3. Any vector of the form $\mathbf{x} = t\begin{bmatrix} 1 \\ -2 \\ 1 \end{bmatrix}$

5. $p_{\mathbf{A}}(\lambda) = \lambda^2 - 2\lambda + 9$, eigenvalues: $1 + 2\sqrt{2}i$, $1 - 2\sqrt{2}i$

7. $p_{\mathbf{A}}(\lambda) = -\lambda^3 + 2\lambda^2 - 3$, eigenvalues: -1, $3/2 + \sqrt{3}/2i$, $3/2 - \sqrt{3}/2i$

9. The characteristic polynomial, $\lambda^2 - 2\lambda + 4$, has no real roots.

11. $\lambda_1 = \lambda_2 = \lambda_3 = 0$

13. All nonzero vectors in $\mathbf{R}^3$

15. $\lambda_1 = 3, \quad \lambda_2 = 5$

19. (a) 2λ (b) $\mathbf{x}$

21. $p_{\mathbf{A}}(\lambda) = p_{\mathbf{B}}(\lambda)$

Exercises 6.2, page 190

1. $\lambda_1 = -8, \quad \lambda_2 = 4, \quad \lambda_3 = 16;$
$\lambda_1 = -6, \quad \lambda_2 = -3, \quad \lambda_3 = 0;$
$\lambda_1 = -\frac{1}{2}, \quad \lambda_2 = 1, \quad \lambda_3 = \frac{1}{4};$
$\lambda_1 = -2, \quad \lambda_2 = 1, \quad \lambda_3 = 4;$
$\lambda_1 = \frac{5}{2}, \quad \lambda_2 = 4, \quad \lambda_3 = \frac{13}{4}$

3. One **5.** One

7. $\lambda_1 = \lambda_2 = -2, \lambda_3 = 7, \left\{\begin{bmatrix} -1 \\ 0 \\ 2 \end{bmatrix}, \begin{bmatrix} 1 \\ 1 \\ 0 \end{bmatrix}, \begin{bmatrix} 2 \\ -2 \\ 1 \end{bmatrix}\right\}$; other basis vectors are possible.

9. No

11. $\lambda_1 = -1, \lambda_2 = 3$, orthonormal basis:

$$\left\{\frac{1}{\sqrt{2}}\begin{bmatrix} 1 \\ -1 \end{bmatrix}, \frac{1}{\sqrt{2}}\begin{bmatrix} 1 \\ 1 \end{bmatrix}\right\}$$

13. $\lambda_1 = -10, \lambda_2 = 5, \lambda_3 = 8$, orthonormal basis:

$$\left\{\frac{1}{\sqrt{6}}\begin{bmatrix} 1 \\ 1 \\ -2 \end{bmatrix}, \frac{1}{\sqrt{3}}\begin{bmatrix} 1 \\ 1 \\ 1 \end{bmatrix}, \frac{1}{\sqrt{2}}\begin{bmatrix} 1 \\ -1 \\ 0 \end{bmatrix}\right\}$$

15.

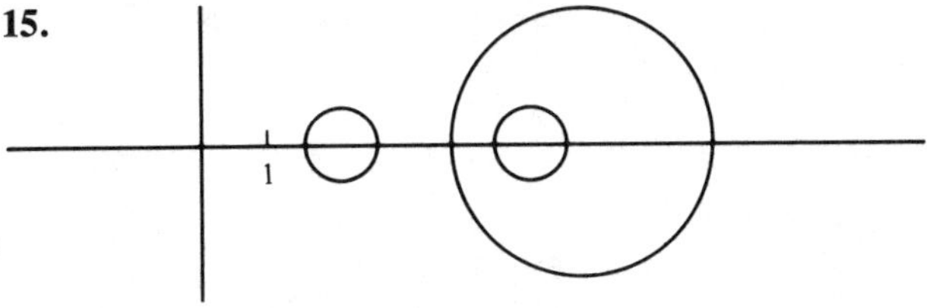

17. All eigenvalues are in the interval [2, 6].

Exercises 6.3, page 198

5. $\mathbf{A}^{-1} = \frac{1}{11}\begin{bmatrix} 3 & 2 \\ -4 & 1 \end{bmatrix}$ **7.** $\mathbf{A}^{-1} = \frac{1}{27}\begin{bmatrix} 18 & -9 & 0 \\ 1 & 1 & 6 \\ 4 & 4 & -3 \end{bmatrix}$

15. $p_A(\lambda) = -\lambda^3 + 13\lambda^2 - 54\lambda + 72$

Exercises 6.4, page 207

1. $\begin{bmatrix} .2 & .75 \\ .8 & .25 \end{bmatrix}, \quad .515$

3. 1 = Geneva, 2 = Beirut, 3 = Cairo $\begin{bmatrix} 0 & 0 & \frac{1}{2} \\ 1 & 0 & \frac{1}{2} \\ 0 & 1 & 0 \end{bmatrix}; \quad \begin{bmatrix} .2 \\ .4 \\ .4 \end{bmatrix}$

7. (a) and (c) are regular; (b) is not regular.

9. Columns: city, sub., rural

$\begin{bmatrix} .76 & .10 & .08 \\ .23 & .88 & .04 \\ .01 & .02 & .88 \end{bmatrix}$ long-term distribution $\begin{bmatrix} .26 \\ .61 \\ .13 \end{bmatrix}$ city, sub., rural

Chapter 7

Exercises 7.1, page 213

5. Yes

9. If $\mathbf{A} \sim \mathbf{B}$, then det $(\mathbf{A})$ = det $(\mathbf{B})$; hence det $(\mathbf{A}) = 0$ if and only if det $(\mathbf{B}) = 0$.

11. det $(\mathbf{A}) = \lambda_1\lambda_2 \cdots \lambda_n \neq 0$, $\mathbf{A} = \mathbf{P}\mathbf{D}^{-1}\mathbf{P}^{-1}$ where $\mathbf{D}^{-1} = \mathbf{diag}\,(\lambda_1^{-1}, \ldots, \lambda_n^{-1})$

Exercises 7.2, page 217

1. $\mathbf{A} = \begin{bmatrix} .75 & .25 \\ .25 & .75 \end{bmatrix}, \quad \mathbf{u}^{(k)} = \mathbf{P}\mathbf{D}^k\mathbf{P}^{-1}\mathbf{u}^{(0)} = .5\begin{bmatrix} 1 \\ 1 \end{bmatrix}$

3. $\mathbf{A}$ does not have two linearly independent eigenvectors.

5. $\mathbf{A}^{101} = \mathbf{A}$

7. (b) $\mathbf{A} = \mathbf{O}$ **9.** Let $\mathbf{A} = \mathbf{PBP}^{-1}$ for suitable $\mathbf{P}$ where $\mathbf{B} = \mathbf{diag}\,(1, 1, -1)$.

11. Each component of $\mathbf{x}^{(k)}$ converges to 0 as $k \to \infty$.

13. (a)
$$\mathbf{A} = \mathbf{Q}\,\mathbf{diag}\,(-2, 4)\mathbf{Q}^{\mathrm{T}}, \text{ where } \mathbf{Q} = \frac{1}{\sqrt{2}}\begin{bmatrix} 1 & 1 \\ -1 & 1 \end{bmatrix}$$

(b)
$$\mathbf{A} = \mathbf{Q}\,\mathbf{diag}\,(3, 6, 6)\mathbf{Q}^{\mathrm{T}}, \text{ where } \mathbf{Q} = \begin{bmatrix} \frac{1}{\sqrt{3}} & \frac{1}{\sqrt{2}} & \frac{1}{\sqrt{2}} \\ \frac{-1}{\sqrt{3}} & 0 & \frac{1}{\sqrt{2}} \\ \frac{1}{\sqrt{3}} & \frac{-1}{\sqrt{2}} & 0 \end{bmatrix}$$

(c)
$$\mathbf{A} = \mathbf{Q}\,\mathbf{diag}\,(-50, -3, 25)\mathbf{Q}^{\mathrm{T}}, \text{ where } \mathbf{Q} = \frac{1}{5}\begin{bmatrix} 3 & 0 & -3 \\ 0 & 5 & 0 \\ 4 & 0 & 4 \end{bmatrix}$$

Exercises 7.3, page 229

1. (a) $-3x^2 + 4xy - 8y^2$ (b) $5uv + 6v^2$ (c) uv (d) $8x_1^2 + 2\sqrt{2}\,x_1x_2 - 4x_2^2$

3. (a) $3x^2 + 2xy + 2y^2$ (b) $3x^2 - 4xy$ (c) $-2xy$
(d) $x^2 + 2y^2 + 6z^2 + 6xy + 8yz - 2xz$

5. (a) $\mathbf{R}_{-\pi/4}$ (b) $\mathbf{R}_{-\pi/4}$ (c) $\mathbf{R}_\theta$, $\theta = \arctan(-2)$ (d) $\mathbf{R}_{-\pi/4}$

7. Note that the discriminant D of the quadratic form $\langle \mathbf{x}, \mathbf{Ax}\rangle$ satisfies $D = -\det(\mathbf{A})$. If $\mathbf{y} = \mathbf{Rx}$, where $\mathbf{R}$ is a rotation, then $\langle \mathbf{x}, \mathbf{Ax}\rangle = \langle \mathbf{R}^{-1}\mathbf{y}, \mathbf{AR}^{-1}\mathbf{y}\rangle = \langle \mathbf{y}, \mathbf{RAR}^{-1}\mathbf{y}\rangle$ and $\det(\mathbf{RAR}^{-1}) = \det(\mathbf{A})$.

9. (a) Hyperbola (b) hyperbola (c) ellipse (d) parabola (e) ellipse
(f) hyperbola

Exercises 7.4, page 235

1. (a) Yes (b) no (c) no (d) no **5.** $\lambda_1 = 1, \lambda_2 = 62$

7. $\mathbf{u}^{(k)} = \mathbf{P\Lambda}^k\mathbf{P}^{-1}\mathbf{u}^{(0)}$, $\mathbf{\Lambda} = \mathbf{diag}\,(-2, 1)$, $\mathbf{P} = \frac{1}{3}\begin{bmatrix} 1 & 2 \\ -1 & 1 \end{bmatrix}$

Exercises 7.5, page 249

1. (a) $x = 4e^{6-3t}$ (b) $u = 3e^{2t}$

3. (a) $\mathbf{x} = -e^{4(t-1)}\begin{bmatrix} 1 \\ -1 \end{bmatrix}$ (b) $\mathbf{x} = .5e^{-2t}\begin{bmatrix} 1 \\ -1 \end{bmatrix} + .5e^{4t}\begin{bmatrix} 1 \\ 1 \end{bmatrix}$

(c)
$$\mathbf{x} = -2e^{-t}\begin{bmatrix} 1 \\ 1 \\ 2 \end{bmatrix} + 2e^{t}\begin{bmatrix} 1 \\ 1 \\ 1 \end{bmatrix} + e^{2t}\begin{bmatrix} 2 \\ 1 \\ 1 \end{bmatrix}$$

7. $e^{\mathbf{D}t} = \mathbf{diag}\,(e^t, 1, e^{-t})$ **9.** (a) $y = \frac{9}{5}e^{1-t} + \frac{1}{5}e^{9(t-1)}$ (b) $x = e^t - 2te^t$

15.
$$\mathbf{x}(t) = \left(-\tfrac{5}{6}e^{2t} + \tfrac{1}{6}e^{-2t}\right)\begin{bmatrix} 1 \\ -1 \end{bmatrix} + \frac{e^{\sqrt{6}\,t} + e^{-\sqrt{6}\,t}}{3}\begin{bmatrix} 1 \\ 2 \end{bmatrix}$$

21.
$$\omega_1 = 1/\sqrt{2},\ \omega_2 = \sqrt{2}, \text{ the modes are } \begin{bmatrix} 1 \\ 1 \end{bmatrix}, \begin{bmatrix} 1 \\ -1 \end{bmatrix}$$

Chapter 8

Exercises 8.1, page 264

1. $\|\mathbf{x}\|_\infty = 3$, $\|\mathbf{x}\| = \sqrt{14}$ **5.** $K_\infty(\mathbf{D}) = 7/2$

9. For $\mathbf{x} = \begin{bmatrix} 1 \\ -1 \\ 1 \end{bmatrix}$ there follows $\|\mathbf{Ax}\|_\infty = 8 = \|\mathbf{A}\|_\infty \|\mathbf{x}\|_\infty$. There are many other choices for $\mathbf{x}$.

13. With partial pivoting:

$$\begin{bmatrix} 2.9999 \\ 0.9999 \\ -0.9994 \end{bmatrix}$$

Without partial pivoting:

$$\begin{bmatrix} 2.9297 \\ 0.8750 \\ -0.9994 \end{bmatrix}$$

Exercises 8.2, page 272

1. $\mathbf{x}^{(4)} = \begin{bmatrix} -1.17 \\ -1.54 \end{bmatrix}$

3. $\mathbf{x}^{(4)} = \begin{bmatrix} 1.18 \\ -.28 \\ -.299 \end{bmatrix}$

5. $\mathbf{x}^{(4)} = \begin{bmatrix} -1.06 \\ -1.52 \end{bmatrix}$

7. $\mathbf{x}^{(4)} = \begin{bmatrix} 1.00 \\ .038 \\ .519 \end{bmatrix}$

9. (a) Is diagonally dominant
(b) Is not diagonally dominant
(c) Is not diagonally dominant

11. $\mathbf{M}^{-1}\mathbf{K} = \begin{bmatrix} 0 & \frac{1}{3} \\ \frac{1}{4} & 0 \end{bmatrix}$, $\rho(\mathbf{M}^{-1}\mathbf{K}) = \dfrac{1}{\sqrt{12}}$

17. $\mathbf{M}^{-1}\mathbf{K} = \begin{bmatrix} 0 & \frac{1}{2} & 0 \\ \frac{1}{2} & 0 & \frac{1}{2} \\ 0 & \frac{1}{2} & 0 \end{bmatrix}$, $\rho(\mathbf{M}^{-1}\mathbf{K}) = \dfrac{1}{\sqrt{2}}$

Exercises 8.3, page 281

1. (a) $\rho(\mathbf{A}) = 12$, $\lambda = 12$

(b) $\rho(\mathbf{A}) = 2$, no dominant eigenvalue
(c) $\rho(\mathbf{A}) = 5$, $\lambda = 5$
(d) $\rho(\mathbf{A}) = 2$, $\lambda = 2$
(e) $\rho(\mathbf{A}) = 2$, no dominant eigenvalue
(f) $\rho(\mathbf{A}) = 6$, $\lambda = -6$

3. $\gamma_3 = .9$, $|(\lambda_2 - 1/\gamma_3)/\lambda_2| = .11$
5. $\gamma_3 = 1/1.75$, $|(\lambda_3 - 1/\gamma_3)/\lambda_3| = .025$
7. $\gamma_4 = .253$ is an approximation to the spectral radius of $\mathbf{M}^{-1}\mathbf{K}$, using $\mathbf{W}^{(0)} = [0 \quad 1 \quad 0]^T$.
9. $\gamma_4 = 1$

Exercises 8.4, page 294

1. $$\begin{bmatrix} 49 & -24 & 33 \\ -24 & 38 & -7 \\ 33 & -7 & 47 \end{bmatrix}\begin{bmatrix} \hat{u}_1 \\ \hat{u}_2 \\ \hat{u}_3 \end{bmatrix} = \begin{bmatrix} -17 \\ 17 \\ 32 \end{bmatrix}$$

3. $$\begin{bmatrix} 4 & 21 \\ 21 & 125 \end{bmatrix}\begin{bmatrix} \hat{u}_1 \\ \hat{u}_2 \end{bmatrix} = \begin{bmatrix} 13 \\ 104 \end{bmatrix}$$

5. The systems of Exercises 1 and 3 are overdetermined.
7. No; $r(\mathbf{A}) = 2$
9. Yes; $r(\mathbf{A}) = 4$
11. $$\mathbf{A}^{\dagger} = \frac{1}{566}\begin{bmatrix} 151 & 13 & 50 \\ -52 & 48 & 54 \end{bmatrix}$$
15. $y = 3.02 + 2.08x$
17. $y = 11.67 - 20.29x + 7.91x^2$

Index